Genome Editing and Biological Weapons

Katherine Paris

Genome Editing and Biological Weapons

Assessing the Risk of Misuse

Springer

Katherine Paris
George Mason University
Arlington, VA, USA

ISBN 978-3-031-21819-4   ISBN 978-3-031-21820-0 (eBook)
https://doi.org/10.1007/978-3-031-21820-0

This Springer imprint is published by the registered company Springer Nature Switzerland AG
The registered company address is: Gewerbestrasse 11, 6330 Cham, Switzerland

Foreword

On November 25, 2018, an obscure Chinese scientist surprised the scientific community by announcing that he had used a new genome editing tool to secretly create the world's first genetically engineered babies. Dr. He Jiankui, based at Southern University of Science and Technology in Shenzhen, China, had used a relatively new technology called CRISPR to modify the genome of two babies, who were born shortly before his announcement, in an attempt to reduce their susceptibility to HIV. Dr. He's revelation dominated the Second International Summit on Human Genome Editing at the University of Hong Kong which started a few days later. The debate over the scientific applications and ethical implications of human germline editing, which had accelerated in recent years as the potential power of CRISPR became more evident, was no longer abstract or hypothetical.

As the global community was wrestling with the profound ethical questions raised by He's experiments, I was preparing for the release of a new report on a different dimension of genome editing: biosecurity. With my colleagues Dr. Jesse Kirkpatrick at George Mason University and Dr. Megan Palmer and Dr. David Relman at Stanford University, we were at the end of a 2-year project to study the risks that genome editing could be misused to cause harm and provide recommendations for strengthening governance of this dual-use technology. Our report, *Editing Biosecurity: Needs and Strategies for Governing Genome Editing*, identified a range of risks posed by genome editing including dual-use risks posed by the creation of new tools and knowledge that could be misused to cause harm, biosafety risks presented by novel types of edited organisms, reckless applications that take advantage of regulatory loopholes, and the threat posed by actors who seek to use this emerging biotechnology for malicious purposes. We also described how the emergence of genome editing had disrupted the biosecurity landscape and challenged our traditional approach to assessing and managing biological risks.

A key take-away from our report was the need to "take the technology seriously" when assessing the risks and benefits of genome editing technologies such as CRISPR. In the conclusion, we wrote, "A thorough, informed, and accessible analysis of a given technology and its social context is crucial to considering the impact that it may have on the security landscape...Absent such detailed analyses, actors may issue pronouncements of risk and impact that are based on unfounded assumptions, which can strain credibility, putting stakeholders in a poor position to assess the veracity of these claims. . . .Often the key actors best equipped to provide analyses of what is possible and plausible are those who work closely with the technology. In addition, those closest to the science and the variety of applications in different domains can help capture key trends. Engaging technologists in sustained dialogue with social scientists and policy makers can illuminate the security-related aspects of these trends that may be less salient at the ground level, while also equipping technologists to appreciate and help undertake governance actions."[1]

In this book, Dr. Katherine Paris delivers the "thorough, informed, and accessible analysis" of genome editing technologies that we called for in our 2018 report. With degrees in biology, biotechnology, and biodefense, Dr. Paris is well equipped to understand the strengths and limitations of the genome editing technologies that she writes about. Just as important, she is able to convey her analysis of these technologies with enough clarity for a lay audience and enough sophistication for a scientific reader. Furthermore, her analysis is enlivened by her personal experience learning how to use CRISPR at a community lab associated with the Do-It-Yourself (DIY) biology movement. Since some of the more alarmist analyses of the dangers posed by genome editing revolve around misuse enabled by the DIY community, her first-hand observations help provide important perspective on the potential for misuse by non-traditional practitioners of the life sciences. As befits a scholar working at the intersection of science and security, Dr. Paris' work is characterized by a rigorous and evidence-based assessment of the risks posed by genome editing technologies. Her analysis of the opportunities presented by genome editing for creating new and improved biological weapons is balanced with an evaluation of the obstacles to achieving those nefarious objectives. In keeping with the philosophy of the Biodefense PhD program from which she graduated (and I direct), Dr. Paris bridges the gap between the scientific and policy communities by offering cogent policy recommendations to prevent and mitigate the misuse of genome editing technologies.

As with recombinant DNA in the 1970s and synthetic biology in the 2000s, the emergence of genome editing technologies, especially CRISPR, has raised hopes and fears about its impact on science, public health, medicine, the economy, and society. Although many of the risks and rewards under discussion today are the same ones featured during previous debates about breakthrough biotechnologies, the current debate needs to take into account not only the distinctive characteristics of genome editing technologies, but also the recent developments in the political, security, economic, and technological domains that influence the development,

[1] Jesse Kirkpatrick, Gregory D. Koblentz, Megan J. Palmer, Edward Perello, David A. Relman, and Sarah W. Denton, *Editing Biosecurity: Needs and Strategies for Governance of Genome Editing* (Arlington, VA and Palo Alto, CA: George Mason University and Stanford University, 2018), 81. https://doi.org/10.13021/mxmm-kn14

diffusion, and applications of this powerful dual-use technology. Dr. Paris has provided the type of rigorous, multidisciplinary analysis that the scientific and policy communities need to navigate the increasingly dynamic and complex global biosecurity landscape ushered in by the advent of genome editing.

Gregory D. Koblentz PhD, MPP
Director, Biodefense Graduate Program,
Schar School of Policy and Government,
George Mason University, Arlington,
VA, USA

May 2022

Acknowledgements

This book was made possible through the support and encouragement of many people. Thank you, Dr. Gregory Koblentz, for inspiring me to research advancements in genome editing technologies and the implications on biodefense and biosecurity. I am grateful for your help throughout the research, writing, and review process. I would also like to thank Dr. James Conant and Dr. Kylene Kehn-Hall for their guidance and helpful feedback. I sincerely appreciate the time that you all took to advise me.

I approached the assessment of the risk of misuse of genome editing technologies through a biosocial frame. This method of study would not have been possible without the hands-on experience and observations that I acquired while working with CRISPR at Genspace, a DIY bio laboratory in Brooklyn, New York. The generous support of the George Mason University Schar School of Policy and Government Student Research Grant made it possible for me to attend the *Genome Editing with CRISPR-Cas9* class. Many thanks to everyone at Genspace for sharing your expertise in CRISPR and DIY bio.

Special thanks to the nonprofit plasmid repository Addgene and its data collection team in Watertown, Massachusetts for assembling and providing me data on the ZFN, TALEN, and CRISPR kits that Addgene receives and sends all over the world. This data made it possible to assess the availability of genome editing technologies more accurately.

Finally, my sincere thanks for the unwavering support that I received from my family, friends, and Constantine.

About the Book

In February 2016, Director of National Intelligence James R. Clapper issued a serious warning over the threat that genome editing poses to the national security of the USA in the annual *Worldwide Threat Assessment of the United States Intelligence Community*. This threat assessment, which was delivered to Congress, cautioned that genome editing could be misused deliberately or unintentionally to create harmful biological agents or products. The explosion in popularity of a genome editing technology called clustered regularly interspaced short palindromic repeat (CRISPR) coincided with the release of the 2016 threat assessment. Unlike earlier genome editing technologies—namely, zinc-finger nuclease (ZFN) and transcription activator-like effector nuclease (TALEN)—it did not take long for CRISPR to spread and become a commonly used technology in laboratories around the world.

Genome editing technologies allow for the precise genetic modification of organisms and cultured cells. While these technologies are used for a variety of beneficial applications from basic research to biotechnology, agriculture, and medicine, experts caution that ZFN, TALEN, and, particularly, CRISPR could be misused to develop new and improved biological weapons. Policymakers are faced with the ultimate challenge of protecting and safeguarding the continued development and use of genome editing for legitimate purposes, while putting in place biodefense and biosecurity strategies to prevent misuse. Prior to determining what type of governance measures would be effective in preventing misuse, it is critical to conduct a technology assessment that considers both the risk of misuse and the level of governability of genome editing technologies.

In *Genome Editing and Biological Weapons: Assessing the Risk of Misuse*, Katherine Paris provides a comprehensive assessment of the actual risk of misuse of ZFN, TALEN, and CRISPR. Paris conducts the assessment of the risk of misuse by applying Jonathan B. Tucker's *Decision Framework for Governance of Dual-Use Technologies*. She investigates the accessibility, ease of misuse, magnitude of potential harm, and imminence of potential misuse of genome editing technologies. To avoid repeating the shortcomings of the past biothreat assessments, Paris applies a biosocial frame, instead of a traditional biotech revolution frame, to better account for the contingencies and complexities involved in biological weapons work. She concludes with an assessment of governability and a tailored set of recommendations that address security concerns. These recommendations are sensitive to the cost-benefit trade-off of regulating genome editing technologies. The book is a must-read for policymakers, defense and security personnel, and intelligence analysts.

Contents

1 Introduction: Rapid Technological Advancements Amid Rising Concerns of Misuse 1
1.1 Calibrating the Threat 2
1.2 Warnings of Misuse 5
1.3 Approach 8
1.4 Key Issues 13
1.5 Organization 14
References 15

2 Background: Genome Editing with Programmable Nucleases 17
2.1 Genes, Genetic Engineering, and Genome Editing 18
2.2 ZFN 21
2.2.1 Zinc-Finger 21
2.2.2 Structure and Mechanism 21
2.3 TALEN 25
2.3.1 TALE 25
2.3.2 Structure and Mechanism 28
2.4 CRISPR 33
2.4.1 CRISPR-Cas System 33
2.4.2 Structure and Mechanism 37
2.4.3 CRISPR Workflow 40
2.5 Comparison of Technologies 42
2.5.1 ZFN, TALEN, and CRISPR 42
2.5.2 Other Technologies 44
2.6 Conclusion 45
References 46

3 Applying Genetic Engineering to Biological Weapons 51
3.1 Threat of Genetic Engineering 52
3.1.1 Framing Assessments 54
3.1.2 Types of Agents 57
3.1.3 Attributes Targeted for Modification 61

3.1.4 Obstacles to Modification . . . 67
3.2 Misuse of Genome Editing Technologies . . . 79
3.2.1 Monitoring for Dual-Use Technologies . . . 79
3.2.2 Applications of Misuse . . . 81
3.3 Conclusion . . . 81
References . . . 82

4 Risk of Misuse Assessment: Part I . . . 87
4.1 Parameter 1: Accessibility . . . 89
4.1.1 Methodology . . . 89
4.1.2 Ease of Acquiring Hardware and Software . . . 90
4.1.3 Cost . . . 98
4.1.4 Dependence on Other Technologies . . . 101
4.2 Parameter 2: Ease of Misuse . . . 105
4.2.1 Methodology . . . 105
4.2.2 Skill and Tacit Knowledge Needed to Use Technology . . . 108
4.2.3 Ways to Acquire Skill and Knowledge . . . 123
4.2.4 Transferability of Skill and Knowledge . . . 128
4.3 Conclusion . . . 132
References . . . 132

5 Risk of Misuse Assessment: Part II . . . 139
5.1 Parameter 3: Magnitude of Potential Harm . . . 141
5.1.1 Methodology . . . 141
5.1.2 Making Pathogens More Dangerous . . . 143
5.1.3 Weaponizing the Human Microbiome . . . 145
5.1.4 Weaponizing Gene Drives . . . 148
5.1.5 Augmenting Humans: Super Soldiers and the General Population . . . 151
5.2 Parameter 4: Imminence of Potential Misuse . . . 155
5.2.1 Methodology . . . 155
5.2.2 Ability to Be Exploited Based on Maturity . . . 158
5.3 Conclusion . . . 173
References . . . 174

6 Raising the Alarm on CRISPR . . . 179
6.1 Overview of the Risk of Misuse Assessment . . . 180
6.1.1 Relative Risk of Misuse . . . 180
6.1.2 Expanding the Pool of Potential Misusers . . . 182
6.1.3 Overcoming Tacit Knowledge and Skill Barriers . . . 185
6.1.4 Benefits and Drawbacks of Kits . . . 186
6.1.5 Constraints on Applying Learning Tools and Experimental Aids . . . 188
6.1.6 Persistence of Technical Barriers to Biological Weapons . . . 191
6.1.7 Technology Maturation and Risk of Misuse . . . 193

6.2 Implications of Misuse for Specific Applications of Genome Editing 194
6.2.1 Making Pathogens More Dangerous 194
6.2.2 Weaponizing the Human Microbiome 197
6.2.3 Weaponizing Gene Drives 199
6.2.4 Augmenting Humans 200
6.3 Conclusion 206
References 207

7 Conclusion: Reducing the Perils from the Misuse of Genome Editing 213
7.1 Assessment of Governability 214
7.2 Strategies to Address Security Concerns 217
7.2.1 Norms Against Misuse 218
7.2.2 Responsible Research Culture 219
7.2.3 Product Screening and Sales Training 222
7.2.4 Financial Loophole Closure 223
7.2.5 Bioengineered Attack Indicators 224
7.2.6 Diagnostics, Vaccines, and Other Medical Countermeasures 224
7.3 The Future of Biodefense and Biosecurity 225
References 227

Appendix A: Providers of Genome Editing Technologies 231

Appendix B: Computational Design Tools 237

Appendix C: JoVE Video Articles 249

Appendix D: Forums to Acquire Skill and Knowledge 259

Appendix E: Company Startups 265

List of Abbreviations and Symbols

A	Adenine
AAV	Adeno-associated viral vector
AD	Activation domain
AI	Artificial intelligence
APHIS	Animal and Plant Health Inspection Service
BHPC	Biotechnologies for Health and Human Performance Council
BP	Base pair
BWC	Biological Weapons Convention
C	Cysteine
C	Cytosine
CAGR	Compound Annual Growth Rate
Cas	CRISPR-associated
CDC	Centers for Disease Control and Prevention
CoDA	Context-dependent assembly
COVID-19	Coronavirus disease 2019
CRD	Central repeat domain
CRISPR	Clustered regularly interspaced short palindromic repeat
crRNA	CRISPR RNA
CTR	C-terminal region
D	Aspartic acid
DARPA	Defense Advanced Research Projects Agency
DASA R&T	Deputy Assistant Secretary of the Army for Research and Technology
DIY	Do-it-yourself
DNA	Deoxyribonucleic acid
DOD	Department of Defense
DNI	Director of National Intelligence
DSB	Double-strand break
DURC	Dual-use research of concern
EMBL	European Molecular Biology Laboratory

EPO	European Patent Office
F	Phenylalanine
FDA	Food and Drug Administration
FLASH	Fast Ligation-based Automatable Solid-phase High-throughput
FY	Fiscal year
G	Glycine
G	Guanine
GDP	Gross domestic product
GFP	Green fluorescent protein
GI	Gastrointestinal
GoF	Gain of function
gRNA	Guide RNA
H	Histidine
HDR	Homology-directed repair
HEPN	Higher eukaryotes and prokaryotes nucleotide-binding
HIV	Human immunodeficiency virus
HMP	Human Microbiome Project
HPV	Human papillomavirus
HSPD	Homeland Security Presidential Directive
I	Isoleucine
IDT	Integrated DNA Technologies
iGEM	International Genetically Engineered Machine
IOM	Institute of Medicine
IPO	Initial Public Offering
IQ	Intelligence quotient
IRB	Institutional Review Board
ISU	Implementation Support Unit
IVT	*In vitro* transcription
JoVE	Journal of Visualized Experiments
K	Lysine
KB	Kilobase
L	Leucine
LIC	Ligation-independent cloning
MERS	Middle East Respiratory Syndrome
MetaHIT	Metagenomics of the Human Intestinal Tract
MGE	Mobile genetic element
MIIS	Monterey Institute of International Studies
MPS	Mucopolysaccharidosis
MTA	Material Transfer Agreement
N	Asparagine
N	Refers to any nitrogenous base (adenine, thymine, cytosine, or guanine)
NASEM	National Academics of Sciences, Engineering, and Medicine
NCBI	National Center for Biotechnology Information

NDU	National Defense University
NEB	New England Biolabs
NHEJ	Non-homologous end joining
NIAID	National Institute of Allergy and Infectious Diseases
NIH	National Institutes of Health
NLS	Nuclear localization signal
NRC	National Research Council
NSABB	National Science Advisory Board for Biosecurity
NSCAI	National Security Commission on Artificial Intelligence
NTR	N-terminal region
Nuc	Nuclease
OPEN	Oligomerized Pool Engineering
PAM	Protospacer adjacent motif
PCAST	President's Council of Advisors on Science and Technology
PCR	Polymerase chain reaction
PCT	Patent Cooperation Treaty
PFS	Protospacer flanking sites
pre-crRNA	pre-CRISPR RNA
PREPARE	PReemptive Expression of Protective Alleles and Response Elements
R	Refers to a purine (adenine or guanine)
R/R&D	Research or research and development
RAC	Research Advisory Committee
RCF	Relative centrifugal force
rDNA	Recombinant DNA
REAL	Restriction enzyme and ligation
RePORT	Research Portfolio Online Reporting Tools
RePORTER	RePORT Expenditures and Results
RMCE	Recombinase-mediated cassette exchange
RNA	Ribonucleic acid
RNAi	RNA interference
RVD	Repeat variable di-residue
SARS	Severe Acute Respiratory Syndrome
SBIR	Small Business Innovation Research
SGD	*Saccharomyces* Genome Database
sgRNA	Single-guide RNA
SHERLOCK	Specific High-Sensitivity Enzymatic Reporter UnLOCKing
STTR	Small Business Technology Transfer
T	Thymine
TALE	Transcription activator-like effector
TALEN	TALE nuclease
TD	Translocation domain
tracrRNA	*Trans*-activating crRNA
UCSF	University of California, San Francisco

USAMRIID	United States Army Medical Research Institute of Infectious Diseases
USCNS/21	United States Commission on National Security/21st Century
USDA	United States Department of Agriculture
UV	Ultraviolet
VACV	Vaccinia virus
WIPO	World Intellectual Property Organization
WMD	Weapon(s) of Mass Destruction
X	Undetermined amino acid
ZFN	Zinc-finger nuclease

List of Figures

Fig. 1.1 Tucker's decision framework for governance of dual-use technologies 9
Fig. 1.2 Relationship between the framework and pathway selected for analysis by this assessment 10
Fig. 2.1 Structural representation of ZFN 22
Fig. 2.2 Structural representation of TALEN 29
Fig. 2.3 CRISPR-Cas-mediated adaptive immunity 35
Fig. 2.4 Structural representation of CRISPR-Cas9 38
Fig. 2.5 General CRISPR experimental workflow 40
Fig. 3.1 Biotech revolution frame vs. biosocial frame 55
Fig. 4.1 Providers of ZFN, TALEN, and CRISPR products and services 91
Fig. 4.2 ZFN, TALEN, and CRISPR units ordered per year from Addgene (2005–2018) 93
Fig. 4.3 ZFN, TALEN, and CRISPR units deposited per year to Addgene (2005–2018) 94
Fig. 4.4 Global distribution of orders for ZFN units from Addgene 96
Fig. 4.5 Global distribution of orders for TALEN units from Addgene 97
Fig. 4.6 Global distribution of orders for CRISPR units from Addgene 98
Fig. 4.7 Software design tools for genome editing (2006–2018) 99
Fig. 4.8 Restrictions to use genome editing software design tools 99
Fig. 4.9 Skill to use genome editing software design tools 115
Fig. 4.10 Guidance on basic lab skills: centrifugation 119
Fig. 4.11 Services offered by providers for ZFN, TALEN, and CRISPR 124
Fig. 4.12 Number of JoVE video articles per year (2012–2018) 126
Fig. 4.13 Number of views of JoVE video articles per year (2012–2018) 127

Fig. 4.14 Forums to acquire skill and knowledge to use genome editing technologies 128
Fig. 4.15 The effectiveness of watching JoVE videos on undergraduate student performance (pre-lab assessment) 129
Fig. 4.16 The effectiveness of watching JoVE videos on undergraduate student performance (post-lab assessment) 130
Fig. 4.17 Success rates of CRISPR knockout for the Genspace DIY bio class and undergraduate class at the University of New Mexico 131
Fig. 5.1 Normal inheritance vs. gene drive inheritance 149
Fig. 5.2 Genome editing technology publications available globally (2002–2018) 159
Fig. 5.3 Global spread of ZFN publications (2002–2018) 160
Fig. 5.4 Global spread of TALEN publications (2002–2018) 161
Fig. 5.5 Global spread of CRISPR publications (2002–2018) 162
Fig. 5.6 NIH funding for genome editing projects (FY 2002–2018) 163
Fig. 5.7 NIH's SBIR and STTR programs investment into genome editing projects with potential for commercialization (2002–2018) 163
Fig. 5.8 Clinical studies involving ZFN, TALEN, and CRISPR 164
Fig. 5.9 Phases of clinical trials involving ZFN, TALEN, and CRISPR 166
Fig. 5.10 Types of conditions studied in clinical studies involving ZFN, TALEN, and CRISPR 168
Fig. 5.11 Number of global patent documents published per year for ZFN, TALEN, and CRISPR 169
Fig. 5.12 Jurisdictions of ZFN, TALEN, and CRISPR patent documents 170
Fig. 5.13 Number of genome editing focused startup companies (1995–2017) 171
Fig. 5.14 Global genome editing and CRISPR market valuations 172
Fig. 6.1 Overview of the level of concern of ZFN, TALEN, and CRISPR 181
Fig. 7.1 Tucker's relationships between parameters and level of governability of dual-use technologies 215

List of Tables

Table 2.1 Summary of ZFN, TALEN, and CRISPR 42
Table 4.1 Overview score for the ease of acquiring genome editing hardware and software 100
Table 4.2 Addgene kit prices 101
Table 4.3 Overview score for the cost of genome editing 101
Table 4.4 Genspace lab CRISPR experimental workflow 103
Table 4.5 Overview score for the dependence of genome editing on other technologies 105
Table 4.6 Summary of skills and knowledge requirements for using CRISPR 112
Table 4.7 Overview score for the skill and tacit knowledge needed to use genome editing technologies 123
Table 4.8 Overview score for the ways to acquire genome editing skill and knowledge 128
Table 4.9 Overview score for the transferability of genome editing skill and knowledge 131
Table 5.1 Overview score for the magnitude of potential harm from making pathogens more dangerous via genome editing 145
Table 5.2 Overview score for the magnitude of potential harm from weaponizing the human microbiome via genome editing 148
Table 5.3 Overview score for the magnitude of potential harm from weaponizing gene drives via genome editing 151
Table 5.4 Overview score for the magnitude of potential harm from augmenting humans via genome editing 155
Table 5.5 Overview score for the ability of genome editing to be exploited based on maturity 173
Table A.1 Providers of ZFN, TALEN, and CRISPR 231
Table B.1 ZFN design tools 237
Table B.2 TALEN design tools 239
Table B.3 CRISPR/Cas9 sgRNA design tools 241
Table C.1 JoVE video articles on ZFN, TALEN, and CRISPR 249

Table D.1 Forums to acquire skills and knowledge to use genome editing technologies 259
Table E.1 Startup companies focused on ZFN, TALEN, or CRISPR 265

Chapter 1
Introduction: Rapid Technological Advancements Amid Rising Concerns of Misuse

In February 2016, Director of National Intelligence (DNI) James R. Clapper issued a serious warning over the threat that genome editing poses to the national security of the United States in the annual *Worldwide Threat Assessment of the United States Intelligence Community*. This threat assessment, which was delivered to Congress, cautioned that genome editing could be misused deliberately or unintentionally to create harmful biological agents or products. This threat was so significant that it was listed alongside other top global proliferation threats, such as the development of weapons of mass destruction by North Korea and the use of chemical weapons in Syria and Iraq.[1] The announcement from DNI Clapper was important because this was the first time that concerns over the potential misuse of genome editing had risen to the head of the Intelligence Community.

The technology that DNI Clapper referred to in 2016 was not developed overnight. In fact, genetic engineering principles were applied to successfully engineer the first organism, a bacterium, in the early-1970s. The former Soviet Union, which had secretly operated one of the largest and most sophisticated state-sponsored biological warfare programs known to date, learned how to harness this power and weaponized the first genetically altered biological threat agent a decade later. Biological threat agents, which are also commonly called biothreats or bioagents, include viruses, bacteria, and toxins derived from biological sources that could be used as a weapon against humans, animals, or plants. For years, limitations in biotechnology have restricted the genetic modifications that actors can make to bioagents. Even as advances in genetic engineering were made in the mid-1990s by the initial development of targetable genome editing technologies, the high cost and high level of skill needed to use these new technologies severely limited their

[1] *Worldwide Threat Assessment of the U.S. Intelligence Community: Statement for the Record before the Senate Armed Services Committee*, 114th Con., 2nd Sess., February 9, 2016 (statement of James R. Clapper, Director of National Intelligence), 6–9.

K. Paris, *Genome Editing and Biological Weapons*,
https://doi.org/10.1007/978-3-031-21820-0_1

use. However, this dramatically changed in the 2010s when rapid advancements in genome editing technologies made it possible to overcome previous constraints.

These advances were a game changer for scientists who work in the life sciences. In particular, a new and powerful genome editing technology, known as clustered regularly interspaced short palindromic repeats (CRISPR), allowed users to more precisely add, delete, or modify multiple genes at the same time and for a fraction of the cost of predecessor technologies. Users could now make genetic modifications that were too difficult, time-consuming, or expensive to do before. This emerging technology was also more versatile, which meant that it could be used in more organisms and for more applications ranging from basic research to biotechnology, agriculture, and medicine. Furthermore, its widespread availability and ease of use allowed for a larger community of users—including those who were nonexperts. Not only was this technology quickly taken up by the traditional type of user, such as scientists who work in academic, commercial, or government laboratories, but it was also accessible to non-traditional individuals and groups working in startup companies or community laboratories, including do-it-yourself (DIY) biologists.

While advances in genome editing technologies overcame many barriers that had once prevented their extensive use, it is important to emphasize that some drawbacks still exist. Simply put, these technologies are not magic wands and cannot do everything. Even so, new variants of the molecular machinery that power genome editing technologies continue to emerge, which gradually reduces their constraints. It is critical for our defenses to understand the current and future capabilities of these technologies and how they might be misused. One of the greatest challenges in biodefense and biosecurity is that new bioagents can emerge or be developed much faster than defenses can be made to counter these threats.[2] This is especially true for genome editing technologies. Thus, one of the objectives of this book is to better understand the current and future threat landscape of genome editing so that we are better positioned to counter these threats.

1.1 Calibrating the Threat

Concerns are mounting over the multiple risk pathways through which genome editing technologies could cause harm. First, there is a risk that research carried out with genome editing technologies could inadvertently produce new knowledge or vulnerabilities that adversaries might exploit. Although there is oversight of dual-use research, which is research that is conducted for legitimate, beneficial purposes but that could be misused to cause harm, the risk is not always apparent. Second, there is a risk that biosafety measures could fail, and genetically modified organisms

[2] *Biological Threats to US National Security: Testimony before the Senate Committee on Armed Services*, 116th Cong., 1st Sess., November 20, 2019 (testimony of Tom Inglesby, MD, Director, Center for Health Security, Johns Hopkins Bloomberg School of Public Health), 3.

could accidentally escape the laboratory. Third, there is a risk of irresponsible behavior. Various motivations, such as notoriety or financial gain, could fuel users to recklessly accelerate research and development timelines, circumvent regulatory systems, rush into field trials, or undertake several other risky activities when using genome editing technologies. Fourth, there is a risk that nefarious users could intentionally misuse genome editing technologies to develop biological weapons.[3] This final risk pathway is especially concerning and is the focus of my book.

A criticism of biological weapons is their perceived lack of utility. A biological weapon is a system that is composed of a payload (bioagent), munition, and dispersal mechanism.[4] The success of these weapons systems is at the mercy of several factors including the selection of bioagent, delivery system, target, and timing of attack with respect to meteorological conditions.[5] Indeed, the questionable utility of these weapons and belief that they had significant unpredictable and uncontrollable consequences factored into President Richard M. Nixon's decision to renounce the United States offensive biological weapons program in 1969. President Nixon's advisors had convinced him that biological weapons were not a reliable or effective strategic deterrent.[6] However, recent advances in genome editing technologies have renewed this debate because of the potential for these technologies to significantly improve the utility of biological weapons. There are scenarios of misuse of genome editing now that are so devastating that it would make recovery very hard. This presents new challenges for strategic stability and nonproliferation.[7]

As genome editing technologies continue to advance, there is no doubt that they *could* be misused to develop biological weapons that have more desirable attributes over their naturally occurring counterparts. Uncertainty exists over what nefarious actors could now achieve using newer genome editing technologies, especially CRISPR, that they could not achieve before. The purpose of my book is to calibrate the threat from the intentional misuse of genome editing technologies and determine the ways in which CRISPR is more worrisome than its predecessor technologies. In the past, we have worried about technologies that could be misused to genetically engineer pathogens to be more dangerous. Not only does CRISPR provide nefarious actors a means to make pathogens more dangerous, but it also has the potential to be misused for completely unprecedented applications that do not involve pathogens.

---

[3] Jesse Kirkpatrick et al., "Editing Biosecurity: Needs and Strategies for Governing Genome Editing," December 2018, 42–48.

[4] Raymond A. Zilinskas, "Gain of Function and the Biological and Toxin Weapons Convention," *Speech at the Herrenhausen Symposium, Dual Use Research on Microbes: Biosafety, Biosecurity, Responsibility*, November 12, 2014.

[5] Gregory D. Koblentz, *Living Weapons: Biological Warfare and International Security* (Ithaca: Cornell University Press, 2009), 17.

[6] Jonathan B. Tucker and Erin R. Mahan, *President Nixon's Decision to Renounce the U.S. Offensive Biological Weapons Program* (Washington, DC: National Defense University Press, 2009), 9–10.

[7] Margaret E. Kosal, "CRISPR and New Genetic-Engineering Techniques: Emerging Challenges to Strategic Stability and Nonproliferation," *The Nonproliferation Review* (September 15, 2021): 1–20, https://doi.org/10.1080/10736700.2020.1879464.

These new applications of misuse include weaponizing gene drives, augmenting humans, and targeting the human microbiome. Gene drives increase the speed that a gene could be spread through a community. While heritable traits normally have a 50% probability of being passed to the next generation, gene drives increase these odds to nearly 100%. This is a dual edge sword because gene drives that could be created to benefit us could be slightly modified to hurt us. Advances in genome editing also enable more precise and complex modifications in humans that could potentially be used to make super soldiers or to confer other desirable traits in the general population. Such changes to humans could be made to gain a social, economic, political, or security advantage. Finally, advances in genome editing make it possible to target the microorganisms that live on and within humans, which are known as the microbiome. While targeting the human microbiome may be easier than targeting humans themselves, our understanding of the human microbiome is limited. Even so, CRISPR has the potential to greatly increase our understanding of the human microbiome and to alleviate several of the technical challenges that once inhibited this pathway of harm.

In addition to concerns over new and existing applications of misuse, there are also fears over the increasing number and types of actors who could potentially misuse genome editing technologies. Traditionally, worries over the misuse of genetic engineering have centered around state-sponsored biological warfare programs that have the high level of knowledge, skills, and resources that are needed to undertake this difficult and complex task. While these groups are still a threat, the affordability, widespread distribution, and ease of use of genome editing technologies open the door for potential misuse by nonstate actors, such as terrorists and lone-wolve, or even actors working out of their kitchen or garage who have more limited means. Another objective of my book is to explore how worrisome the threat is from different types of state and nonstate actors. Furthermore, this book addresses the misconception that genome editing technologies are so easy to use that anyone could misuse them to create a biological weapon.

Oversight of genome editing technologies is complicated because they are so widely disseminated around the world and are commonly used every day for legitimate purposes in biomedical research, human health, industrial biotechnology, and agriculture. As the assessment in Chap. 5 reveals, the genome editing global market is expected to grow from merely $2 billion in 2015 to over $12 billion by 2027. Significant regulations could seriously stifle this growth. It is important not to overstate concerns over misuse since such fears might lead to the stricter regulations of genome editing technologies that would hamper the beneficial use of these tools. Alternatively, it is important not to understate concerns and overlook dire warnings about the dangers of genome editing technologies.

My book concludes with an assessment of governability and recommendations for governance strategy to help policymakers make informed decisions about how to best oversee genome editing technologies, especially CRISPR. It is important to find a balance between the openness of the technology and its research gains, with biosecurity and biodefense concerns. There needs to be an internationally held deep-seated norm against the misuse of CRISPR such that the idea of misuse is unthinkable. The scientific community should continue to foster and emphasize a shared culture of responsibility where scientists are trained to recognize suspicious behaviors and are empowered to speak up. Companies should be advised on how to train sales personnel to identify red flags in CRISPR customers, and governments should continue to invest in research to close gaps in product screening. Similarly, governments should make sure that oversight loopholes on privately funded research are closed. Additionally, governments should work to establish indicators of bioengineered attacks, which will help authorities to promptly detect, respond to, and mitigate attacks. Furthermore, the agent-based list approach to biodefense—which serves as a basis for the development of diagnostics, vaccines, and other medical countermeasures—should be expanded to accommodate genome editing.

1.2 Warnings of Misuse

Five years prior to the 2016 threat assessment report, warnings over the potential misuse of genome editing began to appear on a global scale in background documentation leading up to the Seventh Review Conference of the Biological Weapons Convention (BWC). Biological weapons experts drew attention to advances in technology that enabled more control over the manipulation of biological systems and networks.[8] In particular, worries emerged over genome editing technologies that can be programmed to make targeted cleavages in genetic material. For instance, a change may be made to the genetic code of a human cell that is so precise that only a single letter of DNA is altered.

The BWC Implementation Support Unit (ISU) initially flagged the threat from an older genome editing technology, known as zinc-finger nucleases (ZFN), followed by the threat from two newer genome editing technologies, known as transcription activator-like effector nucleases (TALEN) and clustered regularly interspaced short palindromic repeats (CRISPR).[9] The technical background of these technologies is covered in more detail in Chap. 2. While all three technologies—ZFN, TALEN, and CRISPR—are alike in that they use customizable nucleases for genome editing, they differ in their ability to achieve the modification that the user desires. A nuclease is an enzyme, which is a molecule that catalyzes a chemical reaction, that cleaves

[8] Seventh Review Conference of the States Parties to the Convention on the Prohibition of the Development, Production and Stockpiling of Bacteriological (Biological) and Toxin Weapons and on Their Destruction, U.N. Doc. BWC/CONF.VII INF.3, *New Scientific and Technological Developments Relevant to the Convention* (Geneva, Switzerland, 2011), 6.

[9] Meeting of the States Parties to the Convention on the Prohibition of the Development, Production and Stockpiling of Bacteriological (Biological) and Toxin Weapons and on Their Destruction, U.N. Doc. BWC/MSP/2013/MX/INF.1/Rev.1, *Advances in Science and Technology Related to the Convention* (Geneva, Switzerland, 2013), 7–8.

nucleic acid, such as DNA and RNA.[10] Unlike ZFN and TALEN, CRISPR gained a reputation soon after its debut as a cheaper, faster, and easier to use alternative to its predecessor technologies. CRISPR was hailed as the "biggest game changer" to impact biology since 1985 when the invention of polymerase chain reaction (PCR), a commonly used gene-amplification method, revolutionized genetic engineering.[11]

The explosion in popularity of CRISPR coincided with the release of the 2016 threat assessment report. Subsequently, in November 2016, the President's Council of Advisors on Science and Technology (PCAST) addressed how advances in biotechnology, including genome editing, were making it possible to overcome limitations of older biological engineering technologies that were time-consuming, limited in use, and necessitated complex procedures. According to PCAST, advances in genome editing, specifically CRISPR, were making it possible to target, modify, activate, or repress almost any DNA sequence of interest in many organisms, such as animals, humans, and plants. Genetic engineering that used to take months or years could now be achieved in days or weeks. PCAST cautioned the president about the potential for the intentional misuse of CRISPR.[12]

A series of influential reports followed these concerns over genome editing. In 2017, scholars from George Mason University and Stanford University, along with other subject-matter-experts, began a 2-year study to research important biosecurity issues related to the technologies. This study highlighted the many benefits of genome editing technologies, but it also warned us about the safety and security risks enabled by the technologies. Importantly, this study exposed how advances in genome editing technologies have disrupted the biosecurity threat landscape by creating new applications of misuse that do not necessarily involve pathogens. This presents a problem for current biodefense paradigms that are based on lists of known pathogens.[13]

In 2018, the National Academics of Sciences, Engineering, and Medicine (NASEM) issued a report titled, *Biodefense in the Age of Synthetic Biology,* to assist the United States Department of Defense in addressing security concerns brought on by advances in synthetic biology. Synthetic biology is described as, "a set of concepts, approaches, and tools within biotechnology that enable the modification or creation of biological organisms."[14] CRISPR is one of the many technologies that enable synthetic biology. Overall, the report advises that advances in synthetic biology expands the capabilities that we worry about, increases the range of actors

---

[10] David L. Nelson and Michael M. Cox, *Lehninger Principles of Biochemistry*, 5th edition (New York: W. H. Freeman, 2008), G-5, G-10.

[11] Heidi Ledford, "CRISPR, the Disruptor," *Nature* 522, no. 7554 (June 4, 2015): 20, https://doi.org/10.1038/522020a.

[12] President's Council of Advisors on Science and Technology, "PCAST Letter to the President: Action Needed to Protect Against Biological Attack," November 2016, 2–3.

[13] Kirkpatrick et al., "Editing Biosecurity: Needs and Strategies for Governing Genome Editing."

[14] National Academies of Sciences, Engineering, and Medicine, *Biodefense in the Age of Synthetic Biology* (Washington, DC: The National Academies Press, 2018), 15, https://doi.org/10.17226/24890.

who could misuse these technologies, and reduces the time needed to accomplish these tasks. Furthermore, the report recommends for the United States to expand its biodefense strategies and approaches to accommodate the new capabilities enabled by this rapidly advancing field.[15]

Additionally, during the 2018 BWC Meeting of Experts, concerns over genome editing, particularly CRISPR, were discussed. The State Parties concluded that the risks of synthetic biology had still not yet been "adequately assessed by the international community" and that "security implications of genome editing are uncertain and difficult to predict."[16] Later, during the 2019 BWC Meeting of Experts, a side event was held that strategized how to approach a risk assessment for genome editing. One presentation advised that a "Practical risk assessment will support [an] informed debate about [the] appropriate risk mitigation activities and governance of genome editing technologies."[17] These BWC documents demonstrate that there is a critical need to provide a risk assessment for genome editing.

The importance of understanding the threat from genome editing technologies was underscored in a congressional hearing on traditional threats and emerging challenges to biosecurity. Testimony from Mr. Vayl Oxford, Director of the Defense Threat Reduction Agency, drove home the critical future threat that genome editing and other advancing biotechnologies have on our security. In reflecting on the coronavirus disease 2019 (COVID-19) global pandemic and the subsequent deaths of over 200,000 Americans, Mr. Oxford issued a chilling warning, "We must think about the future threat landscape where gene editing, DNA sequencing, and synthetic biology offer our adversaries the opportunities to capitalize on the observed lessons learned from our [COVID-19] response, and they potentially could launch future attacks with little warning or attribution."[18] This testimony was delivered on the same day that the president and first lady of the United States announced that they tested positive for COVID-19. Mr. Oxford's testimony reminds us of the devastation that an infectious disease can cause, how it can penetrate through to the highest levels of government, and the need to be ready to counter future threats that could be developed by the misuse of genome editing technologies.

---

[15] National Academies of Sciences, Engineering, and Medicine, *Biodefense in the Age of Synthetic Biology*.

[16] Meeting of the States Parties to the Convention on the Prohibition of the Development, Production and Stockpiling of Bacteriological (Biological) and Toxin Weapons and on Their Destruction, U.N. Doc. BWC/MSP/2018/MX.2/3, *Report of the 2018 Meeting of Experts on Review of Developments in the Field of Science and Technology Related to the Convention* (Geneva, 2018), 9.

[17] Mirko Himmel et al., "Strategies for the Risk Analysis of Genome Editing," in *BTWC Meeting of Experts MX2* (Geneva, 2019), accessed November 23, 2021, https://www.unog.ch/80256EDD006B8954/(httpAssets)/DF34C9AC22CDA925C1258455005C515A/$file/BWC_MX2_20190731_MHimmel_compressed.pdf.

[18] *Strengthening Biological Security: Traditional Threats and Emerging Challenges: Testimony before the House of Foreign Affairs Subcommittee on Asia, the Pacific, and Nonproliferation and House Armed Services Subcommittee on Intelligence and Emerging Threats and Capabilities*, 116th Cong., 2nd sess., October 2, 2020 (testimony of Vayl Oxford, DTRA Director).

1.3 Approach

The warnings over the potential dangers of genome editing technologies underscore the need for a thorough assessment to better understand the current and future threat from the intentional misuse of ZFN, TALEN, and CRISPR. My book accomplishes this task by applying a well-established framework developed by Jonathan B. Tucker. Tucker created the *Decision Framework for Governance of Dual-Use Technologies* to determine the risk of misuse of dual-use technologies and assist policymakers in determining appropriate governance measures (Fig. 1.1). The framework is based on the following interconnected processes: (1) technology monitoring, (2) technology assessment, and (3) technology governance. The utility of Tucker's framework has been tested across a variety of case studies of dual-use biological and chemical technologies. Based on empirical evidence from the case studies, Tucker further refined the framework by dropping parameters that had little consequence and adding others that proved to be key elements for assessment. Tucker's framework is both a reliable and appropriate model to use to assess genome editing technologies.

The first part of Tucker's framework calls for technology monitoring. Science and technology fields constantly evolve, and technologies need to be continually examined to detect those with dual-use potential. Furthermore, technologies that have already been flagged as dual use need to be monitored for changes or advances that could impact their governance strategy. One of the best forums to monitor the status quo of dual-use technologies is through the BWC. The BWC provides an international forum for experts and other biodefense stakeholders to meet and discuss enabling technologies at least twice a year. Given the BWC's function, regularity of meetings, and open sharing of documentation, it was possible to trace the evolution of concerns over genome editing technologies through this medium. An overview of the monitoring that was conducted at the BWC for genome editing as a technology with dual-use potential is included in Chap. 3.

Following technology monitoring, the second part of Tucker's framework calls for the systematic assessment of technologies that have dual-use potential. Based on Tucker's framework, the risk of misuse of genome editing technologies in this book is determined through the following parameters: accessibility, ease of misuse, magnitude of potential harm resulting from misuse, and imminence of potential misuse. Instead of approaching this part of the assessment via a traditional pathway that is materials-focused and lacks the social aspects involved in biological weapons development, an alternative pathway known as the biosocial frame is applied. A frame is an explanatory structure used to analyze a problem. The biosocial frame, developed by Kathleen M. Vogel, considers "how the sociotechnical context of biotechnology development infuses and shapes the ways in which materials and

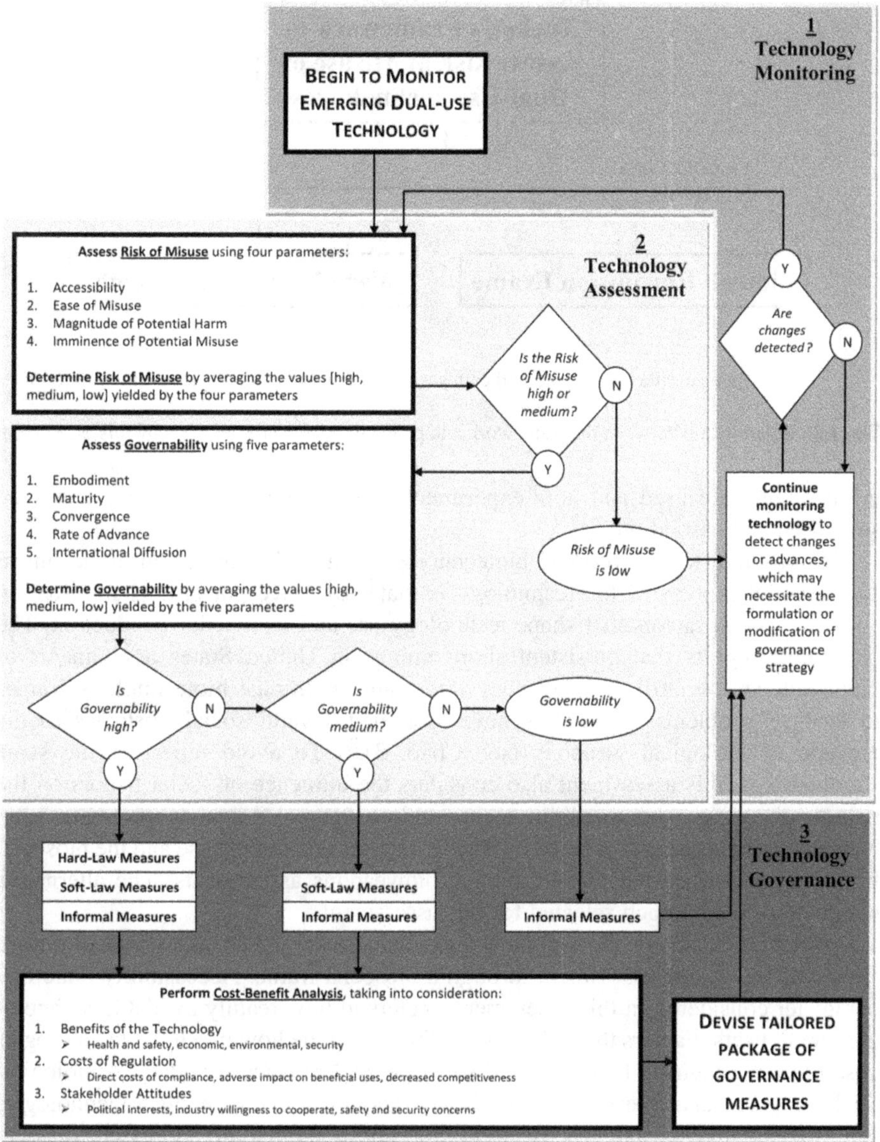

Fig. 1.1 Tucker's decision framework for governance of dual-use technologies. The framework consists of three factors (technology monitoring, technology assessment, and technology governance) and descriptive elements for each of the factors to determine appropriate governance measures for biological and chemical dual-use technologies (Jonathan B. Tucker, ed. foreword by Richard Danzig., *Innovation, Dual Use, and Security,* Fig. 4.1 (p. 69), © 2012 Massachusetts Institute of Technology, by permission of The MIT Press)

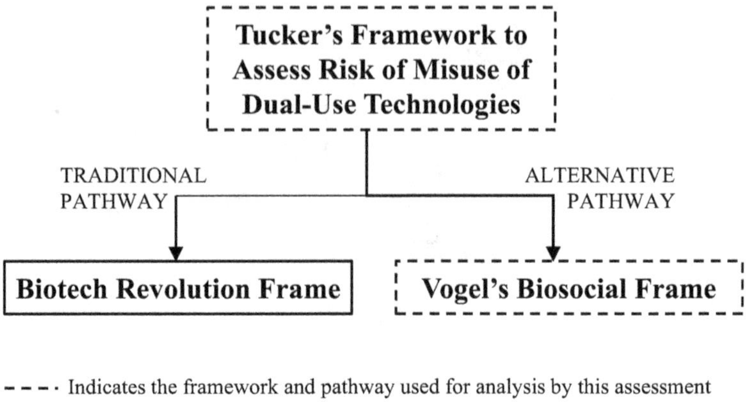

Fig. 1.2 Relationship between the framework and pathway selected for analysis by this assessment

infrastructure are used and how experiments and technological developments are structured and carried out."[19]

The problem with traditional biotechnology frames that center on the technical and material aspects of the technology is that they leave out critical information about the social factors that shape technology use and knowledge production. The literature suggests that persistent shortcomings in United States assessments of biothreats can be attributed to issues with using traditional biotechnology frames to analyze problems.[20] Scholars have identified several social obstacles to the creation of biological weapons (see Chap. 3).[21] To avoid repeating the same shortcomings, this assessment also considers the influence of social factors on the risk of misuse of genome editing technologies. Figure 1.2 depicts the relationship between the framework specified by Tucker to assess risk of misuse and the biosocial frame developed by Vogel to approach bioweapons assessments. The alternative pathway is the approach selected for this assessment.

As mentioned above, the four parameters that Tucker uses to assess risk of misuse are examined in this assessment through a biosocial frame. Accessibility—the first parameter considered in this assessment—refers to how readily available technologies are to users. Factors that affect accessibility include how much the technologies cost, ease of acquiring hardware and software, and reliance on other technologies. Technologies that are very expensive, hard to acquire, and rely on other technologies can limit the number and type of users. Sub questions that are addressed to assess accessibility are as follows: With what ease can hardware and software of genome

[19] Kathleen M. Vogel, *Phantom Menace or Looming Danger? A New Framework for Assessing Bioweapons Threats* (Baltimore: Johns Hopkins University Press, 2013), 68.

[20] Vogel, 4–5.

[21] Sonia Ben Ouagrham-Gormley, *Barriers to Bioweapons: The Challenges of Expertise and Organization for Weapons Development* (Ithaca: Cornell University Press, 2014); Vogel, *Phantom Menace or Looming Danger? A New Framework for Assessing Bioweapons Threats.*

editing technologies be acquired? How much do genome editing technologies cost? What is the dependence of genome editing technologies on other technologies?

Ease of misuse—the second parameter considered in this assessment—takes into consideration whether users who acquire the technology will be successful in using it. Factors that influence ease of misuse include the skill and tacit, or uncodified, knowledge needed to use the technology; ways to acquire skill and knowledge; and transferability of skill and knowledge. Technologies that require more skill and tacit knowledge to use, which can be difficult to transfer between people and locations, may restrict the actual use of the tool. Sub questions that are addressed to assess this ease of misuse are as follows: What skills and tacit knowledge are used to operate CRISPR? What services are offered by providers and forums, both online and in-person, are available to acquire skill and knowledge? What is the transferability of skills and knowledge through these forums?

Magnitude of potential harm—the third parameter considered in this assessment—provides insight into how severe an attack could be that results from the misuse of the technologies. This parameter considers the technology itself. In other words, can the technology be used to do what it is the actor is trying to accomplish? To what extent can the technology do harm over other technologies that exist? Furthermore, this parameter considers the vulnerability of the intended targets. Approximately how many people could be injured or die from an attack? What are the economic costs to mitigate and cleanup an attack? Would such an attack result in other societal effects such as disruption of daily activities or loss of confidence in the government?[22]

Lastly, imminence of potential misuse—the final parameter considered in this assessment—looks at how susceptible technologies are to exploitation based on the level of maturity. Early on during research and development, when technologies are not yet fully formed or developed, the potential for misuse is not very high. Less mature technologies can be more difficult to use because they may need more expertise, and the expertise may be more specialized, arcane, and limited. Furthermore, less mature technologies may have more kinks to work out, the capabilities and limitations of the technologies may not yet be well understood, and the technologies may not be as widely available. However, as technologies become more mature, and reach a point "at which its potential for misuse is more manifest," the risk of misuse increases.[23] Indicators of maturity include the availability of scientific publications, state of clinical studies, patent landscape, and formation of technology-focused startup companies. Sub questions that are addressed to assess this parameter are as follows: What is the availability of scientific publications for each genome editing technology? How long does it take for work that used genome editing to move from the laboratory to clinical trials? How do the patent landscapes differ for the technologies? How long did it take for companies to spring up that were based on the technologies?

---

[22] Tucker, "The Decision Framework," 70.

[23] Tucker, 70.

It should be noted that intent is an important indicator to use to assess whether a nefarious actor will misuse a technology; however, intent is nearly impossible to predict. Tucker does not recommend assessing intent in the decision framework because intent is often only clear when a case is retrospectively examined either by analyzing declassified documents or interviewing key decision makers. Instead, Tucker assumes a nefarious actor has malign intent.[24] Therefore, this assessment assumes that nefarious actors who wish to carry out an attack using a genetically modified biological weapons have malevolent intent.

A mix of qualitative and quantitative research methods were used to conduct this assessment of the risk of misuse of genome editing technology. Accessibility and imminence of potential misuse were mainly studied using quantitative method because these parameters were mostly operationalized using indicators that could be numerically measured. Alternatively, ease of misuse was mainly studied using qualitative methods because this parameter was operationalized by indicators that assessed the ability and knowledge of individuals. In addition to qualitative methods, quantitative methods were also applied to study indicators of the ease of misuse that could be numerically assessed.

Data for this assessment were obtained from a variety of publicly available sources including databases and first-hand laboratory experience. For the first parameter, accessibility, data were mostly obtained through Addgene, a nonprofit provider of the technologies; databases (omicX and Science Exchange); as well as providers' websites and other online publications. For the second parameter, ease of misuse, data were mostly obtained through databases (Science Exchange); first-person observations made during a CRISPR class taken at Genspace, a leading DIY biology community laboratory, located in Brooklyn, New York; the scientific, peer reviewed, video journal called the *Journal of Visualized Experiments* (*JoVE*); and published studies from Clemson University, DeSales University, and University of New Mexico. For the third parameter, magnitude of potential harm, data were obtained through the review of many sources, such as a study from analysts at Gryphon Scientific, LLC on the *Risk and Benefit Analysis of Gain of Function Research* and another report from the DOD Biotechnologies for Health and Human Performance Council (BHPC) on the desirable capabilities for the warfighter.[25] Lastly, for the fourth parameter, imminence of potential misuse, data was mostly obtained from a variety of databases, such as the Web of Science, NIH RePORTER, ClinicalTrials.gov, Lens, Crunchbase, and Owler. The role and utility of the data sources is described in the methodology sections of Chaps. 4 and 5.

[24] Tucker, 71–72.

[25] Peter Emanuel et al., "Cyborg Soldier 2050: Human/Machine Fusion and the Implications for the Future of the DOD," *CCDC CBC-TR-1599,* August 2019; *Risk and Benefit Analysis of Gain of Function Research, Final Report* (Takoma Park, MD: Gryphon Scientific, LLC, 2016).

1.4 Key Issues

In this book, I argue that the relative risk of misuse of CRISPR is high compared to ZFN and TALEN, which have a medium-to-low risk of misuse. Across the board, CRISPR poses the greatest level of concern and is the most worrisome given that it is highly accessible and affordable, is easier to use with ample resources available to help users, and is quickly maturing. Still, there are many technical and social barriers that make it challenging to apply CRISPR to develop biological weapons. As discussed in Chap. 6, the data from this assessment indicate that CRISPR is highly accessible and easier to use than ZFN and TALEN. This increases the pool of potential misusers of the technology and puts CRISPR within the reach of less well-endowed actors.

While it is concerning that the number and types of actors who can use CRISPR has greatly increased, this assessment demonstrates that it still takes skill and tacit knowledge to use the technology. This is unexpected because the literature makes CRISPR seem so easy to use that anyone could misuse it to develop a biological weapon. It is important to clarify that it is easy to do simple, well-understood tasks in well-studied organisms with CRISPR. For instance, in the Genspace DIY bio class on CRISPR that is discussed later in this study, students with little-to-no prior laboratory experience, skills, or understanding of molecular biology were able to follow along with the rest of the class to use CRISPR to knock out a non-complex, well-studied gene (*ade2*) in a well-studied model organism (yeast). However, this does not mean that the same student could apply CRISPR to make a harmful bacterium, such as *Bacillus anthracis*, the causative agent of anthrax, more dangerous.

There are limits to how helpful DIY classes, as well as videos on CRISPR and kits that facilitate CRISPR protocols, are to actors who want to develop biological weapons. There are many technical and social barriers that actors would need to overcome to apply CRISPR to develop a biological weapon. These obstacles would pose only a near-term barrier for states, a medium-term barrier for groups that are skilled, and a longer-term barrier for skilled individuals. It is noted that my assessment confirms that of the joint George Mason-Stanford University study.[26] Other actors with little-to-no laboratory skill, knowledge, or experience would face significant longer-term barriers.

Additionally, this assessment weighs in on the feasibility and plausibility of using CRISPR to achieve certain applications of misuse: altering microbes (pathogens and the human microbiome), weaponizing gene drives, and augmenting humans. Traditional concerns over making pathogens more dangerous are less worrisome because there are already technologies that could do this; however, this is a near-term threat. Weaponizing gene drives is assessed as a mid-term threat. While gene drives are worrisome, investments into defensive monitoring systems could counter this threat. Lastly, augmenting humans via somatic gene-therapies is a mid-term threat given

[26] Kirkpatrick et al., "Editing Biosecurity: Needs and Strategies for Governing Genome Editing," 9.

that it is already in clinical trials, whereas germline gene-therapies is a longer-term threat given the norm against germline genome editing. Applying CRISPR to either type of gene therapy is highly worrisome because it changes the threat landscape.

1.5 Organization

The plan for this book is to assess the risk of misuse of genome editing technologies and then to provide recommendations for ways to prevent nefarious actors from exploiting these technologies to develop biological weapons. Prior to conducting an assessment on the risk of misuse of genome editing technologies, it is important to first understand the fundamental, technical knowledge of the biology of ZFN, TALEN, and CRISPR. Chapter 2 provides a technical overview of genome editing and serves as a foundation for the risk of misuse assessment. Chapter 3 adds to this foundation by tracing the historic buildup of concerns over the threat of genetic engineering that date back to the early 1970s. This chapter identifies the desirable attributes of biological weapons that nefarious actors may want to modify with genetic engineering and the obstacles that they could encounter along the way. Additionally, Chap. 3 addresses the ways in which people worry genome editing technologies could be misused to alter microbes, weaponize gene drives, and augment humans.

The risk of misuse assessment for genome editing technologies is provided in Chaps. 4 and 5. The accessibility, ease of misuse, magnitude of potential harm, and imminence of potential misuse are analyzed in these chapters to determine the relative risk of misuse of ZFN, TALEN, and CRISPR. The risk of misuse assessment reveals several findings about the accessibility, tacit knowledge, and skill needed to use genome editing technologies. Furthermore, the assessment informs us how kits that are designed to make the technologies more user-friendly impact knowledge and skill requirements. It explores the limits of DIY classes and videos for learning how to use the technologies as well as the obstacles that still pose a barrier to biological weapons. These findings, along with an analysis and evaluation of the implications of misuse for specific applications of genome editing, are presented in Chap. 6.

Finally, an assessment of governability and recommendations for policymakers on how to prevent the misuse of genome editing technologies are included in Chap. 7. As a whole, it will be challenging to govern CRISPR because the materials and methods to use it are either ubiquitous or, more specialized, but widely available. Governments need to continue to work toward closing gaps in product screenings, closing loopholes on privately funded research, establishing indicators of bioengineered attacks, and expanding the agent-based list approach to biodefense. Governments will not stop misuse by themselves. Institutions need to implement polices, leaders should reinforce them, and scientists must carry them out. It is more important than ever to instill a deep-seated norm, that is held by scientists around the world, against the misuse of CRISPR and to continue to encourage responsible research conduct by scientists.

References

1. Himmel, Mirko, Eva Oellingrath, Nancy Connell, and Piers Millet. "Strategies for the Risk Analysis of Genome Editing." In BTWC Meeting of Experts MX2. Geneva, 2019. https://www.unog.ch/80256EDD006B8954/(httpAssets)/DF34C9AC22CDA925C1258455005C515A/$file/BWC_MX2_20190731_MHimmel_compressed.pdf.
2. Inglesby. Biological Threats to US National Security: Testimony before the Senate Committee on Armed Services, 116th Cong., 1st Sess., (2019) (testimony of Tom Inglesby, MD, Director, Center for Health Security, Johns Hopkins Bloomberg School of Public Health).
3. Kirkpatrick, Jesse, Gregory D. Koblentz, Megan J. Palmer, Edward Perello, David A. Relman, and Sarah W. Denton. "Editing Biosecurity: Needs and Strategies for Governing Genome Editing," December 2018.
4. Koblentz, Gregory D. Living Weapons: Biological Warfare and International Security. Ithaca: Cornell University Press, 2009.
5. Kosal, Margaret E. "CRISPR and New Genetic-Engineering Techniques: Emerging Challenges to Strategic Stability and Nonproliferation." The Nonproliferation Review, September 15, 2021, 1–20. https://doi.org/10.1080/10736700.2020.1879464.
6. Ledford, Heidi. "CRISPR, the Disruptor." Nature 522, no. 7554 (June 4, 2015): 20. https://doi.org/10.1038/522020a.
7. Meeting of the States Parties to the Convention on the Prohibition of the Development, Production and Stockpiling of Bacteriological (Biological) and Toxin Weapons and on Their Destruction. U.N. Doc. BWC/MSP/2013/MX/INF.1/Rev.1. Advances in Science and Technology Related to the Convention. Geneva, Switzerland, 2013.
8. National Academies of Sciences, Engineering, and Medicine. Biodefense in the Age of Synthetic Biology. Washington, DC: The National Academies Press, 2018. https://doi.org/10.17226/24890.
9. Nelson, David L., and Michael M. Cox. Lehninger Principles of Biochemistry. 5th edition. New York: W. H. Freeman, 2008.
10. Ouagrham-Gormley, Sonia Ben. Barriers to Bioweapons: The Challenges of Expertise and Organization for Weapons Development. Ithaca: Cornell University Press, 2014.
11. Oxford. Strengthening Biological Security: Traditional Threats and Emerging Challenges: Testimony before the House of Foreign Affairs Subcommittee on Asia, the Pacific, and Nonproliferation and House Armed Services Subcommittee on Intelligence and Emerging Threats and Capabilities, 116th Cong., 2nd sess., 2020 (testimony of Vayl Oxford, DTRA Director). Accessed September 19, 2021. https://armedservices.house.gov/2020/10/joint-hearing-strengthening-biological-security-traditional-threats-and-emerging-challenges.
12. President's Council of Advisors on Science and Technology. "PCAST Letter to the President: Action Needed to Protect Against Biological Attack," November 2016.
13. Raymond A. Zilinskas. "Gain of Function and the Biological and Toxin Weapons Convention." Speech at the Herrenhausen Symposium, Dual Use Research on Microbes: Biosafety, Biosecurity, Responsibility, November 12, 2014.
14. "Report of the 2018 Meeting of Experts on Review of Developments in the Field of Science and Technology Related to the Convention." Geneva, 2018. https://undocs.org/BWC/MSP/2018/MX.2/3.
15. Seventh Review Conference of the States Parties to the Convention on the Prohibition of the Development, Production and Stockpiling of Bacteriological (Biological) and Toxin Weapons and on Their Destruction. U.N. Doc. BWC/CONF.VII INF.3. New Scientific and Technological Developments Relevant to the Convention. Geneva, Switzerland, 2011.

16. Tucker, Jonathan B. "The Decision Framework." In Innovation, Dual Use, and Security: Managing the Risks of Emerging Biological and Chemical Technologies, edited by Jonathan B. Tucker, 67–82. Cambridge, Massachusetts: The MIT Press, 2012.
17. Tucker, Jonathan B., and Erin R. Mahan. President Nixon's Decision to Renounce the U.-S. Offensive Biological Weapons Program. Washington, DC: National Defense University Press, 2009.
18. Vogel, Kathleen M. Phantom Menace or Looming Danger? A New Framework for Assessing Bioweapons Threats. Baltimore: Johns Hopkins University Press, 2013.
19. Worldwide Threat Assessment of the U.S. Intelligence Community: Statement for the Record before the Senate Armed Services Committee, 114th Con., 2nd Sess., February 9, 2016 (statement of James R. Clapper, Director of National Intelligence).

Chapter 2
Background: Genome Editing with Programmable Nucleases

In the 1970s, scientists began to directly manipulate the genetic code of organisms. Early approaches to genetic engineering, however, were expensive, cumbersome, hard to direct, and not precise. A key advancement in genome editing came in the mid-1990s when researchers started to use zinc-finger nucleases (ZFN) to modify deoxyribonucleic acid (DNA) near a predetermined site.[1] Nucleases are enzymes that are capable of cleaving nucleic acid.[2] While ZFN allowed scientists to conduct more targeted genome editing than ever before, the technology was costly and difficult to use. Furthermore, a single company, Sangamo, held most of the zinc-finger intellectual property and either owned or licensed most of the zinc-finger patents.[3] This made ZFN inaccessible to many users, especially those in academic and smaller commercial laboratories.

In 2010, researchers described a new genome editing technology called transcription activator-like effector nucleases (TALEN).[4] Like ZFN, TALEN could be programmed by scientists to make double-stranded breaks at specific sites in DNA. While TALEN were less difficult than ZFN to customize to recognize target DNA sequences, there were still challenges using TALEN—TALEN are large, which makes them harder to handle and insert into cells. In 2012, another new genome editing technology called clustered regularly interspaced short palindromic

[1] J Miller, A D McLachlan, and A Klug, "Repetitive Zinc-Binding Domains in the Protein Transcription Factor IIIA from Xenopus Oocytes.," *The EMBO Journal* 4, no. 6 (June 1985): 1609–14; Yang-Gyun Kim, Jooyeun Cha, and Srinivasan Chandrasegaran, "Hybrid Restriction Enzymes: Zinc Finger Fusions to Fok I Cleavage Domain," *Proceedings of the National Academy of Sciences of the United States of America* 93, no. 3 (1996): 1156–60.

[2] Nelson and Cox, *Lehninger Principles of Biochemistry*, G-5, G-10.

[3] Helen Pearson, "Protein Engineering: The Fate of Fingers," *Nature* 455, no. 7210 (September 11, 2008): 161–62, https://doi.org/10.1038/455160a.

[4] Michelle Christian et al., "Targeting DNA Double-Strand Breaks with TAL Effector Nucleases," *Genetics* 186, no. 2 (October 1, 2010): 757–61, https://doi.org/10.1534/genetics.110.120717.

K. Paris, *Genome Editing and Biological Weapons*,
https://doi.org/10.1007/978-3-031-21820-0_2

repeats (CRISPR) emerged. Given the limitations of ZFN and TALEN, it was not surprising that scientists turned to CRISPR as a cheaper, faster, and easier to use alternative.

ZFN, TALEN, and CRISPR are types of programmable nucleases that scientists use for genome editing. Each of these genome editing technologies contains a region that can be programmed to recognize a specific nucleic acid sequence and another region that cleaves the nucleic acid via a nuclease. Once nucleic acid cleavage occurs, nucleic acid breaks are repaired by natural cellular processes, which result in site-specific genome modifications. Repair occurs via one of two pathways: non-homologous end joining (NHEJ) or homology-directed repair (HDR). NHEJ is an error-prone repair process, whereas HDR results in more precise genome editing.[5]

The purpose of the chapter is to provide fundamental technical information on ZFN, TALEN, and CRISPR. This background offers biodefense and biosecurity analysts a foundation for the risk of misuse assessment of genome editing technologies. The chapter begins by describing basic concepts in genes, genetic engineering, and genome editing. Next, the chapter describes the structures and mechanisms of ZFN, TALEN, and CRISPR. Progress in genome editing technologies did not occur overnight. It took years of work leading up to the debut of ZFN, TALEN, and CRISPR. Furthermore, researchers are constantly making refinements to genome editing technologies. The chapter concludes with an analysis of the similarities and differences of ZFN, TALEN, and CRISPR followed by a comparison of these technologies to other methods that are used for targeted genome modification.

2.1 Genes, Genetic Engineering, and Genome Editing

The central dogma of molecular biology is that genetic information flows in cells from DNA to proteins via three cellular processes: replication, transcription, and translation. Replication is the process whereby an identical copy of DNA is made from a template DNA strand. Transcription is the process whereby information encoded in DNA is copied into ribonucleic acid (RNA). Finally, translation is the process whereby information encoded in RNA is decoded into molecules called proteins.[6] Simply put, information flows from DNA to RNA to proteins.

The term *gene* is used by modern biochemists to describe "all the DNA that encodes the primary sequence of some final gene product, which can be either a polypeptide or an RNA with a structural or catalytic function."[7] Informally, a gene is

[5]Hyongbum Kim and Jin-Soo Kim, "A Guide to Genome Engineering with Programmable Nucleases," *Nature Reviews Genetics* 15, no. 5 (May 2014): 321–24, https://doi.org/10.1038/nrg3686.

[6]Nelson and Cox, *Lehninger Principles of Biochemistry*, 945.

[7]Nelson and Cox, 948.

a "basic physical and functional unit of heredity" that is passed from parent to offspring.[8] Genes determine the phenotype, or observable characteristics, of the cell or organism.[9] The collection of all the genetic information in an organism, which includes all the genes, is known as the genome.[10]

The desire for humans to select for desirable phenotypes in organisms goes back thousands of years. For a long time, the only way to achieve this was by purposefully crossing (i.e., selective breeding) plants and animals to produce offspring with the best traits, but this process was labor-intensive and took time. In the 1860s, Gregor Mendel's work on cross-pollinating pea plants led to important discoveries in heredity that laid the foundation for modern genetics. At the beginning of the 1900s, some scientists used a random process, known as induced mutagenesis, to introduce new attributes into DNA. Through induced mutagenesis, radiation and chemicals were applied to DNA, which resulted in uncontrolled and randomly localized changes to the genetic code—this typically harmed the organism.[11]

Aided by X-ray crystallography research from Maurice Wilkins and Rosalind Franklin, James Watson and Francis Crick published the double-stranded structure of DNA in 1953—a discovery that is credited with "paving the way for the 'biotechnology revolution.'"[12] For two decades following this breakthrough, scientists continued to make important discoveries that significantly improved knowledge of fundamental biochemistry, molecular biology, and genetics concepts. These discoveries culminated in the generation of the first recombinant DNA (rDNA) molecule in 1972. The rDNA was made by combining DNA from two different viruses. Shortly afterwards, the first organism (bacteria) was successful genetically engineered in 1973. To accomplish this task, restriction enzymes were used to cleave DNA at specific sites, and DNA ligase was then used to fuse the DNA together.[13]

Genetic engineering is described as "the act of modifying the genetic makeup of an organism."[14] A change in the genetic makeup, known as the genotype, of an organism brings about a change in its phenotype.[15] Unfortunately, early genetic

---

[8] "What Is a Gene?," U.S. National Library of Medicine: Genetics Home Reference, accessed April 13, 2017, https://ghr.nlm.nih.gov/primer/basics/gene.

[9] Nelson and Cox, *Lehninger Principles of Biochemistry*, 947, G-11.

[10] Nelson and Cox, 3, 33.

[11] Elsy Boglioli and Magali Richard, "Rewriting The Book Of Life: A New Era in Precision Gene Editing" (The Boston Consulting Group, September 2015), 3.

[12] "Advances in Biotechnology and Genetic Engineering: Implications for the Development of New Biological Warfare Agents" (Washington, DC: Department of Defense, June 1996), A-1.

[13] Stanley N. Cohen et al., "Construction of Biologically Functional Bacterial Plasmids In Vitro," *Proceedings of the National Academy of Sciences of the United States of America* 70, no. 11 (November 1973): 3240–44; "Advances in Biotechnology and Genetic Engineering," A-2.

[14] "Genetic Engineering," accessed August 7, 2017, https://www.nature.com/subjects/genetic-engineering.

[15] Nelson and Cox, *Lehninger Principles of Biochemistry*, G-6.

engineering technologies were described as cumbersome and hard to direct.[16] There were issues with the new genetic material becoming integrated at random sites as well as low frequencies of successful modifications. While scientists turned to an extrachromosomal DNA approach in which DNA was inserted into non-genomic regions that permitted DNA to operate beyond the context of genomic DNA, the inserted DNA was no longer functional once the cell degraded or expelled the DNA.[17]

Alternatively, a more precise approach has emerged that allows researchers to make changes in the genomes of a variety of organisms by genome editing with programmable nucleases.[18] Programmable nucleases can be directed to induce many different outcomes including gene disruption, gene insertion, gene correction, and chromosomal rearrangement.[19] Genome engineering is described as "genetic engineering applied to genomes."[20] In addition to the genome, genome engineering also encompasses targeted modifications to the genomic contexts, such as epigenetic marks, and outputs, such as transcripts.[21] Genome editing with targetable, or programmable, nucleases is a genome engineering method.[22] It is noted that this assessment refers to "genome editing" instead of "gene editing" because genome editing encompasses modification to non-coding parts of the genome as well as to epigenomes—genome editing is not limited to genes.[23] The three major genome editing technologies are ZFN, TALEN, and CRISPR.

---

[16] Kirkpatrick et al., "Editing Biosecurity: Needs and Strategies for Governing Genome Editing," 29; Edward Perello, *CRISPR Genome Editing: A Technical & Policy Primer* (George Mason University, n.d.), 7, https://editingbiosecurity.org/new-page.

[17] Kirkpatrick et al., "Editing Biosecurity: Needs and Strategies for Governing Genome Editing," 29; Perello, *CRISPR Genome Editing: A Technical & Policy Primer*, 7–8.

[18] Takashi Yamamoto, ed., *Targeted Genome Editing Using Site-Specific Nucleases: ZFNs, TALENs, and the CRISPR/Cas9 System* (Springer, 2015), v.

[19] Kim and Kim, "A Guide to Genome Engineering with Programmable Nucleases," 322.

[20] Peter A. Carr and George M. Church, "Genome Engineering," *Nature Biotechnology* 27, no. 12 (December 2009): 1152, https://doi.org/10.1038/nbt.1590.

[21] Patrick D. Hsu, Eric S. Lander, and Feng Zhang, "Development and Applications of CRISPR-Cas9 for Genome Engineering," *Cell* 157, no. 6 (June 5, 2014): 1262, https://doi.org/10.1016/j.cell.2014.05.010.

[22] Kevin M. Esvelt and Harris H. Wang, "Genome-scale Engineering for Systems and Synthetic Biology," *Molecular Systems Biology* 9, no. 641 (2013): 17, https://doi.org/10.1038/msb.2012.66.

[23] Nuffield Council on Bioethics, *Genome Editing: An Ethical Review* (London: Nuffield Council on Bioethics, 2016), 4.

2.2 ZFN

2.2.1 *Zinc-Finger*

Gene expression in cells is a heavily regulated process. One way that cells control the expression of genes is through the binding of regulatory proteins to DNA. Regulatory proteins identify surface attributes of DNA and bind to target DNA sequences with a higher affinity—about 10^4 to 10^6 times—over non-target DNA sequences. Many regulatory proteins are equipped with specific DNA-binding domains that interact with DNA. Contact with DNA occurs through structural motifs located within the DNA-binding domains of regulatory proteins. A zinc-finger is an example of one such structural motif that binds to DNA. Zinc-fingers are composed of approximately 30 amino acids, which are the building blocks of proteins, arranged in an elongated loop and stabilized at the base by folding around an ion of zinc (Zn^{2+}).[24] Each zinc-finger recognizes about three base pairs of DNA.[25] DNA-binding proteins commonly contain many zinc-fingers because the interaction between one zinc-finger and DNA is usually weak.[26] At a minimum, three consecutive zinc-fingers are required to generate a strong enough binding affinity for DNA.[27]

2.2.2 *Structure and Mechanism*

Zinc-fingers were first described in 1985, and researchers adapted zinc-fingers as a tool to carry out the targeted cleavage of DNA in 1996.[28] A zinc-finger nuclease (ZFN) is an artificially engineered endonuclease that is programed to create a double-stranded break in DNA at a specific location (Fig. 2.1).[29] An endonuclease cleaves an internal region, as opposed to an end, of the nucleic acid.[30] Each monomer, or basic unit, of a ZFN contains a DNA-binding domain at the N-terminal end and DNA-cleavage domain at the C-terminal end; an amino acid

[24] Nelson and Cox, *Lehninger Principles of Biochemistry*, 1121–22.

[25] Dana Carroll, "Genome Engineering with Targetable Nucleases," *Annual Review of Biochemistry* 83, no. 1 (2014): 412, https://doi.org/10.1146/annurev-biochem-060713-035418.

[26] Nelson and Cox, *Lehninger Principles of Biochemistry*, 1122.

[27] Carroll, "Genome Engineering with Targetable Nucleases," 412.

[28] Miller, McLachlan, and Klug, "Repetitive Zinc-Binding Domains in the Protein Transcription Factor IIIA from Xenopus Oocytes."; Kim, Cha, and Chandrasegaran, "Hybrid Restriction Enzymes."

[29] Tom Strachan and Andrew Read, *Human Molecular Genetics, Fourth Edition*, 4 edition (New York: Garland Science, 2010), 654.

[30] Strachan and Read, 723–24.

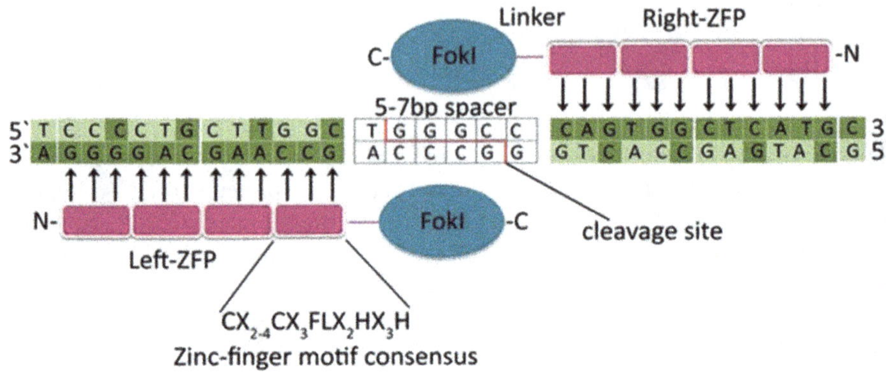

Fig. 2.1 Structural representation of ZFN. Each ZFN monomer contains a DNA-binding domain, here zinc-fingers, and DNA-cleavage domain, here FokI. A linker sequence of amino acids connects the two domains. The N-terminal end of ZFN monomers bind to the 3′ end of DNA. The left and right half-sites are separated by a 5–7 bp spacer. Each zinc-finger module binds to three nucleotides (arrows). The target sequence for the ZFN depicted above is 24 bp. Two FokI nucleases form a heterodimer and induce a double-stranded DNA break at the cleavage site. The consensus motif is the calculated order of amino acids that most frequently appear (Cysteine-C, Histidine-H, Leucine-L, Phenylalanine-F, and Undetermined-X) (Vinita Periwal, "A Comprehensive Overview of Computational Resources to Aid in Precision Genome Editing with Engineered Nucleases," Briefings in Bioinformatics, 18, no. 4 (July 2017): 2, by permission of Oxford University Press)

linker joins the two domains.[31] N-terminus and C-terminus are standard conventions used in biology to differentiate the two ends of a protein. The N-terminus, or amino-terminus, is the side of the protein that contains a free amino group (-NH_2), and the C-terminus, or carboxyl-terminus, is the side of the protein that contains a free carboxyl group (-COOH).[32]

While the DNA-binding domain of ZFN is specially engineered to recognize and bind specific sequences of DNA, the DNA-cleavage domain is constant and has no apparent sequence-specificity. The DNA-binding domain is composed of zinc-fingers, and the DNA-cleavage domain is composed of *Fok*I, which originates from *Flavobacterium okeanokoites*.[33] *Fok*I is a type IIS restriction endonuclease that non-specifically cleaves DNA. The type IIS subgroup is one type of more than 3000 type II restriction endonucleases. This subgroup identifies asymmetric,

[31]Vinita Periwal, "A Comprehensive Overview of Computational Resources to Aid in Precision Genome Editing with Engineered Nucleases," *Briefings in Bioinformatics*, 2, accessed April 6, 2017, https://doi.org/10.1093/bib/bbw052.

[32]Nelson and Cox, *Lehninger Principles of Biochemistry*, 82.

[33]Carroll, "Genome Engineering with Targetable Nucleases," 411; A. A. Chugunova, O. A. Dontsova, and P. V. Sergiev, "Methods of Genome Engineering: A New Era of Molecular Biology," *Biochemistry (Moscow)* 81, no. 7 (July 1, 2016): 665, https://doi.org/10.1134/S0006297916070038; Kim, Cha, and Chandrasegaran, "Hybrid Restriction Enzymes," 1156.

non-palindromic DNA sequences that are 4–7 base pairs long and induces breaks at fixed positions in DNA, which are usually within 20 bp of the recognition site.[34]

In nature, two units of *Fok*I need to dimerize to activate the endonuclease and cleave DNA. Dimerization of *Fok*I is also necessary for genome editing with ZFN. Thus, two monomers of ZFN are only active once they dimerize and bind to so-called "half-sites" on opposite DNA strands. ZFN monomers bind to DNA in a tail-to-tail conformation, which means the N-terminal end of each zinc-finger binds the 3′ end of each DNA strand.[35] 5′ and 3′ are standard conventions used in biology to denote the two ends of DNA. While dimerization does help to improve the accuracy of recognition, ZFN-induced cleavage of DNA still occurs at off-target sites, which causes cytotoxicity and leads to cell death.[36]

One way to reduce cleavage of off-target sites is to improve the activity of the DNA-cleavage domain, *Fok*I. The dimerization of two identical ZFN monomers, termed homodimerization, tends to cause off-target activity. Modification of *Fok*I that induces the dimerization of two different monomers of ZFN, termed heterodimerization, mitigates off-target activity. Additionally, incorrect DNA cleavage occurs when one monomer of ZFN that is correctly bound to DNA interacts with a second monomer of ZFN that is incorrectly bound to DNA. Off-target activity may be minimized by slightly weakening the interaction of *Fok*I dimers, which shifts the focus from the DNA-cleavage domain to the DNA-binding domain. Thus, zinc-fingers that are part of the DNA-binding domain must bind correctly with DNA in order to achieve dimerization of ZFN monomers and ZFN activation.[37]

A second way to decrease off-target activity is to "optimize" the amino acid linker, which joins the two ZFN domains. The length and sequence of the amino acid linker are factors that modulate the activity and specificity of ZFNs.[38] Scholars contend, "to our knowledge no linker has yet been described that provides perfect specificity for a 'spacer' DNA of a single length. The consequence of this ambiguity is that a pair of ZFNs can cut not only at sites with a spacer length of *e.g.*, 6 bp, but also at sites with 5-bp and 7-bp spacers."[39] The *spacer* is the region of DNA that separates the two half-sites between DNA-binding domains, here zinc-fingers. A

[34] Abigail J. Bath et al., "Many Type IIs Restriction Endonucleases Interact with Two Recognition Sites before Cleaving DNA," *Journal of Biological Chemistry* 277, no. 6 (February 8, 2002): 4024, https://doi.org/10.1074/jbc.M108441200.

[35] Carroll, "Genome Engineering with Targetable Nucleases," 412–13; Toni Cathomen and J Keith Joung, "Zinc-Finger Nucleases: The Next Generation Emerges," *Molecular Therapy* 16, no. 7 (July 2008): 1200, https://doi.org/10.1038/mt.2008.114.

[36] Chugunova, Dontsova, and Sergiev, "Methods of Genome Engineering," 665; Cathomen and Keith Joung, "Zinc-Finger Nucleases," 1203.

[37] Cathomen and Keith Joung, "Zinc-Finger Nucleases," 1204.

[38] Eva-Maria Händel, Stephen Alwin, and Toni Cathomen, "Expanding or Restricting the Target Site Repertoire of Zinc-Finger Nucleases: The Inter-Domain Linker as a Major Determinant of Target Site Selectivity," *Molecular Therapy: The Journal of the American Society of Gene Therapy* 17, no. 1 (January 2009): 104, https://doi.org/10.1038/mt.2008.233.

[39] Cathomen and Keith Joung, "Zinc-Finger Nucleases," 1204.

spacer made up of approximately 5–7 bp is considered optimal.[40] Linkers may be optimized via the "systematic analysis of candidate-based linkers" or via the "combinatorial library approach in which both linker composition and length are randomized."[41]

A third way to minimize off-target activity is to enhance the specificity of the DNA-binding domain (i.e., the zinc-fingers) for target DNA.[42] Methods used to engineer the DNA-binding domain of ZFN include modular assembly, context-sensitive selection, and Sangamo's proprietary 2+2 strategy.[43] Using the modular assembly method, zinc-fingers with known DNA-binding specificities are selected from an archive of single zinc-finger "modules." These modules are then connected to each other. Researchers have created zinc-finger modules that are able to identify almost all the 64 nucleotide triplets.[44] A nucleotide, which is the basic unit of DNA, varies depending on its nitrogenous base: adenine, thymine, cytosine, and guanine. Given the four types of nitrogenous bases (adenine, thymine, cytosine, and guanine), a total of 64– or 4^3–combinations of nucleotide triplets are possible.

While the modular assembly strategy is seemingly straightforward, it is difficult to achieve results in practice due to the position and context dependency of zinc-fingers. In other words, a zinc-finger that works in one situation does not necessarily work in a different situation. It is likely that zinc-fingers do not exist as independent "modular" units. Instead, the neighboring zinc-fingers interact and affect each other.[45] Context-sensitive selection methods mitigate issues that arise with the modular assembly method by considering both the position and context of zinc-fingers within the DNA-binding domain of ZFN monomers. Context-sensitive strategies are based on what combinations of zinc-fingers modules work well together. Context-sensitive selection methods consider the position of the zinc-finger within the DNA-binding domain and the interaction between zinc-fingers. The zinc-finger module closest to the N-terminal end of the ZFN monomer occupies position one, and each zinc-finger module thereafter occupies the next highest position.[46] Although context-sensitive selection methods produce "multi-finger domains with high DNA-binding affinities and specificities," many research groups are unable to employ these methods due to a lack of "specialized expertise."[47]

[40] Periwal, "A Comprehensive Overview of Computational Resources to Aid in Precision Genome Editing with Engineered Nucleases," 2; Chugunova, Dontsova, and Sergiev, "Methods of Genome Engineering," 665.

[41] Cathomen and Keith Joung, "Zinc-Finger Nucleases," 1204.

[42] Cathomen and Keith Joung, 1204.

[43] Cathomen and Keith Joung, 1202.

[44] Thomas Gaj, Charles A. Gersbach, and Carlos F. Barbas, "ZFN, TALEN and CRISPR/Cas-Based Methods for Genome Engineering," *Trends in Biotechnology* 31, no. 7 (July 2013): 398, https://doi.org/10.1016/j.tibtech.2013.04.004.

[45] Cathomen and Keith Joung, "Zinc-Finger Nucleases," 1202–3.

[46] Cathomen and Keith Joung, 1202.

[47] Cathomen and Keith Joung, 1202–3.

In addition to modular assembly and context-sensitive selection methods, Sangamo Therapeutics, formally Sangamo Biosciences, developed a proprietary 2+2 strategy to engineer zinc-fingers. While the exact process that Sangamo uses is unknown, it is believed that two of the two-finger units are combined to create a ZFN monomer with a total of four zinc-fingers. The two-finger units are selected from an established, proprietary archive of zinc-fingers with DNA-binding specificities that are known. Following construction of four-finger domains, zinc-fingers undergo additional improvement via a proprietary algorithm. A major drawback of the 2+2 strategy is that key information, namely the archive of zinc-fingers and optimization algorithm, is protected and unknown to the public.[48] Nevertheless, in addition to four-finger domains, Sangamo can create highly specific DNA-binding domains that contain up to six zinc-fingers, which shows the ZFN tool possesses "great potency."[49]

A fourth way to increase specificity is to promote cellular repair through HDR instead of error-prone NHEJ by inducing single-stranded breaks instead of double-stranded breaks in DNA. Nucleases induce double-stranded DNA breaks that are repaired via HDR or NHEJ. NHEJ bring about "uncontrolled and unwanted indels at both target sites and off-target sites."[50] In order to avoid repair through NHEJ, researchers use programmable nickases that create single-stranded DNA breaks and stimulate HDR pathway. Researchers can induce a single-stranded break on both strands of DNA, which in effect produces a double-stranded break. When two separate single-stranded breaks are induced, cellular repair occurs through HDR, which maintains precision as well as editing efficiencies. This is accomplished by using ZFNickase in place of ZFN. While nickases do increase the precision of genome editing, researchers observe a decrease in the efficiency when editing with ZFNickases instead of ZFNs.[51]

2.3 TALEN

2.3.1 TALE

In nature, recall that the process that encodes genetic information from DNA to RNA, termed transcription, is closely regulated by proteins. A transcription activator-like effector (TALE) is a type of protein that binds to DNA and helps to

[48] Cathomen and Keith Joung, 1202–3.

[49] Hiroshi Ochiai and Takashi Yamamoto, "Chapter 1: Genome Editing Using Zinc-Finger Nucleases (ZFNs) and Transcription Activator-Like Effector Nucleases (TALENs)," in *Targeted Genome Editing Using Site-Specific Nucleases: ZFNs, TALENs, and the CRISPR/Cas9 System*, ed. Takashi Yamamoto (Springer, 2015), 14.

[50] Kim and Kim, "A Guide to Genome Engineering with Programmable Nucleases," 330.

[51] Kim and Kim, 330.

initiate the process of transcription. Pathogenic bacteria that belong to the genus *Xanthomonas* use TALE proteins as virulence factors to facilitate infection in crop plants. The bacteria utilize the type III secretion system to inject TALEs into the cytoplasm of plant cells. Once inside, TALEs are transported to the nucleus, which houses the genomic information, of plant cells. Next, TALEs bind to promoter sequences of plant DNA.[52] The promoter is the site on DNA where RNA polymerase, the enzyme that synthesizes RNA from a template strand of DNA or RNA, binds and begins transcription.[53] TALEs "mimic" the transcription factors of eukaryotic cells, which in this case are plant cells.[54] TALEs activate transcription of plant genes that could help the bacterium or initiate host defenses.[55]

A TALE protein contains three main regions: the N-terminal region (NTR), the central repeat domain (CRD), and the C-terminal region (CTR). Located between the NTR and CTR, the CRD confers DNA binding specificity to TALE proteins.[56] The CRD is made up of 1.5–33.5 direct repeat units, also referred to as tandem repeats, of amino acids. At least 6.5 direct repeats are needed to activate the expression of genes in targets. CRDs typically contain 15.5–19.5 direct repeats—TALEs with 17.5 direct repeats are the most common. Direct repeats are usually 34 amino acids long; however, the final repeat—referred to as a half repeat—is only about 20 amino acids long. Altogether, the CRD is approximately 54–909 amino acids long.[57]

As the term "direct" suggests, the amino acid sequence of each repeat unit is very similar. However, variation does occur in each direct repeat unit in the amino acids located at positions 12 and 13, known as the repeat variable di-residue (RVD). The amino acid in position 12 helps to stabilize RVDs, and the amino acid in position 13 binds to DNA.[58] Due to the variation that RVDs afford TALEs, TALEs can distinguish between nucleotides and bind to specific sequences of DNA. Unlike a single module of zinc-finger that recognizes three nucleotides of DNA, each direct

[52]Radek Jankele and Petr Svoboda, "TAL Effectors: Tools for DNA Targeting," *Briefings in Functional Genomics* 13, no. 5 (September 1, 2014): 409, https://doi.org/10.1093/bfgp/elu013.

[53]Nelson and Cox, *Lehninger Principles of Biochemistry*, 1116, G-14.

[54]Sabine Kay et al., "A Bacterial Effector Acts as a Plant Transcription Factor and Induces a Cell Size Regulator," *Science* 318, no. 5850 (October 26, 2007): 648, https://doi.org/10.1126/science.1144956.

[55]Amanda Nga-Sze Mak et al., "The Crystal Structure of TAL Effector PthXo1 Bound to Its DNA Target," *Science (New York, N.Y.)* 335, no. 6069 (February 10, 2012): 716, https://doi.org/10.1126/science.1216211.

[56]Jankele and Svoboda, "TAL Effectors," 410.

[57]Jens Boch and Ulla Bonas, "Xanthomonas AvrBs3 Family-Type III Effectors: Discovery and Function," *Annual Review of Phytopathology* 48, no. 1 (2010): 422, 428, https://doi.org/10.1146/annurev-phyto-080508-081936.

[58]Carroll, "Genome Engineering with Targetable Nucleases," 414; Luke Cuculis et al., "Direct Observation of TALE Protein Dynamics Reveals a Two-State Search Mechanism," *Nature Communications* 6 (June 1, 2015): 2, https://doi.org/10.1038/ncomms8277.

repeat unit of TALE recognizes one nucleotide of DNA.[59] Moreover, while the N-terminal end of zinc-finger modules bind the 3′ end of DNA, the N-terminal end of TALEs bind the 5′ end of DNA.[60]

The DNA binding domain of TALE extends slightly upstream, or towards the N-terminal end, into the NTR. The NTR is approximately 283–290 amino acids long.[61] Importantly, the region of the NTR located adjacent to the CRD mediates binding to thymine—the first nucleotide in the target DNA sequence.[62] For the most part, the literature agrees that the DNA sequence targeted by TALE should begin with thymine.[63] However, artificially constructed TALEs may bind to DNA with a great enough affinity that it is not necessary for target sites to start with thymine. Nevertheless, it is still advisable to construct artificial TALEs that recognize thymine at the beginning of the DNA target sequence.[64] The NTR of TALE also contains the translocation domain (TD), which is needed to secrete and translocate the TALE from the bacterium to the host cell. TALEs are directed to move to the nucleus of the host cell via the nuclear localization signal (NLS), which is located in the CTR. In addition to the NLS, the CTR also contains the transcriptional activation domain (AD) that activates the expression of plant genes.[65] The CTR is approximately 274–297 amino acids long.[66]

In 2009, two research groups—independently of one another—achieved a breakthrough in TALE research when they uncovered the code used by TALE to recognize specific DNA nucleotides.[67] More specifically, the groups discovered that certain amino acid combinations of RVDs recognize specific nucleotides. For instance, the RVD that contains the amino acids asparagine (N) in position 12 and isoleucine (I) in position 13, abbreviated NI, is highly specific with a strong preference for the nucleotide adenine. As an aside, the one-letter symbols for other

[59] Claudio Mussolino and Toni Cathomen, "TALE Nucleases: Tailored Genome Engineering Made Easy," *Current Opinion in Biotechnology*, Tissue, cell and pathway engineering, 23, no. 5 (October 2012): 644, https://doi.org/10.1016/j.copbio.2012.01.013.

[60] Carroll, "Genome Engineering with Targetable Nucleases," 414.

[61] Boch and Bonas, "Xanthomonas AvrBs3 Family-Type III Effectors," 422.

[62] Mussolino and Cathomen, "TALE Nucleases," 410.

[63] Chugunova, Dontsova, and Sergiev, "Methods of Genome Engineering," 669; Gaj, Gersbach, and Barbas, "ZFN, TALEN and CRISPR/Cas-Based Methods for Genome Engineering," 399; Hiroshi Ochiai and Yamamoto, "Chapter 1: Genome Editing Using Zinc-Finger Nucleases (ZFNs) and Transcription Activator-Like Effector Nucleases (TALENs)," 15.

[64] Jankele and Svoboda, "TAL Effectors," 412; Kim and Kim, "A Guide to Genome Engineering with Programmable Nucleases," 325.

[65] Boch and Bonas, "Xanthomonas AvrBs3 Family-Type III Effectors," 431.

[66] Boch and Bonas, 422.

[67] Matthew J. Moscou and Adam J. Bogdanove, "A Simple Cipher Governs DNA Recognition by TAL Effectors," *Science* 326, no. 5959 (December 11, 2009): 1501–1501, https://doi.org/10.1126/science.1178817; Jens Boch et al., "Breaking the Code of DNA Binding Specificity of TAL-Type III Effectors," *Science* 326, no. 5959 (December 11, 2009): 1509–12, https://doi.org/10.1126/science.1178811.

amino acids that are relevant to this discussion are as follows: aspartic acid (D), glycine (G), histidine (H), and lysine (K).[68]

In addition to NI, the RVDs HD, NG, and NK/NH are highly specific with a strong preference for the nucleotides cytosine, thymine, and guanine, respectively. The RVD NN is said to have "degenerated specificity" because it selects for adenine and guanine. The six RVDs mentioned so far—NI, HD, NG, NK, NH, and NN—are known as "standard" RVDs because they are offered in assembly kits for TALE. Alternatively, "non-standard" RVDs are not offered in such kits.[69] Other commonly mentioned RVDs include NS, HG, and N*. NS recognizes adenine, cytosine, and guanine with degenerated specificity and likely sterically excludes thymine; HG recognizes thymine; and N* recognizes cytosine and thymine. Note that N* is part of a direct repeat unit that lacks an amino acid in position 13—the repeat contains a total of 33 amino acids.[70]

The binding affinity and position of RVDs along the CRD influence the activity of TALEs. The binding affinity of RVDs for nucleotides ranges from strong to weak. In other words, certain RVDs bind to DNA stronger than others. In terms of standard RVDs, HD, NG, and NN bind to DNA with a high affinity; NH binds to DNA with an intermediate affinity; and NI and NK bind to DNA with a weak affinity. To clarify, it is possible for RVDs to recognize DNA with a high specificity and bind DNA with a low affinity. For instance, NK is highly specific for guanine, but NK binds to guanine with a weak affinity. The placement of RVDs along the CRD has important implications for off-target activity. TALEs with strong RVDs at the N-terminal end of the CRD are more susceptible to off-target activity. Conversely, TALEs with weak RVDs at the C-terminal end of the CRD are more prone to off-target activity.[71]

2.3.2 Structure and Mechanism

The discovery of the TALE-DNA recognition code allowed researchers to engineer TALEs to create custom DNA-binding proteins. Just as researchers figured out how to repurpose zinc-fingers for genome editing, researchers also determined that TALEs could be fused to a nuclease to make precise modifications in genomes.[72] A transcription activator-like effector nuclease (TALEN) is a genome editing tool that makes targeted double-stranded breaks in DNA (Fig. 2.2). TALENs contain a

[68] Carroll, "Genome Engineering with Targetable Nucleases," 414.

[69] Jankele and Svoboda, "TAL Effectors," 412.

[70] Jankele and Svoboda, 412; Moscou and Bogdanove, "A Simple Cipher Governs DNA Recognition by TAL Effectors," 150; Boch and Bonas, "Xanthomonas AvrBs3 Family-Type III Effectors," 429.

[71] Jankele and Svoboda, "TAL Effectors," 412.

[72] Christian et al., "Targeting DNA Double-Strand Breaks with TAL Effector Nucleases."

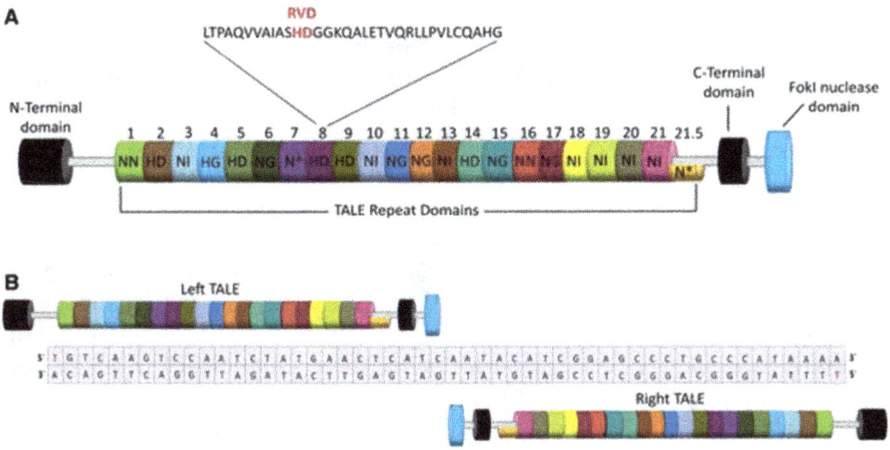

Fig. 2.2 Structural representation of TALEN. (**a**) A TALEN monomer consists of a DNA-binding domain, here TALE, and DNA-cleavage domain, here *Fok*I. The TALE is composed of an N-terminal region (NTR), the central repeat domain (CRD), and the C-terminal region (CTR). The CRD contains a series of direct repeat units that differ in the amino acids located at positions 12 and 13, known as the repeat variable di-residue (RVD). RVDs enable TALEs to distinguish between nucleotides and bind to specific sequences of DNA. The two-letter code for RVDs (e.g., NN, HD, and NI) are depicted above. (**b**) A pair of TALEN monomers bind to target sites located on opposite DNA strands. Dimerization activates the nuclease (i.e., *Fok*I) and allows DNA cleavage to occur (Vinita Periwal, "A Comprehensive Overview of Computational Resources to Aid in Precision Genome Editing with Engineered Nucleases," Briefings in Bioinformatics, 18, no. 4 (July 2017): 3, by permission of Oxford University Press)

DNA-binding domain and DNA-cleavage domain. The DNA-binding domain is composed of TALE proteins, and the DNA-cleavage domain is composed of a nuclease, which is typically *Fok*I. *Fok*I, which makes nonspecific cleavages in DNA, is the same nuclease that researchers fuse to zinc-fingers to create ZFN.[73] A linker of approximately 17–65 amino acids connects the CTR of TALE to *Fok*I.[74] Recall that *Fok*I is only active as a dimer, so two monomers of TALEN must bind to target sites located on opposite DNA strands in order to activate the nuclease (i.e., *Fok*I).[75]

Researchers began to use the natural TALE framework to create TALENs in 2010 and 2011. At first, researchers made minimal changes to TALE and simply swapped the AD in the CTR of TALE for *Fok*I. The success of early experiments that used natural TALE was only "modest" in comparison to later experiments that used truncated TALE. Unlike natural TALE, truncated TALE platforms are shortened on the ends of the protein. TALEN activity increases when the length of the NTR and CTR of truncated TALE is at least 150 amino acids and 20 amino acids,

[73] Chugunova, Dontsova, and Sergiev, "Methods of Genome Engineering," 666.

[74] Jankele and Svoboda, "TAL Effectors," 413.

[75] Mussolino and Cathomen, "TALE Nucleases," 646.

respectively.[76] NTRs that contain 136 amino acids and CTRs that contain 63 amino acids, abbreviated +136/+63, are very common and highly active.[77]

The size of the CTR affects the length of the nucleotide spacer.[78] Recall from the discussion of ZFN that the spacer is the region of DNA that separates the two half-sites between DNA-binding domains, here TALEs. Researchers observed a decrease in cleavage activity of C-terminal truncated TALEN when spacers contain fewer than 10–12 bp.[79] While the length of the spacer varies alongside different TALEN architectures, the range of spacer length generally narrows as the length of the CTR decreases. For instance, untruncated TALEN that contained 288 amino acids in the NTR and 231 amino acids in the CTR had increased cleavage activity for spacers that varied in length from 14 to 33 bp. Similarly, truncated TALEN of the +136/+63 architecture had increased cleavage activity for spacers that ranged in length from 14 to 33 bp. Lastly, truncated TALEN of the +136/+18 architecture had increased cleavage activity for spacers that varied in length from 13 to 16 bp. Importantly, the cleavage activity of +136/+18 TALENs was much higher than that of +136/+63 TALENs for spacers that were 13–15 bp long. In other words, TALENs that contained shorter CTRs were more successful that TALENs that contained longer CTRs in the cleavage of spacers that were the same length.[80]

In order to create functional TALENs, researchers not only need to optimize the NTR and CTR, but they also need to carefully design the CRD. To accomplish this task, researchers consider the DNA target site that TALEN will recognize and bind. As previously mentioned, it is advisable for target sites to start with the nucleotide thymine. The sequence of target sites should be unique—this helps to minimize off-target activity by preventing TALEN from mistakenly binding to sites that are similar in sequence to the target. In general, each TALEN monomer should contain at least 14 direct repeats.[81] Given the high similarity of TALE direct repeat sequences, however, it is technically challenging to construct long arrays.[82] Nevertheless, TALEN monomers are typically designed to recognize the initial thymine and an additional 15–21 bp.[83]

The composition and placement of RVDs within the CRD of TALE is another key factor that influences TALEN activity. At least four RVDs that bind to DNA

---

[76] Mussolino and Cathomen, "TALE Nucleases," 646.

[77] Hiroshi Ochiai and Yamamoto, "Chapter 1: Genome Editing Using Zinc-Finger Nucleases (ZFNs) and Transcription Activator-Like Effector Nucleases (TALENs)," 15.

[78] Hiroshi Ochiai and Yamamoto, 15.

[79] Jeffrey C. Miller et al., "A TALE Nuclease Architecture for Efficient Genome Editing," *Nature Biotechnology* 29, no. 2 (February 2011): 145, https://doi.org/10.1038/nbt.1755.

[80] Michelle L. Christian et al., "Targeting G with TAL Effectors: A Comparison of Activities of TALENs Constructed with NN and NK Repeat Variable Di-Residues," *PLoS ONE* 7, no. 9 (September 24, 2012): 2, 7, https://doi.org/10.1371/journal.pone.0045383.

[81] Jankele and Svoboda, "TAL Effectors," 415–16.

[82] Gaj, Gersbach, and Barbas, "ZFN, TALEN and CRISPR/Cas-Based Methods for Genome Engineering," 399.

[83] Carroll, "Genome Engineering with Targetable Nucleases," 415.

with a high affinity, such as HD, NG, and NN, should be placed evenly along TALE. More than three of the same RVDs should not be placed in a row as this can cause undesirable effects such as deformed folds.[84] In addition to RVDs, which recall make up positions 12 and 13 of direct repeats, variation that occurs in non-RVD positions 4 and 32 is believed to affect the success of TALENs. Although the molecular mechanism is unclear, the success rate as well as the activity of TALENs improve when positions 4 and 32 are varied.[85]

Several variations of Golden Gate assembly are used to construct TALENs. In 2008 and 2009, researchers demonstrated how the Golden Gate cloning strategy was a relatively simple and precise way to generate recombinant plasmids.[86] Recombinant DNA is DNA derived from at least two sources. Typically, DNA fragments are generated by digesting DNA with restriction enzymes, which identify and cleave DNA at certain sequences. Different DNA fragments are then linked together with an enzyme called ligase to form recombinant DNA. Researchers use plasmids as vectors to transfer DNA into cells where DNA is then replicated. Found naturally in bacteria and a few eukaryotes, a plasmid is a circular, double-stranded, extrachromosomal segment of DNA—meaning it exists and replicates independent of the cell's chromosomes. A DNA fragment inserted into a plasmid vector forms recombinant DNA.[87] Golden Gate cloning allows researchers to assemble many DNA fragments in one vector in only one reaction that encompasses both digestion and ligation steps.[88]

Assembly of TALENs based on Golden Gate cloning is regarded as "the easiest method developed to date."[89] In 2011, researchers demonstrated how to use Golden Gate cloning to construct TALENs in 5 days via two steps. First, DNA fragments that encode intermediate direct repeats 1–10 units in length were assembled. Second, the intermediate direct units were linked into a backbone plasmid.[90] Despite its ease of use, Golden Gate cloning "requires considerable hands-on time and quality-control measures" as it takes time to produce and identify properly assembled

[84] Jankele and Svoboda, "TAL Effectors," 416.

[85] Hiroshi Ochiai and Yamamoto, "Chapter 1: Genome Editing Using Zinc-Finger Nucleases (ZFNs) and Transcription Activator-Like Effector Nucleases (TALENs)," 15–16.

[86] Carola Engler, Romy Kandzia, and Sylvestre Marillonnet, "A One Pot, One Step, Precision Cloning Method with High Throughput Capability," *PLoS ONE* 3, no. 11 (November 5, 2008), https://doi.org/10.1371/journal.pone.0003647; Carola Engler et al., "Golden Gate Shuffling: A One-Pot DNA Shuffling Method Based on Type IIs Restriction Enzymes," *PLoS ONE* 4, no. 5 (May 14, 2009), https://doi.org/10.1371/journal.pone.0005553.

[87] Harvey Lodish et al., "DNA Cloning with Plasmid Vectors," 2000, sec. 7.1, accessed June 11, 2017, https://www.ncbi.nlm.nih.gov/books/NBK21498/.

[88] Tomas Cermak et al., "Efficient Design and Assembly of Custom TALEN and Other TAL Effector-Based Constructs for DNA Targeting," *Nucleic Acids Research* 39, no. 12 (July 2011): 2, https://doi.org/10.1093/nar/gkr218.

[89] Hiroshi Ochiai and Yamamoto, "Chapter 1: Genome Editing Using Zinc-Finger Nucleases (ZFNs) and Transcription Activator-Like Effector Nucleases (TALENs)," 16.

[90] Cermak et al., "Efficient Design and Assembly of Custom TALEN and Other TAL Effector-Based Constructs for DNA Targeting," 2.

TALENs.[91] As an alternative to assembly based on cloning, assembly based on polymerase chain reaction (PCR) "requires fewer plasmids and a shorter handling time."[92] Note that PCR is a process used to amplify desired sequences of DNA.[93] Despite the benefits, researchers warn that PCR-based assembly can erroneously induce "mutations or recombined repeats."[94]

In addition to Golden Gate assembly methods, researchers can construct TALENs via restriction enzyme and ligation (REAL) assembly. The REAL method assembles TALENs through a series of ligation steps that fuse together DNA fragments, generated via restriction enzymes, that code for TALE repeats. In the first ligation step, two TALE repeats are linked together. In each ligation step thereafter, the set of TALE repeats from the previous step are ligated together. DNA fragments that code for the NTR and CTR are added to the ends along with the final half repeat unit, and the TALE repeat array is cloned into a plasmid vector.[95] While REAL is acclaimed as the "simplest method" used to construct TALENs, REAL is more time consuming than other approaches due to numerous steps.[96] Nevertheless, a slightly modified approach, known as REAL-Fast, uses pre-assembled plasmids that encode 2–4 direct TALE repeat units, which speeds up the process by eliminating the initial 1–2 ligation steps.[97]

Another method, known as the Fast Ligation-based Automatable Solid-phase High-throughput (FLASH) system, uses the same pre-assembled direct repeats units as REAL-Fast in combination with automation via robotic liquid-handling instruments to accelerate assembly of TALE repeats. With FLASH, TALE repeat units are ligated together in a single direction to an immobilized DNA fragment. FLASH allows researchers to quickly assemble numerous DNA fragments that encode TALE repeat arrays.[98] In addition to FLASH, high-throughput assembly is also achieved

[91] Jonathan L. Schmid-Burgk et al., "A Ligation-Independent Cloning Technique for High-Throughput Assembly of Transcription Activator-like Effector Genes," *Nature Biotechnology* 31, no. 1 (2013): 76, https://doi.org/10.1038/nbt.2460.

[92] Hiroshi Ochiai and Yamamoto, "Chapter 1: Genome Editing Using Zinc-Finger Nucleases (ZFNs) and Transcription Activator-Like Effector Nucleases (TALENs)," 17.

[93] Nelson and Cox, *Lehninger Principles of Biochemistry*, 317.

[94] Cermak et al., "Efficient Design and Assembly of Custom TALEN and Other TAL Effector-Based Constructs for DNA Targeting," 2.

[95] Jeffry D. Sander et al., "Targeted Gene Disruption in Somatic Zebrafish Cells Using Engineered TALENs," *Nature Biotechnology* 29, no. 8 (August 2011): fig. 9 (supplementary).

[96] Hiroshi Ochiai and Yamamoto, "Chapter 1: Genome Editing Using Zinc-Finger Nucleases (ZFNs) and Transcription Activator-Like Effector Nucleases (TALENs)," 18.

[97] Deepak Reyon et al., "Current Protocols in Molecular Biology Engineering Designer Transcription Activator-Like Effector Nucleases (TALENs)," *Current Protocols in Molecular Biology / Edited by Frederick M. Ausubel ... [et Al.]* 0 12 (October 2012): 2, https://doi.org/10.1002/0471142727.mb1215s100; Hiroshi Ochiai and Yamamoto, "Chapter 1: Genome Editing Using Zinc-Finger Nucleases (ZFNs) and Transcription Activator-Like Effector Nucleases (TALENs)," 17–18.

[98] Deepak Reyon et al., "FLASH Assembly of TALENs for High-Throughput Genome Editing," *Nature Biotechnology* 30, no. 5 (2012): 463, https://doi.org/10.1038/nbt.2170; Hiroshi Ochiai and

through ligation-independent cloning (LIC). Researchers boast that LIC can be achieved with "simple pipetting step…and, moreover, it can be practiced with standard laboratory equipment."[99] Besides the methods described above, custom designed TALENs are also available to purchase from companies.[100]

2.4 CRISPR

2.4.1 *CRISPR-Cas System*

To defend against invading mobile genetic elements (MGE), such as bacteriophages and plasmids, many bacteria and archaea are equipped with CRISPR-based molecular defense systems. These adaptive immune systems consist of RNA and protein components. The RNA component is transcribed from genetic loci formed from clustered regularly interspaced short palindromic repeats (CRISPR). The protein component is transcribed and translated from CRISPR-associated (Cas) genes. Together, these components make up CRISPR-Cas systems, which, as mentioned above, protect microbes.[101]

CRISPR-Cas systems provide defense in three stages: adaptation, expression, and interference. First, adaptation occurs when a fragment of foreign nucleic acid, termed protospacer, is captured and incorporated into a new sequence, termed spacer, within CRISPR loci.[102] Spacers are non-identical sequences that interspace, or come between, identical sequences, known as direct repeats. Spacers and direct repeats together make up CRISPR loci. About half of all CRISPR-containing genomes have two or more CRISPR loci.[103] Spacers are approximately 21–72 bp in length, and direct repeats are approximately 21–48 bp in length. A given loci

---

Yamamoto, "Chapter 1: Genome Editing Using Zinc-Finger Nucleases (ZFNs) and Transcription Activator-Like Effector Nucleases (TALENs)," 18.

[99] Schmid-Burgk et al., "A Ligation-Independent Cloning Technique for High-Throughput Assembly of Transcription Activator-like Effector Genes," 81.

[100] Gaj, Gersbach, and Barbas, "ZFN, TALEN and CRISPR/Cas-Based Methods for Genome Engineering," 399.

[101] Philippe Horvath and Rodolphe Barrangou, "CRISPR/Cas, the Immune System of Bacteria and Archaea," *Science* 327, no. 5962 (January 8, 2010): 167, https://doi.org/10.1126/science.1179555; Prarthana Mohanraju et al., "Diverse Evolutionary Roots and Mechanistic Variations of the CRISPR-Cas Systems," *Science (New York, N.Y.)* 353, no. 6299 (August 5, 2016): aad5147-1, https://doi.org/10.1126/science.aad5147; Michael P. Terns and Rebecca M. Terns, "CRISPR-Based Adaptive Immune Systems," *Current Opinion in Microbiology* 14, no. 3 (June 2011): 321, https://doi.org/10.1016/j.mib.2011.03.005.

[102] Mohanraju et al., "Diverse Evolutionary Roots and Mechanistic Variations of the CRISPR-Cas Systems," aad5147-1.

[103] Hélène Deveau, Josiane E. Garneau, and Sylvain Moineau, "CRISPR/Cas System and Its Role in Phage-Bacteria Interactions," *Annual Review of Microbiology* 64, no. 1 (2010): 476–77, https://doi.org/10.1146/annurev.micro.112408.134123.

contains as many as 375 direct repeats interspaced by 374 spacers; however, many loci have fewer than 50 direct repeats interspaced by 49 spacers.[104] A leader sequence approximately 20–534 bp in length is usually found upstream of the first direct repeat. The promoter is situated within the leader sequence. Additionally, *cas* genes are commonly located upstream or downstream of the CRISPR loci.[105]

After the adaptation of spacers, expression of CRISPR loci occurs followed by interference of invading DNA. During expression, CRISPR loci are transcribed into pre-CRISPR RNA (pre-crRNA) and processed into CRISPR RNA (crRNA). crRNA assemble with at least one Cas protein to form a crRNA-Cas effector complex. Finally, during interference, crRNA—which recall includes transcribed spacers—directs the complex to invading nucleic acid sequences (i.e., protospacers) that are complementary to spacers. Upon finding a match and binding to foreign nucleic acid, Cas nucleases degrade invading nucleic acid.[106] The process of CRISPR-Cas-mediated adaptive immunity is illustrated in Fig. 2.3.

A short, 2–7 bp, sequence, known as the protospacer adjacent motif (PAM), is located next to the protospacer and helps the host cell to discriminate between itself (i.e., self) and invading DNA (i.e., non-self). More specifically, PAM is present on foreign DNA, but PAM is not incorporated into the host's CRISPR loci.[107] PAM ensures the host does not mistakenly target spacers that are part of the host's CRISPR loci. Note that PAM plays a role in CRISPR-Cas systems that target double-stranded DNA and is not necessarily present in all types of CRISPR-Cas systems—this is described in more detail below.[108]

Scientists broadly categorize CRISPR-Cas systems into two classes: class 1 and class 2. Class 1 contains types I, III, and IV systems. Class 2 contains types II, V, and VI systems. Classification is based on the Cas protein(s) used during the interference stage: class 1 systems have multiple genes that encode many Cas proteins, whereas class 2 systems usually have one gene that encodes a single Cas protein. Class 2 systems are used for genome editing because of the simplicity of the its crRNA-Cas protein effector complex, which consist of a single Cas protein.[109] Type II, V, and VI systems use Cas9, Cas12, and Cas13 effectors, respectively.[110] While there are several type II sub-types, only a few have been well-characterized: Cas9 from

[104] Deveau, Garneau, and Moineau, 476–77; Horvath and Barrangou, "CRISPR/Cas, the Immune System of Bacteria and Archaea," 167–68.

[105] Deveau, Garneau, and Moineau, "CRISPR/Cas System and Its Role in Phage-Bacteria Interactions," 476–77.

[106] Mohanraju et al., "Diverse Evolutionary Roots and Mechanistic Variations of the CRISPR-Cas Systems," aad5147-1.

[107] Mohanraju et al., aad5147-4.

[108] Karthik Murugan et al., "The Revolution Continues: Newly Discovered Systems Expand the CRISPR-Cas Toolkit," *Molecular Cell* 68, no. 1 (October 5, 2017): 21, https://doi.org/10.1016/j.molcel.2017.09.007.

[109] Mohanraju et al., "Diverse Evolutionary Roots and Mechanistic Variations of the CRISPR-Cas Systems," 556.

[110] Mohanraju et al., aad5147-3; Murugan et al., "The Revolution Continues," 15.

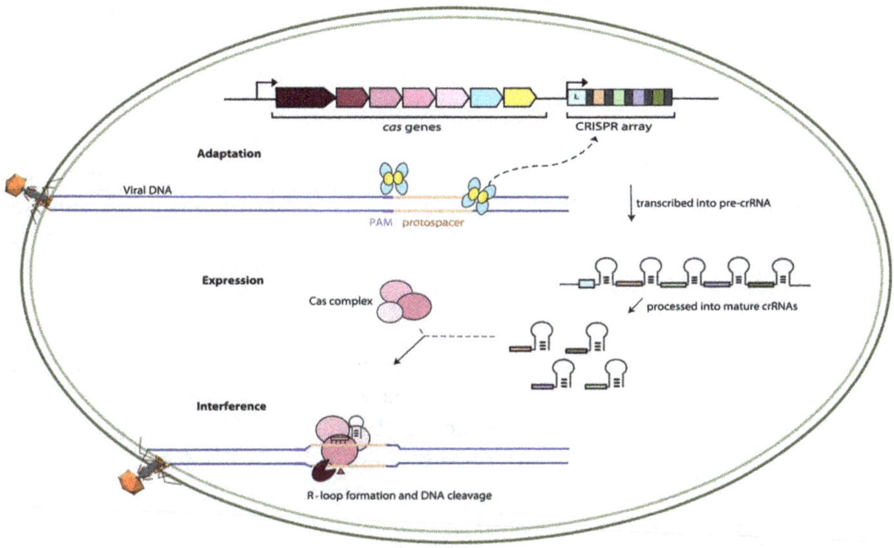

Fig. 2.3 CRISPR-Cas-mediated adaptive immunity. The CRISPR-Cas adaptive immune system provides defense in three stages via adaptation, expression, and interference. During adaptation, foreign nucleic acid, termed protospacer is acquired and incorporated into a new sequence, termed spacer, within CRISPR loci. The Cas1 and Cas2 proteins assist with selecting and processing foreign nucleic acid. The protospacer adjacent motif (PAM) is present on foreign nucleic acid to help discriminate between self and non-self, but PAM is not incorporated into CRIPSR loci. During expression, CRISPR loci are transcribed into pre-CRISPR RNA (pre-crRNA) and processed into CRISPR RNA (crRNA). crRNA assemble with at least one Cas protein to form a crRNA-Cas effector complex. During interference, crRNA directs the complex to invading nucleic acid sequences (i.e., protospacers) that are complementary to spacers. Upon finding a match and binding to foreign nucleic acid, Cas nucleases degrade invading nucleic acid (From Prarthana Mohanraju et al., "Diverse Evolutionary Roots and Mechanistic Variations of the CRISPR-Cas Systems," *Science (New York, N.Y.)* 353, no. 6299 (August 5, 2016): aad5147-2, https://doi.org/10.1126/science.aad5147. Reprinted with permission from AAAS)

Streptococcus pyogenes (abbreviated SpCas9 or SpyCas9), Cas12a (formerly Cpf1), Cas12b (formerly C2c1), and Cas13a (formerly C2c2).[111]

A few notable differences occur among class 2 systems including whether nucleases cleave DNA or RNA and the composition of Cas effector domains. While Cas9 and Cas12 target double-stranded DNA, Cas13 targets single-stranded RNA.[112] Cas9 cleaves DNA via two domains, HNH and RuvC, which respectively cleave the target strand and non-target strand.[113] The HNH domain is defined as, "an endonuclease domain named for characteristic histidine [H] and asparagine

[111]Fuguo Jiang and Jennifer A. Doudna, "CRISPR-Cas9 Structures and Mechanisms," *Annual Review of Biophysics* 46 (2017): 509, 521, 523, https://doi.org/10.1146/annurev-biophys-062215-010822; Murugan et al., "The Revolution Continues," 16–17, 19.

[112]Murugan et al., "The Revolution Continues," 16–17, 18–19.

[113]Murugan et al., 16, 19.

[N] residues."[114] The RuvC domain is defined as, "an endonuclease domain named for an *E. coli* protein involved in DNA repair."[115] Unlike Cas9, Cas12 effectors do not have an HNH domain. Cas12 contains the RuvC domain and another domain known as the nuclease (Nuc) domain. It is believed, however, that RuvC is solely responsible for the cleavage of both DNA strands.[116] Conversely, Cas13 does not have the RuvC and HNH or Nuc domains. Instead, Cas13 have two higher eukaryotes and prokaryotes nucleotide-binding (HEPN) ribonuclease domains that function in RNA cleavage.[117] Note that a ribonuclease, known as RNase, is an enzyme that degrades RNA.[118]

Class 2 systems also differ from each other in terms of the mechanism used to process pre-crRNA. As mentioned above, CRISPR loci are transcribed into pre-CRISPR RNA (pre-crRNA) and processed into target-specific CRISPR RNA (crRNA). In type II and some sub-types of type V systems, *trans*-activating crRNA (tracrRNA) helps in pre-crRNA processing. crRNA and tracrRNA are both needed to activate Cas9, Cas12b, and Cas12e.[119] In these systems, tracrRNA and crRNA form a complex with Cas that is involved in target recognition.[120] However, Cas12a—and possibility Cas12c and 12d—do not use tracrRNA.[121] Alternatively, in type VI systems, Cas13 assists in pre-crRNA processing.[122]

Finally, class II systems differ with respect to PAM. Type II and type V systems are similar in that Cas effectors in both systems locate PAM and then locally unwind DNA, which allows crRNA to base pair with the protospacer.[123] Note that Cas effectors from different species recognize unique PAM sequences. For instance, SpCas9 detects 5′-NGG-3′ PAM (N represents any nitrogenous base: adenine, thymine, cytosine, or guanine), Cas9 from *Staphylococcus aureus* (SaCas9) detects 5′-NNGRRT-3′ PAM (R represents adenine or guanine), and Cas9 from *Neisseria meningiditis* (NmCas9 or NmeCas9) detects 5′-NNNNGATT-3′ PAM.[124] Type VI

---

[114] "Genome Editing Glossary," New England BioLabs, n.d., accessed October 27, 2019, https://www.neb.com/applications/genome-editing/genome-editing-glossary.

[115] "Genome Editing Glossary."

[116] Murugan et al., "The Revolution Continues," 17, 19.

[117] Murugan et al., 19, 21.

[118] "Ribonuclease," Science Direct, accessed March 12, 2018, https://www.sciencedirect.com/topics/biochemistry-genetics-and-molecular-biology/ribonuclease.

[119] Murugan et al., "The Revolution Continues," 16–17.

[120] Mohanraju et al., "Diverse Evolutionary Roots and Mechanistic Variations of the CRISPR-Cas Systems," aad5147-2.

[121] Murugan et al., "The Revolution Continues," 17.

[122] Murugan et al., 19.

[123] Murugan et al., 16, 18.

[124] Jiang and Doudna, "CRISPR-Cas9 Structures and Mechanisms," 524; F Ann Ran et al., "Genome Engineering Using the CRISPR-Cas9 System," *Nature Protocols* 8, no. 11 (November 2013): 2282, https://doi.org/10.1038/nprot.2013.143.

systems, however, use protospacer flanking sites (PFS) in place of PAM to differentially recognize target sequences.[125]

2.4.2 Structure and Mechanism

Just as researchers determined how to use ZFN and TALEN to carry out targeted genome modifications, they also determined how to do the same with CRISPR-Cas systems. There are several pioneers of CRISPR.[126] Two significant studies published in 2012 set the stage for the reprogramming of CRISPR-Cas systems as genome editing tools.[127] In one study, researchers determined the mode of action of Cas9, a CRISPR-associated protein, which, as mentioned above, is an RNA-guided endonuclease that generates double-stranded DNA cleavages via the HNH and RuvC domains.[128] In the other study, researchers also characterized the mode of action of Cas9 and showed how to simplify the CRISPR-Cas system by combining the crRNA:tracr-RNA complex into a single-guide RNA (sgRNA). More specifically, researchers demonstrated how Cas9 could be programmed with sgRNA to cut double-stranded DNA.[129] In 2013, researchers used Cas9 complexed with either crRNA:tracr-RNA or sgRNA to conduct targeted genome editing in mammalian cells—namely in human cells.[130]

As described above, researchers first adapted the type II CRISPR-Cas system, which uses Cas9, for genome editing. In 2014, the type II system derived from *S. pyogenes* was "the most commonly used and extensively characterized system"

[125] Murugan et al., "The Revolution Continues," 21.

[126] Heidi Ledford and Ewen Callaway, "Pioneers of Revolutionary CRISPR Gene Editing Win Chemistry Nobel," *Nature* 586, no. 7829 (October 7, 2020): 346–47, https://doi.org/10.1038/d41586-020-02765-9.

[127] Rodolphe Barrangou and Philippe Horvath, "A Decade of Discovery: CRISPR Functions and Applications," *Nature Microbiology* 2, no. 17092 (June 5, 2017): 3–4, https://doi.org/10.1038/nmicrobiol.2017.92.

[128] Giedrius Gasiunas et al., "Cas9–CrRNA Ribonucleoprotein Complex Mediates Specific DNA Cleavage for Adaptive Immunity in Bacteria," *Proceedings of the National Academy of Sciences of the United States of America* 109, no. 39 (September 25, 2012): E2579–86, https://doi.org/10.1073/pnas.1208507109.

[129] Martin Jinek et al., "A Programmable Dual-RNA-Guided DNA Endonuclease in Adaptive Bacterial Immunity," *Science* 337, no. 6096 (August 17, 2012): 816, 819–20, https://doi.org/10.1126/science.1225829.

[130] Le Cong et al., "Multiplex Genome Engineering Using CRISPR/Cas Systems," *Science* 339, no. 6121 (February 15, 2013): 819–23, https://doi.org/10.1126/science.1231143; Prashant Mali et al., "RNA-Guided Human Genome Engineering via Cas9," *Science* 339, no. 6121 (February 15, 2013): 823–26, https://doi.org/10.1126/science.1232033; Martin Jinek et al., "RNA-Programmed Genome Editing in Human Cells," *ELife* 2 (January 29, 2013): e00471, https://doi.org/10.7554/eLife.00471.

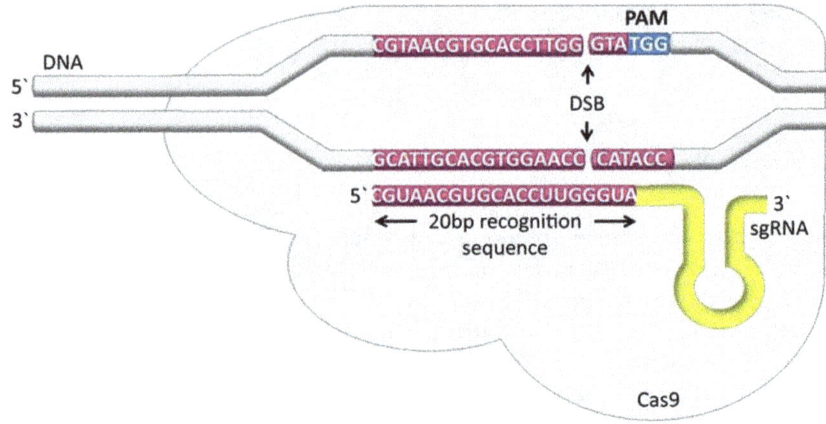

Fig. 2.4 Structural representation of CRISPR-Cas9. CRISPR-Cas9 engineered for genome editing contains a DNA-binding domain, here sgRNA, and DNA-cleavage domain, here Cas9. The 5′ end of sgRNA is designed to recognize a specific target sequence on DNA. The PAM sequence, here 5′-TGG-3′, is situated adjacent to the target sequence. Cas9 induces a double-strand break (DSB) in DNA three base pairs upstream of the PAM sequence (Vinita Periwal, "A Comprehensive Overview of Computational Resources to Aid in Precision Genome Editing with Engineered Nucleases," Briefings in Bioinformatics, 18, no. 4 (July 2017): 4, by permission of Oxford University Press)

for genome editing.[131] The tool is composed of the Cas9 nuclease, in this case SpCas9, and guide RNA (gRNA). gRNA is either a duplexed crRNA:tracrRNA or, more commonly, a fusion of the two RNA strands—known as single-guide RNA (sgRNA). Researchers design the first 20 nucleotides of sgRNA (i.e., those on the 5′ end) so that this region is complementary to the target sequence. Note that the custom designed sequence on sgRNA is akin to the protospacer on crRNA. A structural representation of CRISPR-Cas9 is illustrated in Fig. 2.4.

PAM requirements restrict the sites that researchers can target. Since SpCas9 recognizes a 5′-NGG-3′ PAM sequence, where N is any nitrogenous base, target sites need to be situated upstream of this specific PAM sequence. Note that SpCas9 recognizes other PAM sequences, such as 5′-NAG-3′ PAM, but with reduced efficiency.[132] The 5′-NGG-3′ and 5′-NAG-3′ motifs occur approximately every 8 bp and 4 bp, respectively. Thus, eligible CRISPR-Cas9 target sequences occur every 8 bp to 4 bp depending on the PAM sequence.[133] The target constraint posed by PAM, however, is alleviated as more Cas nucleases that recognize a greater range

[131]Jeffry D. Sander and J. Keith Joung, "CRISPR-Cas Systems for Editing, Regulating and Targeting Genomes," *Nature Biotechnology* 32, no. 4 (April 2014): 348, https://doi.org/10.1038/nbt.2842.

[132]Sander and Joung, "CRISPR-Cas Systems for Editing, Regulating and Targeting Genomes," 348.

[133]Chugunova, Dontsova, and Sergiev, "Methods of Genome Engineering," 668, 670; Kim and Kim, "A Guide to Genome Engineering with Programmable Nucleases," 327.

of PAM are engineered or discovered. There are several variants of SpCas9 that recognize different PAM sequences: D1135E, VQR, EQR, and VRER variants of SpCas9 respectively recognize 5′-NGG-3′ PAM, 5′-NGA-3′ PAM, 5′-NGAG-3′ PAM, and 5′-NGCG-3′ PAM sequences.[134] In addition to SpCas9 derived from *S. pyogenes*, numerous Cas9 equivalents from other bacterial species—such as the aforementioned SaCas9 and NmCas9—have been characterized.[135]

There are also Cas nucleases from other CRISPR-Cas systems, specifically those from types V and VI, that are used for genetic engineering. In 2015, researchers described how Cas12a, which is part of the type V CRISPR-Cas system, recognizes PAM sequences abundant in thymines. For instance, *Francisella novicida* (FnCas12a, formerly FnCpf1) recognizes a 5′-TTN-3′ PAM sequence.[136] To activate Cas12a in nature, tracrRNA is not needed—only a crRNA about 44 nucleotides long is required.[137] As opposed to genome editing with Cas9, which uses an sgRNA about 100 nucleotides long, researchers can program a shorter, and therefore less expensive, gRNA to use with Cas12a.[138] Genome editing with Cas12a and its gRNA could overcome challenges associated with using an adeno-associated viral vector (AAV), which has a packaging limit of approximately 4.5 kb, to deliver SpCas9 and its gRNA, which are approximately 4.2 kb.[139] Since Cas12a and its gRNA are both relatively small, they are an "attractive" option to use with viral delivery systems.[140] Lastly, unlike Cas9 that generates blunt ends during DNA cleavage, Cas12a generates sticky ends on DNA that enable the directional insertion of genes.[141]

The PAM requirement is further alleviated by genome editing with Cas13, which is part of the type VI CRISPR-Cas system. Recall that type VI systems do not use PAM. Furthermore, unlike Cas9 and Cas12 that target and cleave DNA, Cas13 targets and cleaves RNA. While Cas9 can be programmed to target RNA, an additional segment of DNA that is complementary to the PAM sequence is needed. Since Cas13 only needs crRNA that is complementary to the target RNA, it is

[134] Benjamin P. Kleinstiver et al., "Engineered CRISPR-Cas9 Nucleases with Altered PAM Specificities," *Nature* 523, no. 7561 (July 23, 2015): 481–83, https://doi.org/10.1038/nature14592; Xixun Hu et al., "Increasing the Efficiency of CRISPR-Cas9-VQR Precise Genome Editing in Rice," *Plant Biotechnology Journal* 16, no. 1 (January 1, 2018): 292, https://doi.org/10.1111/pbi.12771.

[135] Jiang and Doudna, "CRISPR-Cas9 Structures and Mechanisms," 521–24.

[136] Bernd Zetsche et al., "Cpf1 Is a Single RNA-Guided Endonuclease of a Class 2 CRISPR-Cas System," *Cell* 163, no. 3 (October 22, 2015): 760, https://doi.org/10.1016/j.cell.2015.09.038.

[137] Murugan et al., "The Revolution Continues," 17.

[138] Murugan et al., 22; Mary Gearing, "Chapter 2 – What Is CRISPR?: Cpf1: A Cas9 Homolog," in *CRISPR 101: A Desktop Resource*, ed. AddGene, 2nd ed., 2015, 28.

[139] Gearing, "Chapter 2 – What Is CRISPR?: Cpf1: A Cas9 Homolog," 28; Mary Gearing, "Chapter 4 – CRISPR Expression and Delivery Methods: Using SaCAS9 for Easier AAV-Based CRISPR Delivery," in *CRISPR 101: A Desktop Resource*, ed. AddGene, 2nd ed., 2015, 152.

[140] Murugan et al., "The Revolution Continues," 22.

[141] Mohanraju et al., "Diverse Evolutionary Roots and Mechanistic Variations of the CRISPR-Cas Systems," aad5147-9; Murugan et al., "The Revolution Continues," 22.

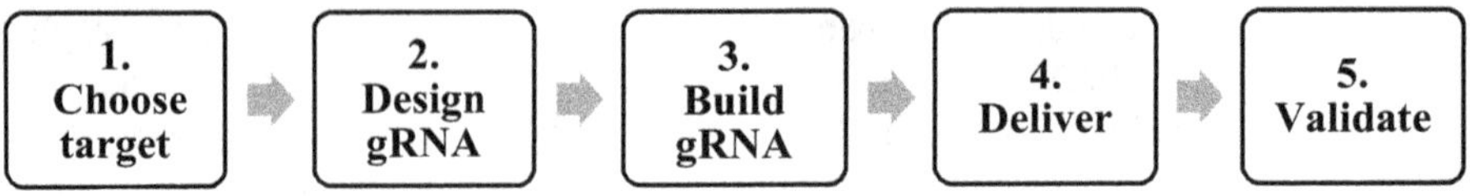

Fig. 2.5 General CRISPR experimental workflow

regarded as a "simpler system" to use for RNA-based genetic engineering applications.[142]

Scientists continue to devise ways to increase the specificity and decrease off-target effects of genome editing with CRISPR-based systems by making slight modifications to the Cas nuclease. As described at the end of the section on the composition and construction of ZFN, researchers can use nickases to induce single-stranded cuts. Cas9 can be converted into a nickase by mutating either the HNH or RuvC, which are responsible for cleavage. When both the HNH and RuvC domains are mutated, it results in a catalytically inactive (i.e., dead) Cas9, termed dCas9. dCas9 can be used during transcriptional regulation to either activate (CRISPRa) or interfere (CRISPRi) with gene expression.[143] Additionally, dCas9 fused with *Fok*I (termed fCas9) in complex with sgRNA (FokI-dCas9:sgRNA or fCas9:sgRNA) has approximately 140-fold greater specificity over wild-type Cas9. Furthermore, two monomers of fCas9 provided "at least fourfold" greater specificity over paired nickases.[144]

2.4.3 CRISPR Workflow

The CRISPR workflow includes the following steps: choose target, design gRNA, synthesize gRNA, deliver CRISPR components (i.e., Cas effector and gRNA), and validate. The CRISPR workflow, as depicted in Fig. 2.5, is based on journal articles and guidance documents from CRISPR suppliers as well as a working paper that was prepared in support of the 2018 *Editing Biosecurity* report.[145] To begin an

---

[142] Murugan et al., "The Revolution Continues," 23.

[143] Barrangou and Horvath, "A Decade of Discovery," 6.

[144] John P. Guilinger, David B. Thompson, and David R. Liu, "Fusion of Catalytically Inactive Cas9 to FokI Nuclease Improves the Specificity of Genome Modification," *Nature Biotechnology* 32, no. 6 (June 2014): 577, https://doi.org/10.1038/nbt.2909.

[145] Joel McDade, "Chapter 3 – Using CRISPR in Your Experiments: Planning Your CRISPR Experiment," in *CRISPR 101: A Desktop Resource*, ed. AddGene, 2nd ed., 2015, 36–42; Genewiz, "CRISPR Workflow," accessed March 26, 2018, https://web.genewiz.com/hubfs/CRISPR_Workflow.pdf; F. Ann Ran et al., "Genome Engineering Using the CRISPR-Cas9 System," *Nature Protocols* 8, no. 11 (November 2013): 2281–2308, https://doi.org/10.1038/nprot.2013.143; Rongxue Peng, Guigao Lin, and Jinming Li, "Potential Pitfalls of CRISPR/Cas9-Mediated Genome Editing," *FEBS Journal* 283, no. 7 (April 1, 2016): 1222, https://doi.org/10.1111/febs.13586; "CRISPR/Cas9 Genome Editing Tools," Takara Bio USA, accessed January 30, 2018, http://www.

experiment, researchers select an organism and decide on the genetic manipulation they want to achieve. CRISPR can be used to target DNA or RNA and to achieve several outcomes such as precision editing, temporary activation or repression of gene expression, and permanent gene knockout.[146] Based on this information, researchers identify the target genetic element that they need to modify to achieve the experimental objective.

After researchers choose the target, they design gRNA complementary to the target site. Computational tools, such as CRISPRdirect, are available to guide the design process. Next, researchers generate gRNAs. gRNAs can be made "as part of a plasmid or vector via cloning; through a process known as *in vitro* transcription (IVT) or chemically synthesized as RNA oligonucleotides."[147] Researchers select the best format based on several factors such as the amount of labor required, off-target effects, efficiency, and consistency.[148]

Once gRNA is designed and synthesized, CRISPR components, including the Cas effector and gRNA, are delivered into target cells. The method of delivery depends on the experimental system. Examples of delivery methods include lipid transfection, electroporation, and microinjection.[149] The Cas effector can be delivered in several different formats, including as DNA (e.g., plasmid DNA), RNA, and protein. Note that efficiency, cost, and specificity vary among delivery formats.[150] Finally, it is important to confirm the presence of the genetic modification and figure out the efficiency of CRISPR events. Researchers often use genetic analysis to validate experimental results.[151]

clontech.com/US/Products/Genome_Editing/CRISPR_Cas9/Resources/CRISPR_Cas9_Workflow; Mirus Bio LLC, "CRISPR/Cas9 Genome Editing Brochure," 2017, accessed February 18, 2018, https://www.mirusbio.com/assets/web_documents/crispr-cas9-genome-editing-brochure-pdna-mrna-cas9-grna-rnp.pdf; Synthego, "How to Perform Successful CRISPR Experiments: A Step-by-Step Guide," December 15, 2016, accessed February 18, 2018, https://www.biolegio.com/media/4148/how-to-perform-successful-crispr-experiments.pdf; Cornelia Hampe, "CRISPR/Cas9 Gene Editing Discover How to Speed up Your Workflow with Guide-ItTM!," http://separations.co.za/wp-content/uploads/2016/07/Takara-Clontech-CRISPR-Cas9-Gene-Editing-in-5-easy-steps.pdf; Perello, *CRISPR Genome Editing: A Technical & Policy Primer*, 6.

[146] McDade, "Chapter 3 – Using CRISPR in Your Experiments: Planning Your CRISPR Experiment," 37–38.

[147] Synthego, "How to Perform Successful CRISPR Experiments: A Step-by-Step Guide," 9.

[148] Synthego, 9–11.

[149] Synthego, 13.

[150] Mirus Bio LLC, "CRISPR/Cas9 Genome Editing Brochure," 4.

[151] Synthego, "How to Perform Successful CRISPR Experiments: A Step-by-Step Guide," 16.

Table 2.1 Summary of ZFN, TALEN, and CRISPR

	ZFN	TALEN	CRISPR
Components	Zinc-fingers/*Fok*I	TALE/*Fok*I	gRNA/Cas variants
Design and build of constructs	Difficult	Difficult to moderate	Easy
Multiplexibility	Difficult	Difficult	Possible
Targeting	Nearly unlimited	Some constraints	Some constraints
Efficiency of modifications	Low	High	High

2.5 Comparison of Technologies

2.5.1 ZFN, TALEN, and CRISPR

An ideal genome editing technology is one that is easy to design and use. Furthermore, an ideal genome editing technology should be able to recognize the desired portion of the genome that the scientist wants to modify (i.e., it is precise), should reliably make the desired modification (i.e., it is efficient), and should not mistakenly make modifications elsewhere (i.e., it is specific). While there are benefits and trade-off for each of the genome editing technologies considered in this assessment, CRISPR is regarded as the overall "simplest and most convenient in use."[152] This section highlights the similarities and differences of ZFN, TALEN, and CRISPR (Table 2.1).

ZFN is described as "difficult to use" because scientists need to design and build unique protein constructs custom-made to recognize each new target DNA sequence. Although ZFN emerged as a technology over 20 years ago, which gave it time to be "developed and improved," it is still considered a "time-consuming process."[153] Not only is this process hard and time-consuming, but also genome editing with ZFN is finicky and is not consistently successful.[154] In relation to ZFN, scientists consider TALEN constructs easier to customize for each new target DNA sequence. Nevertheless, TALENs are physically big constructs, which makes them more difficult to handle and harder to insert into target cells.[155] Moreover, the sequence of the DNA binding domain of TALENs is highly repetitive, which increases the technical challenge when cloning.[156]

Experts maintain that a major benefit of the CRISPR system is the ease with which the genetically engineered construct is generated. In contrast to ZFNs and TALENs, the CRISPR system uses RNA instead of proteins to recognize target

[152] Chugunova, Dontsova, and Sergiev, "Methods of Genome Engineering," 670.

[153] Chugunova, Dontsova, and Sergiev, "Methods of Genome Engineering," 666, 671.

[154] Susan Young, "Genome Surgery," *MIT Technology Review*, April 2014, 57.

[155] Young, 57.

[156] Gaj, Gersbach, and Barbas, "ZFN, TALEN and CRISPR/Cas-Based Methods for Genome Engineering," 399.

DNA sequences. RNA is known to "be made routinely in the lab," and the RNA required for the CRISPR system can be "quickly synthesize[d]" by a laboratory technician.[157] While two constructs are needed for ZFNs and TALENs to cleave double-stranded DNA, only one construct of the CRISPR system is needed to achieve the same outcome. Accordingly, several ZFN and TALEN monomers—each pair programmed to dimerize at a different target site—need to simultaneously be introduced into the cell to edit multiple target sites at once. This is challenging and results in mismatched nuclease dimers. Since CRISPR does not dimerize, it is easier to make many changes in a single reaction. The efficiency of multiple modifications is high for CRISPR-Cas9 and low for both TALEN and ZFN.[158]

An aspect of construct generation that is difficult for programmable nucleases is the design of the DNA binding domain. ZFN is limited by the frequency of eligible target sites. The frequency of ZFN target sites varies in the literature from approximately one every 100 bp or 500 bp.[159] Moreover, target sites should be rich in guanines and composed of triplets containing guanine followed by any two nucleotides (abbreviated 5′-GNN-3′).[160] When scholars compare ZFNs to TALENs, they conclude the "targetability of TALENs will prove superior to the public ZFN platforms."[161] Depending on the source, candidate TALEN target sites occur approximately one per base pair or every 35 base pairs.[162] Eligible TALEN target sites occur more frequently than ZFN target sites. This is the case because RVDs allow TALENs to distinguish between nucleotides and bind to specific sequences of DNA. CRISPR also has target site limitations—namely those imposed by PAM. This targeting constraint of CRISPR, however, is alleviated as more variants of Cas effectors that recognize a broader range of PAM sequences are engineered or discovered.

After the initial wave of CRISPR optimism, researchers advised "this technology was not as easy as once assumed."[163] A key disadvantage of the CRISPR system, which also results from genome editing with ZFNs and TALENs, is the generation of unintended off-target mutations.[164] Off-target effects, or off-target activity, are caused when nucleases cut sequences of nucleic acid that resemble the sequence targeted by the nuclease. Repair of these breaks is often error-prone and results in small insertions and/or deletions that cause mutation or inactivation of genes. The

157 Young, "Genome Surgery," 57.

158 Chugunova, Dontsova, and Sergiev, 668–669.

159 Mussolino and Cathomen, "TALE Nucleases," 646; Kim and Kim, "A Guide to Genome Engineering with Programmable Nucleases," 325.

160 Chugunova, Dontsova, and Sergiev, "Methods of Genome Engineering," 670.

161 Cermak et al., "Efficient Design and Assembly of Custom TALEN and Other TAL Effector-Based Constructs for DNA Targeting," 10.

162 Cermak et al., 1, 10; Kim and Kim, "A Guide to Genome Engineering with Programmable Nucleases," 330.

163 Peng, Lin, and Li, "Potential Pitfalls of CRISPR/Cas9-Mediated Genome Editing," 1221.

164 Chugunova, Dontsova, and Sergiev, "Methods of Genome Engineering," 668.

likelihood of gross chromosomal rearrangements, such as deletions, inversions, and translocations, increase with multiple cleavages.[165] Unintended cuts in exonic or regulatory DNA sequences are especially problematic and can interfere with gene expression and cellular function.[166] Nevertheless, as described in the earlier sections of this chapter, researchers continue to develop methods to mitigate off-target mutations.

2.5.2 *Other Technologies*

Targeted genome-modification technologies, such as those that use recombinases and recombineering, have high potential to make large-scale modifications or many small-scale modifications to genomes; however, they are limited by some or all of the following: targeting efficiency, programmability, cost, and use across different organisms.[167] Recombinases are described as "DNA-binding enzymes that catalyze highly specific and efficient DNA splicing reactions between two sites."[168] Examples include λ integrase, φC31 integrase, Cre, Flp, and recombinase-mediated cassette exchange (RMCE). Typically, recombinases have limited programmability because they induce site-specific recombination only at certain recognition sites. When these target sites are not present, researchers need to introduce the sites—this process is considered "too laborious" for many laboratories.[169]

Recombination-mediated genetic engineering, known as recombineering, is a targeted genome-modification tool, based on homologous recombination, used for insertions, deletions, or other modifications.[170] Recombineering drives homologous recombination, independent of restriction enzyme sites, by using machinery derived from bacteriophage. Note that restriction enzymes limit programmability as cuts can only be made at restriction enzyme sites. Recombineering examples include λ-Red, RecET, Redβ, and RecT. Recombineering is highly programmable; however, there are issues with efficiency. Nevertheless, efficiency generally increases when recombineering with single-stranded DNA substrates instead of double-stranded

[165] Jiing-Kuan Yee, "Off-Target Effects of Engineered Nucleases," *The FEBS Journal* 283, no. 17 (September 2016): 3241, https://doi.org/10.1111/febs.13760.

[166] Ciaran M Lee et al., "Nuclease Target Site Selection for Maximizing On-Target Activity and Minimizing Off-Target Effects in Genome Editing," *Molecular Therapy* 24, no. 3 (March 2016): 483, https://doi.org/10.1038/mt.2016.1.

[167] Esvelt and Wang, "Genome-scale Engineering for Systems and Synthetic Biology," 4.

[168] Esvelt and Wang, 4.

[169] Esvelt and Wang, 5.

[170] Shyam K. Sharan et al., "Recombineering: A Homologous Recombination-Based Method of Genetic Engineering," *Nature Protocols* 4, no. 2 (2009): 206, https://doi.org/10.1038/nprot.2008.227.

DNA substrates.[171] Even so, recombineering is designed to work in bacteria, so research is still needed to expand use of the technology to more organisms.[172]

In addition to the technologies mentioned above, it is also important to describe another technology—RNA interference (RNAi)—because of its prominent use as a tool to learn about gene function. To study gene function, researchers often reduce or knockout gene expression and study the downstream effect. While homologous recombination has been used to target genes for inactivation, this method is described as having "low efficiency," "time-consuming and labor-insensitive selection/screening strategies," and "adverse mutagenic effects."[173] When RNAi was discovered, it was acclaimed as a "magic bullet" that could "target any gene."[174] It should be noted that RNAi is based on RNA, which made it very flexible and easy to use. Although RNAi is rapid and inexpensive, it suffers from issues, such as incomplete knockdown and off-target effects, which limit its application.[175] ZFN, TALEN, and, more recently, CRISPR, are recognized as powerful alternatives to RNAi.[176]

2.6 Conclusion

It is important to understand how genome editing technologies work in order to assess their potential for misuse. Scientists determined how to harness the zinc-finger, TALE, and CRISPR-Cas system from nature and utilize them for genome editing in the laboratory. The overview of the structures and mechanisms of genome editing technologies that were presented in this chapter elucidates why scientists often prefer to use CRISPR over ZFN and TALEN. While there are pros and cons to using each of the technologies, the benefits of CRISPR generally outweigh the drawbacks. The qualities that make CRISPR desirable to use, however, are the same qualities that potentially could be exploited by nefarious actors to develop biological weapons. Overall, the fundamental technical information on ZFN,

[171]Esvelt and Wang, "Genome-scale Engineering for Systems and Synthetic Biology," 4, 6; Beth Kenkel, "Lambda Red: A Homologous Recombination-Based Technique for Genetic Engineering," Addgene, December 15, 2016, http://blog.addgene.org/lambda-red-a-homologous-recombination-based-technique-for-genetic-engineering.

[172]Esvelt and Wang, "Genome-scale Engineering for Systems and Synthetic Biology," 6.

[173]Gaj, Gersbach, and Barbas, "ZFN, TALEN and CRISPR/Cas-Based Methods for Genome Engineering," 397.

[174]Michael Boettcher and Michael T. McManus, "Choosing the Right Tool for the Job: RNAi, TALEN, or CRISPR," *Molecular Cell* 58, no. 4 (May 21, 2015): 575, https://doi.org/10.1016/j.molcel.2015.04.028.

[175]Gaj, Gersbach, and Barbas, "ZFN, TALEN and CRISPR/Cas-Based Methods for Genome Engineering," 397.

[176]Boettcher and McManus, "Choosing the Right Tool for the Job"; Gaj, Gersbach, and Barbas, "ZFN, TALEN and CRISPR/Cas-Based Methods for Genome Engineering."

TALEN, and CRISPR that was included in this chapter provides a foundation for the risk of misuse assessment of genome editing technologies.

As demonstrated in the chapter, a lot of progress has been made in genetic engineering since the 1970s. Undoubtedly, concerns that genetic engineering could be misused to develop biological weapons have emerged alongside these advancements. Chapter 3 contains a review of the longstanding debate over the threat posed by genetic engineering. This includes an outline of the types of attributes of agents that nefarious actors may want to target for modification as well as the obstacles that would make it difficult to apply genetic engineering to develop biological weapons. Additionally, Chap. 3 goes into the more current fears that have emerged over the potential misuse of ZFN, TALEN, and CRISPR.

References

1. "Advances in Biotechnology and Genetic Engineering: Implications for the Development of New Biological Warfare Agents." Washington, DC: Department of Defense, June 1996.
2. Barrangou, Rodolphe, and Philippe Horvath. "A Decade of Discovery: CRISPR Functions and Applications." *Nature Microbiology* 2, no. 17092 (June 5, 2017). https://doi.org/10.1038/nmicrobiol.2017.92.
3. Bath, Abigail J., Susan E. Milsom, Niall A. Gormley, and Stephen E. Halford. "Many Type IIs Restriction Endonucleases Interact with Two Recognition Sites before Cleaving DNA." *Journal of Biological Chemistry* 277, no. 6 (February 8, 2002): 4024–33. https://doi.org/10.1074/jbc.M108441200.
4. Boch, Jens, and Ulla Bonas. "Xanthomonas AvrBs3 Family-Type III Effectors: Discovery and Function." *Annual Review of Phytopathology* 48, no. 1 (2010): 419–36. https://doi.org/10.1146/annurev-phyto-080508-081936.
5. Boch, Jens, Heidi Scholze, Sebastian Schornack, Angelika Landgraf, Simone Hahn, Sabine Kay, Thomas Lahaye, Anja Nickstadt, and Ulla Bonas. "Breaking the Code of DNA Binding Specificity of TAL-Type III Effectors." *Science* 326, no. 5959 (December 11, 2009): 1509–12. https://doi.org/10.1126/science.1178811.
6. Boettcher, Michael, and Michael T. McManus. "Choosing the Right Tool for the Job: RNAi, TALEN, or CRISPR." *Molecular Cell* 58, no. 4 (May 21, 2015): 575–85. https://doi.org/10.1016/j.molcel.2015.04.028.
7. Boglioli, Elsy, and Magali Richard. "Rewriting The Book Of Life: A New Era in Precision Gene Editing." The Boston Consulting Group, September 2015.
8. Carr, Peter A., and George M. Church. "Genome Engineering." *Nature Biotechnology* 27, no. 12 (December 2009): 1151–62. https://doi.org/10.1038/nbt.1590.
9. Carroll, Dana. "Genome Engineering with Targetable Nucleases." *Annual Review of Biochemistry* 83, no. 1 (2014): 409–39. https://doi.org/10.1146/annurev-biochem-060713-035418.
10. Cathomen, Toni, and J Keith Joung. "Zinc-Finger Nucleases: The Next Generation Emerges." *Molecular Therapy* 16, no. 7 (July 2008): 1200–1207. https://doi.org/10.1038/mt.2008.114.
11. Cermak, Tomas, Erin L. Doyle, Michelle Christian, Li Wang, Yong Zhang, Clarice Schmidt, Joshua A. Baller, Nikunj V. Somia, Adam J. Bogdanove, and Daniel F. Voytas. "Efficient Design and Assembly of Custom TALEN and Other TAL Effector-Based Constructs for DNA Targeting." *Nucleic Acids Research* 39, no. 12 (July 2011): e82. https://doi.org/10.1093/nar/gkr218.
12. Christian, Michelle, Tomas Cermak, Erin L. Doyle, Clarice Schmidt, Feng Zhang, Aaron Hummel, Adam J. Bogdanove, and Daniel F. Voytas. "Targeting DNA Double-Strand Breaks

with TAL Effector Nucleases." *Genetics* 186, no. 2 (October 1, 2010): 757–61. https://doi.org/10.1534/genetics.110.120717.
13. Christian, Michelle L., Zachary L. Demorest, Colby G. Starker, Mark J. Osborn, Michael D. Nyquist, Yong Zhang, Daniel F. Carlson, Philip Bradley, Adam J. Bogdanove, and Daniel F. Voytas. "Targeting G with TAL Effectors: A Comparison of Activities of TALENs Constructed with NN and NK Repeat Variable Di-Residues." *PLoS ONE* 7, no. 9 (September 24, 2012). https://doi.org/10.1371/journal.pone.0045383.
14. Chugunova, A. A., O. A. Dontsova, and P. V. Sergiev. "Methods of Genome Engineering: A New Era of Molecular Biology." *Biochemistry (Moscow)* 81, no. 7 (July 1, 2016): 662–77. https://doi.org/10.1134/S0006297916070038.
15. Cohen, Stanley N., Annie C. Y. Chang, Herbert W. Boyer, and Robert B. Helling. "Construction of Biologically Functional Bacterial Plasmids In Vitro." *Proceedings of the National Academy of Sciences of the United States of America* 70, no. 11 (November 1973): 3240–44.
16. Cong, Le, F. Ann Ran, David Cox, Shuailiang Lin, Robert Barretto, Naomi Habib, Patrick D. Hsu, et al. "Multiplex Genome Engineering Using CRISPR/Cas Systems." *Science* 339, no. 6121 (February 15, 2013): 819–23. https://doi.org/10.1126/science.1231143.
17. Takara Bio USA. "CRISPR/Cas9 Genome Editing Tools." Accessed January 30, 2018. http://www.clontech.com/US/Products/Genome_Editing/CRISPR_Cas9/Resources/CRISPR_Cas9_Workflow.
18. Cuculis, Luke, Zhanar Abil, Huimin Zhao, and Charles M. Schroeder. "Direct Observation of TALE Protein Dynamics Reveals a Two-State Search Mechanism." *Nature Communications* 6 (June 1, 2015): 7277. https://doi.org/10.1038/ncomms8277.
19. Deveau, Hélène, Josiane E. Garneau, and Sylvain Moineau. "CRISPR/Cas System and Its Role in Phage-Bacteria Interactions." *Annual Review of Microbiology* 64, no. 1 (2010): 475–93. https://doi.org/10.1146/annurev.micro.112408.134123.
20. Engler, Carola, Ramona Gruetzner, Romy Kandzia, and Sylvestre Marillonnet. "Golden Gate Shuffling: A One-Pot DNA Shuffling Method Based on Type IIs Restriction Enzymes." *PLoS ONE* 4, no. 5 (May 14, 2009). https://doi.org/10.1371/journal.pone.0005553.
21. Engler, Carola, Romy Kandzia, and Sylvestre Marillonnet. "A One Pot, One Step, Precision Cloning Method with High Throughput Capability." *PLoS ONE* 3, no. 11 (November 5, 2008). https://doi.org/10.1371/journal.pone.0003647.
22. Esvelt, Kevin M., and Harris H. Wang. "Genome-scale Engineering for Systems and Synthetic Biology." *Molecular Systems Biology* 9, no. 641 (2013): 17. https://doi.org/10.1038/msb.2012.66.
23. Gaj, Thomas, Charles A. Gersbach, and Carlos F. Barbas. "ZFN, TALEN and CRISPR/Cas-Based Methods for Genome Engineering." *Trends in Biotechnology* 31, no. 7 (July 2013): 397–405. https://doi.org/10.1016/j.tibtech.2013.04.004.
24. Gasiunas, Giedrius, Rodolphe Barrangou, Philippe Horvath, and Virginijus Siksnys. "Cas9–CrRNA Ribonucleoprotein Complex Mediates Specific DNA Cleavage for Adaptive Immunity in Bacteria." *Proceedings of the National Academy of Sciences of the United States of America* 109, no. 39 (September 25, 2012): E2579–86. https://doi.org/10.1073/pnas.1208507109.
25. Gearing, Mary. "Chapter 2 – What Is CRISPR?: Cpf1: A Cas9 Homolog." In *CRISPR 101: A Desktop Resource*, edited by AddGene, 2nd ed., 27–28, 2015.
26. ———. "Chapter 4 – CRISPR Expression and Delivery Methods: Using SaCAS9 for Easier AAV-Based CRISPR Delivery." In *CRISPR 101: A Desktop Resource*, edited by AddGene, 2nd ed., 152–54, 2015.
27. "Genetic Engineering." Accessed August 7, 2017. https://www.nature.com/subjects/genetic-engineering.
28. Genewiz. "CRISPR Workflow." Accessed March 26, 2018. https://web.genewiz.com/hubfs/CRISPR_Workflow.pdf.
29. New England BioLabs. "Genome Editing Glossary," n.d. https://www.neb.com/applications/genome-editing/genome-editing-glossary.

30. Guilinger, John P., David B. Thompson, and David R. Liu. "Fusion of Catalytically Inactive Cas9 to FokI Nuclease Improves the Specificity of Genome Modification." *Nature Biotechnology* 32, no. 6 (June 2014): 577–82. https://doi.org/10.1038/nbt.2909.
31. Hampe, Cornelia. "CRISPR/Cas9 Gene Editing Discover How to Speed up Your Workflow with Guide-ItTM!" Presented at the Takara Clontech, 2016. http://separations.co.za/wp-content/uploads/2016/07/Takara-Clontech-CRISPR-Cas9-Gene-Editing-in-5-easy-steps.pdf.
32. Händel, Eva-Maria, Stephen Alwin, and Toni Cathomen. "Expanding or Restricting the Target Site Repertoire of Zinc-Finger Nucleases: The Inter-Domain Linker as a Major Determinant of Target Site Selectivity." *Molecular Therapy: The Journal of the American Society of Gene Therapy* 17, no. 1 (January 2009): 104–11. https://doi.org/10.1038/mt.2008.233.
33. Hiroshi Ochiai, and Takashi Yamamoto. "Chapter 1: Genome Editing Using Zinc-Finger Nucleases (ZFNs) and Transcription Activator-Like Effector Nucleases (TALENs)." In *Targeted Genome Editing Using Site-Specific Nucleases: ZFNs, TALENs, and the CRISPR/Cas9 System*, edited by Takashi Yamamoto, 3–24. Springer, 2015.
34. Horvath, Philippe, and Rodolphe Barrangou. "CRISPR/Cas, the Immune System of Bacteria and Archaea." *Science* 327, no. 5962 (January 8, 2010): 167–70. https://doi.org/10.1126/science.1179555.
35. Hsu, Patrick D., Eric S. Lander, and Feng Zhang. "Development and Applications of CRISPR-Cas9 for Genome Engineering." *Cell* 157, no. 6 (June 5, 2014): 1262–78. https://doi.org/10.1016/j.cell.2014.05.010.
36. Hu, Xixun, Xiangbing Meng, Qing Liu, Jiayang Li, and Kejian Wang. "Increasing the Efficiency of CRISPR-Cas9-VQR Precise Genome Editing in Rice." *Plant Biotechnology Journal* 16, no. 1 (January 1, 2018): 292–97. https://doi.org/10.1111/pbi.12771.
37. Jankele, Radek, and Petr Svoboda. "TAL Effectors: Tools for DNA Targeting." *Briefings in Functional Genomics* 13, no. 5 (September 1, 2014): 409–19. https://doi.org/10.1093/bfgp/elu013.
38. Jiang, Fuguo, and Jennifer A. Doudna. "CRISPR-Cas9 Structures and Mechanisms." *Annual Review of Biophysics* 46 (2017): 505–29. https://doi.org/10.1146/annurev-biophys-062215-010822.
39. Jinek, Martin, Krzysztof Chylinski, Ines Fonfara, Michael Hauer, Jennifer A. Doudna, and Emmanuelle Charpentier. "A Programmable Dual-RNA-Guided DNA Endonuclease in Adaptive Bacterial Immunity." *Science* 337, no. 6096 (August 17, 2012): 816–21. https://doi.org/10.1126/science.1225829.
40. Jinek, Martin, Alexandra East, Aaron Cheng, Steven Lin, Enbo Ma, and Jennifer Doudna. "RNA-Programmed Genome Editing in Human Cells." *ELife* 2 (January 29, 2013): e00471. https://doi.org/10.7554/eLife.00471.
41. Kay, Sabine, Simone Hahn, Eric Marois, Gerd Hause, and Ulla Bonas. "A Bacterial Effector Acts as a Plant Transcription Factor and Induces a Cell Size Regulator." *Science* 318, no. 5850 (October 26, 2007): 648–51. https://doi.org/10.1126/science.1144956.
42. Kenkel, Beth. "Lambda Red: A Homologous Recombination-Based Technique for Genetic Engineering." Addgene, December 15, 2016. http://blog.addgene.org/lambda-red-a-homologous-recombination-based-technique-for-genetic-engineering.
43. Kim, Hyongbum, and Jin-Soo Kim. "A Guide to Genome Engineering with Programmable Nucleases." *Nature Reviews Genetics* 15, no. 5 (May 2014): 321–34. https://doi.org/10.1038/nrg3686.
44. Kim, Yang-Gyun, Jooyeun Cha, and Srinivasan Chandrasegaran. "Hybrid Restriction Enzymes: Zinc Finger Fusions to Fok I Cleavage Domain." *Proceedings of the National Academy of Sciences of the United States of America* 93, no. 3 (1996): 1156–60.
45. Kirkpatrick, Jesse, Gregory D. Koblentz, Megan J. Palmer, Edward Perello, David A. Relman, and Sarah W. Denton. "Editing Biosecurity: Needs and Strategies for Governing Genome Editing," December 2018.
46. Kleinstiver, Benjamin P., Michelle S. Prew, Shengdar Q. Tsai, Ved V. Topkar, Nhu T. Nguyen, Zongli Zheng, Andrew P. W. Gonzales, et al. "Engineered CRISPR-Cas9 Nucleases with

Altered PAM Specificities." *Nature* 523, no. 7561 (July 23, 2015): 481–85. https://doi.org/10.1038/nature14592.
47. Ledford, Heidi, and Ewen Callaway. "Pioneers of Revolutionary CRISPR Gene Editing Win Chemistry Nobel." *Nature* 586, no. 7829 (October 7, 2020): 346–47. https://doi.org/10.1038/d41586-020-02765-9.
48. Lee, Ciaran M, Thomas J Cradick, Eli J Fine, and Gang Bao. "Nuclease Target Site Selection for Maximizing On-Target Activity and Minimizing Off-Target Effects in Genome Editing." *Molecular Therapy* 24, no. 3 (March 2016): 475–87. https://doi.org/10.1038/mt.2016.1.
49. Lodish, Harvey, Arnold Berk, S. Lawrence Zipursky, Paul Matsudaira, David Baltimore, and James Darnell. "DNA Cloning with Plasmid Vectors," 2000. https://www.ncbi.nlm.nih.gov/books/NBK21498/.
50. Mak, Amanda Nga-Sze, Philip Bradley, Raul A. Cernadas, Adam J. Bogdanove, and Barry L. Stoddard. "The Crystal Structure of TAL Effector PthXo1 Bound to Its DNA Target." *Science (New York, N.Y.)* 335, no. 6069 (February 10, 2012): 716–19. https://doi.org/10.1126/science.1216211.
51. Mali, Prashant, Luhan Yang, Kevin M. Esvelt, John Aach, Marc Guell, James E. DiCarlo, Julie E. Norville, and George M. Church. "RNA-Guided Human Genome Engineering via Cas9." *Science* 339, no. 6121 (February 15, 2013): 823–26. https://doi.org/10.1126/science.1232033.
52. McDade, Joel. "Chapter 3 – Using CRISPR in Your Experiments: Planning Your CRISPR Experiment." In *CRISPR 101: A Desktop Resource*, edited by AddGene, 2nd ed., 36–42, 2015.
53. Miller, J, A D McLachlan, and A Klug. "Repetitive Zinc-Binding Domains in the Protein Transcription Factor IIIA from Xenopus Oocytes." *The EMBO Journal* 4, no. 6 (June 1985): 1609–14.
54. Miller, Jeffrey C., Siyuan Tan, Guijuan Qiao, Kyle A. Barlow, Jianbin Wang, Danny F. Xia, Xiangdong Meng, et al. "A TALE Nuclease Architecture for Efficient Genome Editing." *Nature Biotechnology* 29, no. 2 (February 2011): 143–48. https://doi.org/10.1038/nbt.1755.
55. Mirus Bio LLC. "CRISPR/Cas9 Genome Editing Brochure," 2017. https://www.mirusbio.com/assets/web_documents/crispr-cas9-genome-editing-brochure-pdna-mrna-cas9-grna-rnp.pdf.
56. Mohanraju, Prarthana, Kira S. Makarova, Bernd Zetsche, Feng Zhang, Eugene V. Koonin, and John van der Oost. "Diverse Evolutionary Roots and Mechanistic Variations of the CRISPR-Cas Systems." *Science (New York, N.Y.)* 353, no. 6299 (August 5, 2016): aad5147. https://doi.org/10.1126/science.aad5147.
57. Moscou, Matthew J., and Adam J. Bogdanove. "A Simple Cipher Governs DNA Recognition by TAL Effectors." *Science* 326, no. 5959 (December 11, 2009): 1501–1501. https://doi.org/10.1126/science.1178817.
58. Murugan, Karthik, Kesavan Babu, Ramya Sundaresan, Rakhi Rajan, and Dipali G. Sashital. "The Revolution Continues: Newly Discovered Systems Expand the CRISPR-Cas Toolkit." *Molecular Cell* 68, no. 1 (October 5, 2017): 15–25. https://doi.org/10.1016/j.molcel.2017.09.007.
59. Mussolino, Claudio, and Toni Cathomen. "TALE Nucleases: Tailored Genome Engineering Made Easy." *Current Opinion in Biotechnology*, Tissue, cell and pathway engineering, 23, no. 5 (October 2012): 644–50. https://doi.org/10.1016/j.copbio.2012.01.013.
60. Nelson, David L., and Michael M. Cox. *Lehninger Principles of Biochemistry*. 5th edition. New York: W. H. Freeman, 2008.
61. Nuffield Council on Bioethics. *Genome Editing: An Ethical Review*. London: Nuffield Council on Bioethics, 2016.
62. Pearson, Helen. "Protein Engineering: The Fate of Fingers." *Nature* 455, no. 7210 (September 11, 2008): 160–64. https://doi.org/10.1038/455160a.
63. Peng, Rongxue, Guigao Lin, and Jinming Li. "Potential Pitfalls of CRISPR/Cas9-Mediated Genome Editing." *FEBS Journal* 283, no. 7 (April 1, 2016): 1218–31. https://doi.org/10.1111/febs.13586.
64. Perello, Edward. *CRISPR Genome Editing: A Technical & Policy Primer*. George Mason University, n.d. https://editingbiosecurity.org/new-page.

65. Periwal, Vinita. "A Comprehensive Overview of Computational Resources to Aid in Precision Genome Editing with Engineered Nucleases." *Briefings in Bioinformatics*. Accessed April 6, 2017. https://doi.org/10.1093/bib/bbw052.
66. Ran, F Ann, Patrick D Hsu, Jason Wright, Vineeta Agarwala, David A Scott, and Feng Zhang. "Genome Engineering Using the CRISPR-Cas9 System." *Nature Protocols* 8, no. 11 (November 2013): 2281–2308. https://doi.org/10.1038/nprot.2013.143.
67. Ran, F. Ann, Patrick D. Hsu, Jason Wright, Vineeta Agarwala, David A. Scott, and Feng Zhang. "Genome Engineering Using the CRISPR-Cas9 System." *Nature Protocols* 8, no. 11 (November 2013): 2281–2308. https://doi.org/10.1038/nprot.2013.143.
68. Reyon, Deepak, Cyd Khayter, Maureen R. Regan, J. Keith Joung, and Jeffry D. Sander. "Current Protocols in Molecular Biology Engineering Designer Transcription Activator-Like Effector Nucleases (TALENs)." *Current Protocols in Molecular Biology / Edited by Frederick M. Ausubel ... [et Al.]* 0 12 (October 2012): Unit-12.15. https://doi.org/10.1002/0471142727.mb1215s100.
69. Reyon, Deepak, Shengdar Q. Tsai, Cyd Khayter, Jennifer A. Foden, Jeffry D. Sander, and J. Keith Joung. "FLASH Assembly of TALENs for High-Throughput Genome Editing." *Nature Biotechnology* 30, no. 5 (2012): 460–65. https://doi.org/10.1038/nbt.2170.
70. Science Direct. "Ribonuclease." Accessed March 12, 2018. https://www.sciencedirect.com/topics/biochemistry-genetics-and-molecular-biology/ribonuclease.
71. Sander, Jeffry D., Lindsay Cade, Cyd Khayter, Deepak Reyon, Randall T. Peterson, J. Keith Joung, and Jing-Ruey J. Yeh. "Targeted Gene Disruption in Somatic Zebrafish Cells Using Engineered TALENs." *Nature Biotechnology* 29, no. 8 (August 2011): 697–98.
72. Sander, Jeffry D., and J. Keith Joung. "CRISPR-Cas Systems for Editing, Regulating and Targeting Genomes." *Nature Biotechnology* 32, no. 4 (April 2014): 347–55. https://doi.org/10.1038/nbt.2842.
73. Schmid-Burgk, Jonathan L., Tobias Schmidt, Vera Kaiser, Klara Höning, and Veit Hornung. "A Ligation-Independent Cloning Technique for High-Throughput Assembly of Transcription Activator-like Effector Genes." *Nature Biotechnology* 31, no. 1 (2013): 76–81. https://doi.org/10.1038/nbt.2460.
74. Sharan, Shyam K., Lynn C. Thomason, Sergey G. Kuznetsov, and Donald L. Court. "Recombineering: A Homologous Recombination-Based Method of Genetic Engineering." *Nature Protocols* 4, no. 2 (2009): 206–23. https://doi.org/10.1038/nprot.2008.227.
75. Strachan, Tom, and Andrew Read. *Human Molecular Genetics, Fourth Edition*. 4 edition. New York: Garland Science, 2010.
76. Synthego. "How to Perform Successful CRISPR Experiments: A Step-by-Step Guide," December 15, 2016. https://www.biolegio.com/media/4148/how-to-perform-successful-crispr-experiments.pdf.
77. Terns, Michael P., and Rebecca M. Terns. "CRISPR-Based Adaptive Immune Systems." *Current Opinion in Microbiology* 14, no. 3 (June 2011): 321–27. https://doi.org/10.1016/j.mib.2011.03.005.
78. U.S. National Library of Medicine: Genetics Home Reference. "What Is a Gene?" Accessed April 13, 2017. https://ghr.nlm.nih.gov/primer/basics/gene.
79. Yamamoto, Takashi, ed. *Targeted Genome Editing Using Site-Specific Nucleases: ZFNs, TALENs, and the CRISPR/Cas9 System*. Springer, 2015.
80. Yee, Jiing-Kuan. "Off-Target Effects of Engineered Nucleases." *The FEBS Journal* 283, no. 17 (September 2016): 3239–48. https://doi.org/10.1111/febs.13760.
81. Young, Susan. "Genome Surgery." *MIT Technology Review*, April 2014.
82. Zetsche, Bernd, Jonathan S. Gootenberg, Omar O. Abudayyeh, Ian M. Slaymaker, Kira S. Makarova, Patrick Essletzbichler, Sara E. Volz, et al. "Cpf1 Is a Single RNA-Guided Endonuclease of a Class 2 CRISPR-Cas System." *Cell* 163, no. 3 (October 22, 2015): 759–71. https://doi.org/10.1016/j.cell.2015.09.038.
83. Sigma-Aldrich. "Zinc Finger Nuclease – FAQS." Accessed November 16, 2018. https://www.sigmaaldrich.com/life-science/zinc-finger-nuclease-technology/learning-center/zfn-faqs.html.

Chapter 3
Applying Genetic Engineering to Biological Weapons

It may seem like the debate over the threat posed by genetic engineering on the creation of biological weapons is new given the recent surge of concerns over the potential misuse of genome editing technologies, specifically CRISPR; however, it is a decades-old debate. Nobel laureate Joshua Lederberg, who is recognized as one of the earliest scientists to weigh in on the debate, expressed worries even a few years before the first organism was successfully genetically engineered.[1] In 1971, Lederberg cautioned, "the potential undoubtedly exists for the design and development of infective agents against which no credible defense is possible, through the genetic and chemical manipulation of these agents."[2]

For over 50 years, many reports have warned the world about the threat of genetic engineering. The first part of this chapter reviews older concerns over how genetic engineering could be misused to make pathogens more dangerous. Next, these reports are analyzed to determine the ways they exhibit elements of the biotech revolution frame—an explanatory structure applied to investigate a problem based largely on technical elements of knowledge production and use.[3] To provide context for the types of changes that could be achieved through genetic engineering, this part includes a review of the differences between non-genetically engineered agents and genetically engineered agents. Afterwards, it examines a list of the desirable attributes of biological weapons that people worry could be targeted for modification by genetic engineering. Finally, it reviews the types of technical and social obstacles that exist that can prevent genetic engineers from achieving modifications, and

[1]Robert P. Kadlec and Alan P. Zelicoff, "Implications of the Biotechnology Revolution for Weapons Development and Arms Control," in *Biological Warfare: Modern Offense and Defense*, ed. Raymond A. Zilinskas (Boulder, CO: Lynne Rienner Publishers, 2000), 11.

[2]Joshua Lederberg, "Views: Biological Warfare: A Global Threat: Will Biomedical Knowledge Be Dedicated to Peaceful Uses by International Agreement or Do We Face a New Weapons Spiral?," *American Scientist* 59, no. 2 (1971): 196.

[3]Vogel, *Phantom Menace or Looming Danger? A New Framework for Assessing Bioweapons Threats*.

K. Paris, *Genome Editing and Biological Weapons*,
https://doi.org/10.1007/978-3-031-21820-0_3

introduces an alternative explanatory structure known as the biosocial frame. The biosocial frame considers technical elements of bioweapons assessment within a broader social context that includes intangible factors that might have an impact on bioweapons development.

The second part of this chapter includes a review of the more recent and specific concerns that have emerged in the literature over the misuse of genome editing. Beginning with an analysis of official meeting documents from the BWC in 2011, the second part includes an overview of the monitoring that took place in which ZFN, TALEN, and CRISPR were flagged as technologies with dual-use potential. Note that this review of monitoring is in line with Jonathan Tucker's decision framework for technology governance, which first calls for technologies with dual-use potential to be detected via technology monitoring and then assessed for their potential risk of misuse. Lastly, the chapter lays out the specific fears that people have over how genome editing could be misused: altering microbes (pathogens and the human microbiome), weaponizing gene drives, and augmenting humans (super soldiers and general population).

3.1 Threat of Genetic Engineering

Although not known for about another decade, in 1983, the former Soviet Union weaponized the first genetically altered agent, which was a hypervirulent version of the agent that causes tularemia (*Francisella tularensis*), through its covert offensive program.[4] The former Soviet Union secretly operated one of the largest and most sophisticated state-sponsored biological warfare programs known to date. This program is believed to have started in the 1920s and ramped up following the signature and ratification of the BWC by the Soviet Union in 1972 and 1975, respectively. The offensive component of this program is said to have ceased in 1992–1993.[5]

Several reports in the late 1990s and 2000s addressed concerns over the potential for the misuse of genetic engineering. In 1999, the United States Commission on National Security/21st Century (USCNS/21), also known as the Hart-Rudman Commission, described the proliferation threat from former Soviet Union biological weapons scientists and warned about the design and development of more dangerous

[4]Janet R. Gilsdorf and Raymond A. Zilinskas, "New Considerations in Infectious Disease Outbreaks: The Threat of Genetically Modified Microbes," *Clinical Infectious Diseases: An Official Publication of the Infectious Diseases Society of America* 40, no. 8 (April 15, 2005): 1160, https://doi.org/10.1086/428843.

[5]Raymond A. Zilinskas, *The Soviet Biological Weapons Program and Its Legacy in Today's Russia*, CSWMD Occasional Paper 11 (Washington, DC: National Defense University Press, 2016), 44.

genetically engineered agents.[6] In 2003, it was suggested that the future threat from artificially engineered advanced agents indefinitely increased alongside advancements in biotechnology.[7] In 2004, the Fink Report—named after Gerald R. Fink, chair of the expert panel—suggested it was difficult to engineer agents better than what is available in nature—yet, the report predicted the use of new genetic technologies would likely lead to "the creation of 'designer diseases' and pathogens with increased military utility."[8]

The conventional agent versus engineered agent debate was further addressed in 2006 by the Institute of Medicine (IOM) and the National Research Council (NRC) Committee on Advances in Technology and the Prevention of Their Application to Next Generation Biowarfare Threats. Ultimately, the IOM/NRC report concluded it was possible to create pathogens more virulent than those found in nature.[9] The IOM/NRC report explained how pathogens are "built" and "tested" in nature for "hundreds of thousands of years," so the pathogens that exist in nature are the "most 'successful'" in terms of the ability to "survive within, on, or near human populations over time."[10] However, organisms that are successful in terms of survival are not necessarily the most pathogenic or virulent organisms as it may not be in the best interest of a pathogen to risk die off by causing severe disease and quickly killing off the host population.[11] Finally, in 2008, the Commission on the Prevention of Weapons of Mass Destruction (WMD) Proliferation and Terrorism, known as the Graham-Talent WMD Commission, advised the more genetic engineering capabilities move around the world, "the greater the potential that terrorists will use them to develop biological weapons."[12]

---

[6]The United States Commission on National Security/21st Century, *New World Coming: American Security in the 21st Century, Supporting Research and Analysis, The Phase I Report on the Emerging Global Security Environment for the First Quarter of the 21st Century*, 1999, 51.

[7]James B. Petro, Theodore R. Plasse, and Jack A. McNulty, "Biotechnology: Impact on Biological Warfare and Biodefense," *Biosecurity and Bioterrorism: Biodefense Strategy, Practice, and Science* 1, no. 3 (September 1, 2003): 162, https://doi.org/10.1089/153871303769201815.

[8]National Research Council (US) Committee on Research Standards and Practices to Prevent the Destructive Application of Biotechnology, *Biotechnology Research in an Age of Terrorism* (Washington, DC: National Academies Press, 2004), 25.

[9]Institute of Medicine and National Research Council, *Globalization, Biosecurity, and the Future of the Life Sciences* (Washington, DC: National Academies Press, 2006), 44, https://doi.org/10.17226/11567.

[10]Institute of Medicine and National Research Council, 39–40.

[11]Institute of Medicine and National Research Council, 40–44.

[12]Commission on the Prevention of WMD Proliferation and Terrorism, "World at Risk: The Report of the Commission on the Prevention of WMD Proliferation and Terrorism" (New York: Vintage Books, 2008), 23.

3.1.1 Framing Assessments

There are many ways that the reports from the late 1990s and 2000s, which expressed concerns over the potential for the misuse of genetic engineering, exhibit features of what Kathleen Vogel terms the biotech revolution frame. Frames are applied by analysts to understand, explain, and solve problems. An analyst's studies and experience influence how they frame their work—information that falls within the frame is deemed relevant and considered in assessments, while information that falls outside of the frame is "marginalized, ignored, or suppressed."[13] It is important to consider the way an assessment is framed because different framings can generate very different conclusions.[14]

Molecular biology as well as genetic engineering are said to have revolutionized biotechnology. The discovery that DNA houses genetic information as well as the discovery of the structure of DNA and understanding of the genetic code paved the way for the first successful recombination experiment in the 1970s. After this time, the development of genetic engineering quickly took off.[15] The biotechnology revolution is described as the third significant technology wave that had a major impact on shaping chemical and biological warfare. Prior to the biotechnology revolution, the use of commercial chemical compounds during World War I and synthesis of highly lethal chemical nerve agents in the 1940s marked the beginning of the first and second technology waves, respectively. Unlike the previous technology waves, advancements in genetic engineering during the third wave gave scientists tools to manipulate the fundamental molecular material that gives rise to all life.[16]

Figure 3.1 presents key elements of the biotech revolution frame including: "(1) codified knowledge; (2) material end products; (3) the accessibility of biological materials; (4) the economic drivers of technology; (5) the future; (6) the globalization and diffusion of technology; and (7) fixed technological trajectories."[17] These seven elements of the biotech revolution frame are explained below, and the elements of an alternative frame, known as the biosocial frame, are discussed later in this chapter.

First, the biotech revolution frame focuses on codified, or explicit, knowledge—knowledge that can be written down (e.g., laboratory manuals) and transferred

---

[13] Vogel, *Phantom Menace or Looming Danger? A New Framework for Assessing Bioweapons Threats*, 13, 17.

[14] Vogel, 11.

[15] Kathryn Nixdorff, Jens Brauburger, and Dorte Hahlbohm, "The Biotechnology Revolution: The Science and Applications," in *Verification of the Biological and Toxin Weapons Convention*, vol. 32, NATO ASI Series, 1: Disarmament Technologies, 2000, 78, https://doi.org/10.1007/978-94-017-3643-5_3.

[16] "Advances in Biotechnology and Genetic Engineering," 4.

[17] Vogel, *Phantom Menace or Looming Danger? A New Framework for Assessing Bioweapons Threats*, 20.

Biotech revolution frame	Biosocial frame
Focus on material aspects of biotechnology • Materials, equipment • Physical infrastructure • Codified knowledge • Simplicity • Successful end products	Focus on sociotechnical aspects of biotechnology • Materials and equipment with emphasis on use in communities of scientific and technological practice • Socio-organizational infrastructure • Tacit knowledge • Complexity • Problems, troubleshooting, and failures on the way to the end product
Temporal and spatial orientation • Future-looking • Assumes ease of diffusion, routinization • Assumes increasing pace of growth and diffusion of biotechnology • Globalization	Temporal and spatial orientation • Future not privileged • Assumes challenges in diffusion, routinization • Assumes bottlenecks can affect pace of growth and diffusion of biotechnology • Local specificity
Linear, exponential technological and bioweapons threat trajectories	Varying technological and bioweapons threat trajectories
Increasing pace and momentum	Varying pace and momentum
Technological imperatives	Interpretive flexibility
Assumes potentially disruptive character of biotechnology increases the bioweapons threat	Assumes potentially disruptive character of biotechnologies is modulated by the sociotechnical elements listed above; does not necessarily increase the bioweapons threat

Fig. 3.1 Biotech revolution frame vs. biosocial frame (Kathleen M. Vogel, *Phantom Menace or Looming Danger? A New Framework for Assessing Bioweapons Threats*. pp. 68, Table O.1. © 2013 Johns Hopkins University Press. Reprinted with permission of The Johns Hopkins University Press)

through impersonal means (e.g., publications)—but does not always distinguish it from the know-how it takes to apply this knowledge.[18] For instance, a user who obtains a publication from a lab that genetically engineered *B. anthracis* to resist antibiotics does not guarantee that the user will have the know-how to replicate this work. Second, the biotech revolution frame highlights material end products—implying the work it took to achieve these products is straight-forward without uncertainties prompting the need to troubleshoot. In this frame, advancements in biotechnologies are viewed as becoming more "black-boxed"—meaning there are

[18] Vogel, 20–23.

"predictable inputs and outputs, such that the inner workings of the device need not be known for it to be used."[19]

Third, the biotech revolution frame emphasizes the accessibility of biological material, focusing on the ease of obtaining reagents and equipment, without considering inherent limitations of the material or how local contextual factors may impede use. Fourth, the biotech revolution frame focuses on the economy as a driver biotechnology. In particular, the large pharmaceutical industry, which is worth billions of dollars and spans the world, is thought to preside over the life sciences and set the trajectory of technology development—yet these claims do not address the issues that biotech and pharma have had when trying to get new biotechnologies to work for specific purposes. Fifth, the biotech revolution frame looks to the future and emphasizes the increasing threat from emerging biotechnologies—this approach marginalizes the potential threat from the misuse of less advanced, older technologies to make bioweapons.[20]

Sixth, the biotech revolution frame highlights the global and diffuse nature of biotechnologies. The technologies transcend the borders of countries. This spatial orientation, however, does not factor in the challenges that can occur when taking up biological material and equipment at a local level. Seventh, the biotech revolution frame assumes the trajectories of technologies and the bioweapons threat as fixed, or rather, as new biotechnologies emerge, the bioweapons threat increases.[21] The biotech revolution frame views technological developments as "a set of cumulative technical results, each with its own momentum or trajectory, largely independent and autonomous from social influence or control."[22] This frame also emphasizes technological imperatives, or rather the assumption that "technological developments, once set in motion, are unstoppable, unavoidable, and inevitable."[23]

The biotech revolution frame has been the main frame used for biological weapons assessments and policymaking in the United States; however, as Vogel points out, there have been "shortcomings in U.S. government and nongovernment assessments of biological weapons threats."[24] In certain instances in the aforementioned reports, there could have been additional consideration given to the feasibility of getting and using the technologies. For instance, the Fink Report and Graham-Talent WMD Commission emphasize how biological science, knowledge, and technology are borderless.[25] But, focusing on the "globalization and diffusion of

[19] Vogel, 23.

[20] Vogel, 24–28.

[21] Vogel, 28–31.

[22] Vogel, 19.

[23] Vogel, 20.

[24] Vogel, 4, 14.

[25] National Research Council (US) Committee on Research Standards and Practices to Prevent the Destructive Application of Biotechnology, *Fink Report*, 18; Commission on the Prevention of WMD Proliferation and Terrorism, "Graham-Talent WMD Commission," 23; Vogel, *Phantom Menace or Looming Danger? A New Framework for Assessing Bioweapons Threats*, 28.

biological materials and equipment" does not necessarily capture "the complicated ways in which individuals, groups, and states take up and use materials and equipment."[26]

Moreover, the Fink report emphasized the information needed to genetically engineer agents is available online, and the equipment and expertise to apply the information to make novel agents is globally available.[27] Yet, as Vogel suggests, concentrating on the accessibility of biological materials, supplies, and infrastructure marginalizes "the social and material conditions required for such biotech equipment to work in different contexts (e.g., an outpost in Afghanistan versus an academic laboratory)."[28] Additionally, the laboratory equipment may be used to perform simple tasks but may require additional "techniques, equipment, materials, and complex skills" to accomplish more difficult tasks.[29] Furthermore, in emphasizing the belief that the future is privileged and new technologies will bring about new threats, the Fink Report as well as the report from Petro and colleagues marginalize the past. This feature of the biotech revolution frame "ignores how older, cruder biotechnology techniques could be appropriated for bioweapons purposes."[30]

Ultimately, Vogel developed the "biosocial frame," as an alternative to the biotech revolution. The biosocial frame better captures the contingencies and complexities involved in biological weapons work that are omitted from the biotech revolution frame. Key elements of the biosocial frame, as well as a discussion of why the biosocial frame is a better description of the current situation as compared to the biotech revolution, are considered later in this chapter.

3.1.2 Types of Agents

The literature broadly groups biological warfare agents into four categories (only the last two of which are genetically engineered): traditional agents, emerging agents, genetically modified traditional agents, and advanced agents. Within the context of biodefense, an agent, also referred to as a biological agent or bioagent, is used to describe "any product created using biological components that may be intended to cause harm."[31] An agent could be "a pathogen, a toxin, or even a biological

[26] Vogel, *Phantom Menace or Looming Danger? A New Framework for Assessing Bioweapons Threats*, 28.

[27] National Research Council (US) Committee on Research Standards and Practices to Prevent the Destructive Application of Biotechnology, *Fink Report*, 23.

[28] Vogel, *Phantom Menace or Looming Danger? A New Framework for Assessing Bioweapons Threats*, 24–25.

[29] Vogel, 25.

[30] Vogel, 25–28.

[31] National Academies of Sciences, Engineering, and Medicine, *A Proposed Framework for Identifying Potential Biodefense Vulnerabilities Posed by Synthetic Biology: Interim Report* (The National Academies Press, 2017), 6, https://doi.org/10.17226/24832.

component."[32] A pathogen, or infectious agent, is defined as "a biological agent that causes disease or illness to its host."[33] This comparison of the agents in this section reveals the desirable attributes of a good biological weapon and provides context for the objectives that bioweaponeers who apply genetic engineering might try to achieve.

3.1.2.1 Non-genetically Engineered

Traditional Agents

Prior to discussing the threat of genetic engineering on the development of more desirable agents, it is necessary to first identify and define traditional agents of concern. Traditional, also referred to as classical, biological warfare agents are described as "those that have been developed and weaponized in the course of past BW [i.e., biological warfare] programs."[34] Traditional agents are naturally occurring biological agents and their toxins that were selected based on attributes such as "toxicity, stability, and ease of production" and limited by factors such as "environmental stability, infectious dose, time to effect, clinical progression, and lethality."[35] Examples of traditional agents include *Bacillus anthracis*, *Clostridium botulinum* toxin, and *Yersinia pestis*, which are the pathogens that cause anthrax, botulism, and plague, respectively.

There are several lists that specify ideal qualities of traditional agents. For instance, the 1964 United States Department of Army and Air Force list of requirements for biological agents, as cited in a June 1996 United States Department of Defense publication, identified five qualities of traditional agents. These qualities focused on the reliability to produce a desired effect (e.g., death); the large-scale ability to be manufactured; the stability during production, storage, and transportation; the ability to be efficiently disseminated; and the stability after dissemination.[36] In 1999, the Working Group on Civilian Biodefense, organized by the Center for Civilian Biodefense Studies at the Johns Hopkins University Bloomberg School of Public Health, published a more comprehensive list of nine notable features of traditional agents. In addition to the characteristics already listed, the new list also addressed the potential for transmission between people, low infective dose, limited vaccine availability, agent availability, and previous research and development of

[32] National Academies of Sciences, Engineering, and Medicine, 6.

[33] "Reference Terms: Pathogen," ScienceDaily, accessed August 17, 2017, https://www.sciencedaily.com/terms/pathogen.htm.

[34] Kadlec and Zelicoff, "Implications of the Biotechnology Revolution for Weapons Development and Arms Control," 16.

[35] Petro, Plasse, and McNulty, "Biotechnology," 162.

[36] "Advances in Biotechnology and Genetic Engineering," 3.

the agent as a bioweapon. Additionally, this new list also cited the potential for the outbreak to promote public and healthcare worker anxiety.[37]

While traditional agents have many desirable qualities, they also have limitations. Issues, including "delayed action, uncertain dispersal pattern (to the extent that wind and weather cannot be predicted), and indiscriminate effects," limit the effectiveness of biological weapons on the battlefield.[38] For instance, *B. anthracis* forms a hardy spore that can persist in the environment for decades; however, an area contaminated by these spores would be off limits to both sides of combatants.[39] Genetic engineering could be used to modify the persistence of the agent once disseminated or to create a therapeutic to protect the aggressor's forces from the agent—thus creating a genetically engineered agent that better fits the needs of the user. Attributes targeted for modification are described later in this chapter.

Emerging Agents

In addition to traditional agents, there are other naturally occurring agents emerging in the environment that could be used as biological weapons. Homeland Security Presidential Directive (HSPD)-18 describes emerging agents as "previously unrecognized pathogens that might be naturally occurring and present a serious risk to human populations, such as the virus responsible for Severe Acute Respiratory Syndrome (SARS)."[40] The United States Army Medical Research Institute of Infectious Diseases (USAMRIID) also identifies Middle East Respiratory Syndrome (MERS)-coronavirus, a virus closely related to the virus that causes SARS, as another example of an emerging agent.[41]

3.1.2.2 Genetically Engineered

Genetically Modified Traditional Agents

Genetically modified traditional agents, also known as enhanced agents, are the first of two types of genetically engineered agents examined in the literature. These agents are described as traditional agents and toxins that are genetically engineered. Genetically modified traditional biological warfare agents are "closely related to the

[37] Philip E. Coyle, "Biological Warfare and Bioterrorism: How Do They Differ from Other WMD Threats?," in *Global Biosecurity: Threats and Responses*, ed. Peter Katona, John P. Sullivan, and Michael D. Intriligator (Routledge, 2010), 35.

[38] Jonathan Tucker, "The Future of Biological Warfare," in *The Proliferation of Advanced Weaponry: Technology, Motivations, and Responses*, ed. Eric H. Arnett and W. Thomas Wander (Washington, DC: American Association for the Advancement of Science, 1992), 55.

[39] Tucker, 55–56.

[40] Administration of George W. Bush, "Homeland Security Presidential Directive/HSPD-18—Medical Countermeasures Against Weapons of Mass Destruction" (GPO, January 31, 2007), 110.

[41] Army Medical Research Institute for Infectious Diseases, *Usamriid's Medical Management of Biological Casualties Handbook*, ed. Mark R. Withers, 8th ed. (Government Printing Office, 2014), 148.

parent agent at the genetic level."[42] Enhancements may be made to traditional agents to "enhance their ability to harm" through modification in attributes such as "antibiotic resistance, increased aerosol stability, or heightened pathogenesis."[43] For instance, bioweaponeers may seek to genetically engineer *B. anthracis* to make the bacteria resistant to common therapeutic drugs used for treatment such as the antibiotics ciprofloxacin and doxycycline.[44] As mentioned earlier, attributes targeted for modification are described later in this chapter.

Advanced Agents

Advancements in biotechnology and improvements in biomedical research make it possible to develop new types of biological agents termed advanced biological warfare agents. Advanced agents are described as "novel pathogens or other materials of biological nature that have been artificially engineered in the laboratory to bypass traditional countermeasures or produce a more severe or otherwise enhanced spectrum of disease."[45] As opposed to focusing on the agent itself, development of advanced agents keeps the target organism in mind. In particular, concerns are raised over the development of bioregulators as biological warfare agents. Recall that bioregulators are "chemicals normally produced in the human body that control communication between cells and that play a crucial role in governing the nervous, endocrine, and immune systems."[46] According to Petro and colleagues, "advanced BW [i.e., biological warfare] agents will be able to target specific biological systems, such as the cardiovascular, immunological, neurological, and gastrointestinal systems."[47] Matthew Meselson, a molecular biologist, describes the possibility of manipulating processes such as "cognition, development, reproduction, and inheritance."[48]

In addition to targeting human biological systems, advanced agents could be engineered to target the processes of plants and animals.[49] For instance, agents could be developed to target drug-producing plants—such an application would help law

[42] Petro, Plasse, and McNulty, "Biotechnology," 162.

[43] Petro, Plasse, and McNulty, 162; Administration of George W. Bush, "Homeland Security Presidential Directive/HSPD–18—Medical Countermeasures Against Weapons of Mass Destruction," 110; Army Medical Research Institute for Infectious Diseases, *Usamriid's Medical Management of Biological Casualties Handbook*, 148.

[44] The U.S. National Response Team, "NRT Quick Reference Guide: Bacillus Anthracis (Causes the Disease Anthrax)," September 2011.

[45] Administration of George W. Bush, "Homeland Security Presidential Directive/HSPD–18—Medical Countermeasures Against Weapons of Mass Destruction," 110. This description of advanced agents is also cited in Army Medical Research Institute for Infectious Diseases, *Usamriid's Medical Management of Biological Casualties Handbook*, 148.

[46] Koblentz, *Living Weapons*, 10.

[47] Petro, Plasse, and McNulty, "Biotechnology," 162.

[48] Matthew Meselson, "The Problem of Biological Weapons," *Bulletin of the American Academy of Arts and Sciences* 52, no. 5 (1999): 49, https://doi.org/10.2307/3824391.

[49] Petro, Plasse, and McNulty, "Biotechnology," 162.

enforcement.[50] Plants could also be genetically engineered to overcome challenges associated with biological weapons production. Specifically, transgenic plants could function as "bioproduction reactors" that are engineered to synthesize protein-based bioregulators or toxins. In this case, standard mechanical equipment used to produce biological weapons would not be needed. Furthermore, transgenic insects, such as mosquitoes, could be engineered to synthesize and secrete a protein-based bioregulator or toxin in the saliva transmitted to hosts during the feeding process.[51] Other types of agents, such as "non-lethal" agents that degrade plastic or metal materials, could be developed for military application.[52]

3.1.3 Attributes Targeted for Modification

Several scholars and studies have identified characteristics of agents that could be modified via genetic engineering for hostile purposes. At least as early as the mid-1980s, biodefense experts began to compile lists of the desirable features of agents.[53] Since this time, many publications have further added to, and refined, these lists.[54] Attributes that could be targeted for modification include increased virulence; resistance to drugs and preventatives; protection of own forces; ability to evade host immunological response; ability to avoid detection and diagnostics; enhanced environmental resistance; enhanced production, storage, and dissemination; increased vector efficiency; ability to control agents; and ability to alter host tropism. The following subsection elaborates on the qualities of these attributes.

---

[50] Jan van Aken and Edward Hammond, "Genetic Engineering and Biological Weapons," *EMBO Reports* 4, no. Suppl 1 (2003): S59–60, https://doi.org/10.1038/sj.embor.embor860.

[51] Petro, Plasse, and McNulty, "Biotechnology," 163.

[52] van Aken and Hammond, "Genetic Engineering and Biological Weapons," S59–60.

[53] Kadlec and Zelicoff, "Implications of the Biotechnology Revolution for Weapons Development and Arms Control," 16.

[54] "Advances in Biotechnology and Genetic Engineering"; Steven M. Block, "Living Nightmares: Biological Threats Enabled by Molecular Biology," in *The New Terror: Facing the Threat of Biological and Chemical Weapons*, ed. Sidney D. Drell, Abraham D. Sofaer, and George D. Wilson (Stanford, Calif: Hoover Institution Press, 1999); Tucker, "The Future of Biological Warfare," 53–73; Raymond A. Zilinskas, "Possible Terrorist Use of Modern Biotechnology Techniques" (Possible Terrorist Use of Modern Biotechnology Techniques, Istituto Diplomatico "Mario Toscano" Villa Madama, Rome, Italy, 2000), 2; National Research Council (US) Committee on Research Standards and Practices to Prevent the Destructive Application of Biotechnology, *Fink Report*, 114–15; National Academies of Sciences, Engineering, and Medicine, *Biodefense in the Age of Synthetic Biology*, 57–60, 63–66.

3.1.3.1 Increased Virulence

Virulence refers to the severity of disease in infected hosts. Disease symptoms may be severe and lead to death.[55] Highly virulent pathogens typically act quickly and cause severe disease. Some virulence factors cause local damage to the host by secreting enzymes or other molecules near the resident pathogen. Other virulence factors travel farther away from the resident pathogen via the host's circulatory or lymphatic systems and cause distant damage. Still, other virulence factors evade the host's immune system or use substances made by the host for the benefit of the pathogen.[56] Bioweaponeers may seek to engineer a pathogen with enhanced virulence or cause an avirulent pathogen to become virulent.[57] Finally, it is noted that virulence differs from pathogenicity. Pathogenicity refers to the ability of a pathogen to cause disease in the host.

3.1.3.2 Resistance to Drugs and Preventatives

Resistance describes "the ability of a microorganism to defeat the actions of therapeutic drugs, such as antibiotics, and preventives, such as vaccines."[58] Genetic modifications could be made to agents so that therapeutic drugs and vaccines used to treat and prevent infections are no longer effective. Bioweaponeers could develop strains of bacteria, viruses, and fungi that are respectively resistant to antibiotics, antivirals, and fungicides or antifungals.[59] Additionally, a pathogen such as *Variola major*, which causes the disease smallpox, could be engineered to make it resistant to vaccines that offer protection against the agent. For instance, the pathogen could be engineered to resist the live, replicating vaccinia virus vaccine ACAM2000, which is a smallpox vaccine that is licensed by the FDA.[60] It is noted that replicating vaccinia virus "can lead to severe side effects, especially in people with atopic dermatitis or other forms of eczema, as well as in people with immune deficiencies."[61] Likewise,

---

[55] Stephen R. Thomas and Joseph S. Elkinton, "Pathogenicity and Virulence," *Journal of Invertebrate Pathology* 85, no. 3 (March 1, 2004): 146–51, https://doi.org/10.1016/j.jip.2004.01.006.

[56] Zilinskas, "Possible Terrorist Use of Modern Biotechnology Techniques," 5.

[57] National Research Council (US) Committee on Research Standards and Practices to Prevent the Destructive Application of Biotechnology, *Fink Report*, 114.

[58] Zilinskas, "Possible Terrorist Use of Modern Biotechnology Techniques," 3.

[59] Zilinskas, 4; National Research Council (US) Committee on Research Standards and Practices to Prevent the Destructive Application of Biotechnology, *Fink Report*, 114.

[60] National Research Council (US) Committee on Research Standards and Practices to Prevent the Destructive Application of Biotechnology, *Fink Report*, 114; Center for Biologics Evaluation and Research, "Questions about Vaccines – Smallpox," WebContent, accessed July 8, 2018, https://www.fda.gov/BiologicsBloodVaccines/Vaccines/QuestionsaboutVaccines/ucm070429.htm.

[61] Axel Lambert de Rouvroit and Erik D. Heegaard, "Total Costs Associated with Replicating and Non-Replicating Smallpox Vaccines," *Global Security: Health, Science and Policy* 1, no. 1 (January 1, 2016): 3, https://doi.org/10.1080/23793406.2016.1171944.

the pathogen could be engineered to resist Jynneos Smallpox and Monkeypox Vaccine, which is a different live, non-replicating vaccinia virus vaccine that was recently approved by the FDA to prevent smallpox and monkeypox.[62]

3.1.3.3 Protection of Own Forces

Prior to releasing a genetically engineered agent, an aggressor could develop a highly specific vaccine to immunize the aggressor's own forces against the agent. This would confer an advantage to immunized offensive troops because defensive forces would be vulnerable to infection.[63] Likewise, modifications could be made to allow for the safer handling and deployment of agents by those who use them.[64]

3.1.3.4 Ability to Evade Host Immunological Response

An agent could be modified so that it circumvents immunological response of the host. For instance, bioweaponeers may engineer an agent to secrete enzymes, so the modified agent can evade certain antibodies, such as Immunoglobulin A secreted from the host's skin cells.[65] Antigenic markers of agents could be modified, so individuals with acquired immunity against the agent, which could be acquired from immunization or prior infection, would be vulnerable again to infection by the newly engineered agent.[66] Antigenicity refers to "the ability of a foreign material (antigen) to bind to, or interact with, the products of the final cell-mediated response such as B-cell or T-cell receptors."[67] Antigenic determinants, also known as epitopes, are the portion of the antigen that interacts with the immune cells.[68] When an individual is exposed to an antigenic marker, such as through a vaccine, the individual develops immunity against the agent. Genetic engineering could be used to modify the antigenic marker, so individuals no longer have protection against the agent.

---

[62] "FDA Approves First Live, Non-Replicating Vaccine to Prevent Smallpox and Monkeypox," FDA, September 24, 2019, http://www.fda.gov/news-events/press-announcements/fda-approves-first-live-non-replicating-vaccine-prevent-smallpox-and-monkeypox.

[63] Tucker, "The Future of Biological Warfare," 64.

[64] Block, "Living Nightmares: Biological Threats Enabled by Molecular Biology," 46.

[65] Zilinskas, "Possible Terrorist Use of Modern Biotechnology Techniques," 4; Kadlec and Zelicoff, "Implications of the Biotechnology Revolution for Weapons Development and Arms Control," 20–21.

[66] Zilinskas, "Possible Terrorist Use of Modern Biotechnology Techniques," 7–8.

[67] Jonathan M. Fishman, Katherine Wiles, and Kathryn J. Wood, "Chapter 8: The Acquired Immune System Response to Biomaterials, Including Both Naturally Occurring and Synthetic Biomaterials," in *Host Response to Biomaterials* (Elsevier Inc., 2015), 151–52.

[68] Fishman, Wiles, and Wood, 152.

3.1.3.5 Ability to Avoid Detection and Diagnostics

An agent could be modified so that it evades existing environmental and clinical detection and identification methods.[69] This would make it difficult for the defender to know what medical countermeasures, including drugs and vaccines, to administer to victims. The medical response would probably be "delayed, sub-optimal, or erroneous."[70] In addition to modifying an agent's antigenic structure to evade the host's immune response, genetic engineering could be used to modify antigenic markers, so antibody-based environmental and clinical detection tests are no longer effective.[71]

3.1.3.6 Enhanced Environmental Resistance

Recall that biological weapons are derived from or made up of living organisms. Most traditional agents are fragile and are susceptible to degradation from ambient conditions such as high temperature, ultraviolet (UV) light, high relative humidity, and air movement.[72] Bioweaponeers may seek to genetically engineer more stable agents capable of better withstanding environmental stresses.[73] For instance, bacteria are generally less susceptible than viruses to UV light.[74] Additionally, Gram positive bacteria generally hold up better in the environment than Gram negative bacteria.[75] Thus, bioweaponeers may want to genetically engineer viruses to resist UV light or genetically engineer bacterial cell walls to look like Gram positive bacteria.[76]

---

[69] National Research Council (US) Committee on Research Standards and Practices to Prevent the Destructive Application of Biotechnology, *Fink Report*, 115; Zilinskas, "Possible Terrorist Use of Modern Biotechnology Techniques," 7.

[70] Zilinskas, "Possible Terrorist Use of Modern Biotechnology Techniques," 7.

[71] Kadlec and Zelicoff, "Implications of the Biotechnology Revolution for Weapons Development and Arms Control," 20–21; Zilinskas, "Possible Terrorist Use of Modern Biotechnology Techniques," 7–8.

[72] Kadlec and Zelicoff, "Implications of the Biotechnology Revolution for Weapons Development and Arms Control," 19–20; National Academies of Sciences, Engineering, and Medicine, *Biodefense in the Age of Synthetic Biology*, 60.

[73] National Research Council (US) Committee on Research Standards and Practices to Prevent the Destructive Application of Biotechnology, *Fink Report*, 115.

[74] Reynolds M. Salerno et al., "A BW Risk Assessment: Historical and Technical Perspectives," *The Nonproliferation Review* 11, no. 3 (2004): 24, https://doi.org/10.1080/10736700408436977.

[75] National Academies of Sciences, Engineering, and Medicine, *Biodefense in the Age of Synthetic Biology*, 66.

[76] National Academies of Sciences, Engineering, and Medicine, 66.

3.1.3.7 Enhanced Production, Storage, and Dissemination

Bioweaponeers traditionally face challenges due to limited production capacity and issues with delivery and dispersal.[77] Issues with stability during storage, both before and after dissemination, compound production and dissemination challenges.[78] Agents have a finite shelf-life. Although the shelf-life of many agents can be extended 3–6 months through refrigeration or several months to years via a process known as freeze-drying, or lyophilization, attributes of agents, such as virulence, can subside over time.[79] To reduce the likelihood that stored agents will need to be discarded and replaced with fresh agents, bioweaponeers may seek to engineer agents to stabilize these attributes. To increase the likelihood that agents will survive longer once released, bioweaponeers may seek to engineer agents to better withstand the physical and chemical stresses in the environment through the enhancement of genes that enable resistance to desiccation, UV radiation, and decontamination measures.[80]

3.1.3.8 Increased Vector Efficiency

A vector refers to "an organism that carries a pathogen to a host."[81] For instance, mosquitoes are vectors that carry the malarial parasites of the genus *Plasmodium* to humans.[82] A bioweaponeer may wish to modify vector competence to enhance the transmission of disease.[83] Vector competence is used to describe the ability of the vector "to become infected, to support the development or replication of a pathogen, and to transmit" the pathogen to the host.[84] Competent vectors spread the pathogen whereas incompetent vectors do not. Intrinsic, or rather genetic, factors influence the relationship between the vector and pathogen—and ultimately the transmission of the pathogen to the host.[85] The genes of vectors or pathogens, or both, could be

[77] Kadlec and Zelicoff, "Implications of the Biotechnology Revolution for Weapons Development and Arms Control," 21.

[78] Salerno et al., "A BW Risk Assessment: Historical and Technical Perspectives," 40–48.

[79] Tucker, "The Future of Biological Warfare," 67–68.

[80] Raymond A. Zilinskas, ed., *Biological Warfare: Modern Offense and Defense*, 1 edition (Boulder, CO: Lynne Rienner Pub, 2000), 3.

[81] Salerno et al., "A BW Risk Assessment: Historical and Technical Perspectives," 45.

[82] Salerno et al., 45.

[83] National Research Council (US) Committee on Research Standards and Practices to Prevent the Destructive Application of Biotechnology, *Fink Report*, 114.

[84] Claudia Fortuna et al., "Experimental Studies on Comparison of the Vector Competence of Four Italian Culex Pipiens Populations for West Nile Virus," *Parasites & Vectors* 8 (September 17, 2015): 2, https://doi.org/10.1186/s13071-015-1067-z.

[85] Brenda T. Beerntsen, Anthony A. James, and Bruce M. Christensen, "Genetics of Mosquito Vector Competence," *Microbiology and Molecular Biology Reviews* 64, no. 1 (March 2000): 116.

modified to improve their compatibility—thus increasing the likelihood that diseases will be transmitted to hosts.

3.1.3.9 Ability to Control Agents

Genetic engineering may allow bioweaponeers to limit the persistence of modified agents. Bioweaponeers could deliberately cause senescence, so agents "die on cue."[86] For instance, agents could be modified with suicide genes that program the agent to die after a certain number of generations or to activate only under certain environmental conditions.[87] This would allow more control over the manner in which contagious agents are used including the timeframe of attack.[88] Additionally, bioweaponeers could purposefully cause agents to go into a viable but nonculturable state, so it is dormant and undetectable. The aggressor could then supply a signal to induce the agent to be active and pathogenic.[89]

Additionally, bioweaponeers may seek to control the speed of action of agents. The incubation period refers to the time between exposure to the pathogen and the appearance of symptoms of the illness. Bioweaponeers could use genetic engineering to reduce the amount of time it takes for symptoms to begin. If the onset of symptoms began within hours, instead of days, of exposure, then the opponent would be impacted much faster. Thus, agents may "become more usable on the tactical battlefield."[90] For instance, bioweaponeers may be able to use a chemical or biological trigger to make a virus pathogenic much faster than normal.[91]

3.1.3.10 Ability to Alter Host Tropism

Host specificity describes the tendency of a pathogen to prefer a certain host. Bioweaponeers may wish to change the host range of an agent and make a nonzoonotic agent into a zoonotic agent.[92] The host range of pathogens could be expanded by modifying tropism to increase the odds that the agent will become established in the population. Tropism is described as the capacity of the pathogen

---

[86] Zilinskas, "Possible Terrorist Use of Modern Biotechnology Techniques," 8.

[87] Tucker, "The Future of Biological Warfare," 64.

[88] Zilinskas, "Possible Terrorist Use of Modern Biotechnology Techniques," 9.

[89] Charles Lucey, "Summary of Conference Papers," in *Emerging Technologies: Recommendations for Counter-Terrorism* (Institute for Security Technology Studies Dartmouth College, 2001), 358.

[90] Tucker, "The Future of Biological Warfare," 63.

[91] National Academies of Sciences, Engineering, and Medicine, *Biodefense in the Age of Synthetic Biology*, 60.

[92] National Research Council (US) Committee on Research Standards and Practices to Prevent the Destructive Application of Biotechnology, *Fink Report*, 114.

"to infect or damage specific cells, tissues, or species."[93] For instance, three mutations are needed in the hemagglutinin gene to alter the tropism of avian influenza H7N9 from avian to human.[94]

Alternatively, bioweaponeers may seek to diminish the pathogen's ability to impact populations that are not the target population. Moreover, bioweaponeers may seek to enhance the pathogen's ability to target a specific population.[95] More specifically, there are concerns over the possibility of creating "ethnic weapons" that target certain racial groups. Development of such weapons, however, is considered unlikely since human populations are not "genetically homogenous enough to be uniquely susceptible to infection by a given pathogen."[96] Use of ethnic weapons may result in significant collateral damage as unintended populations would likely be impacted because "genetic variation in susceptibility to disease is likely to be greater within ethnic groups than between them."[97]

3.1.4 Obstacles to Modification

Although there are several attributes that could be targeted for modification, users could run into technical, social, organization, and management issues when trying to genetically engineer agents. The section first reviews key technical challenges that users encounter during genetic modifications, which includes a lack of information due to the current state of understanding in the scientific field, complex modifications that are beyond the level that can be achieved, unwanted or unplanned effects following the modification, the inability to achieve expression of the modification, and properties that are inherent to the organism. Following the review of technical obstacles, this section considers the social barriers that could add to the existing complexity of modifications, which includes knowledge and skill as well as foreign technical assistance. Lastly, this section considers organization and management factors.

[93] National Academies of Sciences, Engineering, and Medicine, *Biodefense in the Age of Synthetic Biology*, 57.

[94] National Academies of Sciences, Engineering, and Medicine, 57, 63.

[95] Zilinskas, "Possible Terrorist Use of Modern Biotechnology Techniques," 6.

[96] Tucker, "The Future of Biological Warfare," 64–65.

[97] C. Dennis, "The Bugs of War," *Nature* 411, no. 6835 (May 17, 2001): 233, https://doi.org/10.1038/35077161.

3.1.4.1 Technical

State of Scientific Knowledge

Although the tools needed to make a modification may exist, users can be held back by the current state of scientific knowledge.[98] Users need to understand what it is they are trying to manipulate. Ellen Jorgensen, a molecular biologist and leader in DIY biology, advises, "We don't really know that much about cells. They are still kind of black boxes."[99] For instance, *E. coli* is a relatively simple and well-studied model organism. While scientists understand the function of most of the genes in *E. coli*, "much of its [i.e., *E. coli*] overall biology remains mysterious."[100] Jorgensen says, "we don't really know that much about how to make a particular thing happen by changing particular spots in the genome."[101] To illustrate this point, Jorgensen explains how scientists are a "long way away from" determining how to genetically engineer a pig to grow an extra leg or set of wings.[102]

In terms of humans, a "limited understanding of the microbiome," "limited understanding of complex immune processes," and "lack of knowledge about regulation of human gene expression" make it challenging to modify the human microbiome, human immune system, and human genome, respectively.[103] It is noted that there are concerns that the human microbiome, which is described as "the microorganisms that live on and within us, especially those associated with the gut, oral cavity, nasopharyngeal space, and skin," could be genetically engineered to develop a biological weapon.[104] These concerns are analyzed in depth in Chap. 5 and evaluated in Chap. 6.

Level of Difficulty of Modification

The level of difficulty varies depending on the modification. Simple cut-and-paste tasks to insert one gene into an organism are more straightforward than "introducing networks of genes and other elements under regulated control to rewire the internal workings' of the organism."[105] A 2018 National Academies of Sciences, Engineering, and Medicine (NASEM) report on synthetic biology cites "engineering complex

[98] Paul Voosen, "Synthetic Biology Comes Down to Earth," *The Chronicle of Higher Education*, March 4, 2013, http://www.chronicle.com/article/Synthetic-Biology-Comes-Down/137587.

[99] Ellen Jorgensen, *What You Need to Know about CRISPR* (Banff, Canada: TEDSummit, 2016), pt. 6:20.

[100] Voosen, "Synthetic Biology Comes Down to Earth."

[101] Jorgensen, *What You Need to Know about CRISPR*, 8:00.

[102] Jorgensen, 8:10–8:18.

[103] National Academies of Sciences, Engineering, and Medicine, *Biodefense in the Age of Synthetic Biology*, 106.

[104] National Academies of Sciences, Engineering, and Medicine, 71.

[105] Voosen, "Synthetic Biology Comes Down to Earth."

viral traits" and "engineering complex bacterial traits" as significant bottlenecks or barriers to creating more dangerous viruses and bacteria, respectively.[106]

Unwanted or Unplanned Effects

When an agent is genetically engineered to modify a certain attribute, other unwanted or unplanned effects may occur. In other words, bioweaponeers may try to enhance attributes, such as antibiotic resistance, but may unintentionally diminish other qualities, such as virulence and hardiness.[107] The former Soviet biological warfare program experienced this issue when trying to modify *F. tularensis* and other bacteria to resist antibiotics. As Raymond Zilinskas explains, the Soviets were able to transfer antibiotic resistance genes to *F. tularensis*; however, the modified agent lost its virulence and was no longer pathogenic.[108]

While the Soviets eventually developed *F. tularensis* that resisted antibiotics and maintained pathogenicity, they still had trouble when trying to make other bacteria strains, including *B. anthracis, B. mallei, B. pseudomallei, and Y. pestis*, resistant to 10 antibiotics—most only resisted seven or eight antibiotics. Issues with stability arose when Soviet scientists introduced constructs engineered with antibiotic resistance into the host cell because these new constructs frequently replaced former constructs.[109] However, the extent that the findings can be generalized to the present day are complicated by advances in biotechnology. As Zilinskas acknowledges:

> The failed attempt by Russian scientists to develop strains of bacterial pathogens resistant to 10 antibiotics demonstrates that the more antibiotic resistance one attempts to imbue an organism with, the larger the technical barriers that have to be overcome. It must be noted, however, that the Russian work on this subject was done before 1992. Since that time there have been remarkable advances in the biosciences so what was not doable by the Russians probably is possible today.[110]

This acknowledgement suggests that issues with unwanted or unplanned effects may be mitigated with current biotechnologies.

Expression of the Modification

Bioweaponeers may have trouble getting the modification to work in the target organism. Transfer of virulence factors between organisms is not always successful. For instance, scientists successfully transferred the toxin gene from *Listeria monocytogenes*, a pathogenic bacterium, to *Bacillus subtilis*, a non-pathogenic bacterium. Once transfected, *B. subtilis* could produce the toxin when grown in

---

[106]National Academies of Sciences, Engineering, and Medicine, *Biodefense in the Age of Synthetic Biology*, 7.

[107]Zilinskas, "Possible Terrorist Use of Modern Biotechnology Techniques," 4, 6, 8.

[108]Milton Leitenberg and Raymond A. Zilinskas, *The Soviet Biological Weapons Program: A History* (Cambridge, Massachusetts: Harvard University Press, 2012), 185.

[109]Leitenberg and Zilinskas, 187–88.

[110]Leitenberg and Zilinskas, 779.

mammalian cell culture; however, the engineered *B. subtilis* was avirulent when delivered into mice.[111]

Properties Inherent to the Organism
Properties inherent to the agent, such as the form of genetic information (DNA or RNA) and antigenic structure, can present their own set of technical challenges. For instance, it is considered more difficult to genetically engineer viruses whose genomic information is in the form of RNA instead of DNA. DNA is inherently more stable than RNA and more "amenable to the genetic splicing techniques that have made recombinant DNA technology as versatile."[112] In terms of RNA viruses, positive-sense single-stranded RNA viruses are considered the most straightforward to engineer. The RNA of these viruses is "the same sense as the messenger RNA that encodes the viral proteins," so these viruses are intrinsically infectious when inserted into cells.[113] It is noted that many RNA viruses can now be produced from plasmid (s) (one or multiple) systems (reverse genetics), which allow them to be modified much more readily.

Bioweaponeers may encounter more challenges when modifying the antigenic structures of bacteria than when modifying those of viruses. Bacteria typically have "an array of different antigenic structures on their cell surfaces," whereas the antigenic composition of viruses is usually not as complex.[114] Thus, several antigenic structures of bacteria may need modified to achieve immune or detection evasion. However, despite the relative simplicity of viruses, these pathogens may not be able to tolerate much modification before function is lost.[115]

3.1.4.2 Social

This section marks a transition point in the literature review from considering the technical obstacles to modification to other obstacles to modification including social, organization, and management. Scholars, such as Sonia Ben Ouagrham-Gormley and Kathleen M. Vogel, have identified social factors that are barriers to the development of biological weapons.[116] As the review of the literature reveals, bioweaponeers should have the knowledge and skill necessary to apply genetic engineering to agents; however, general biological knowledge is not the same as specialized biological weapons knowledge that can be tough to obtain, use, and transfer to others. To better account for the social context of biological weapons

[111] Nixdorff, Brauburger, and Hahlbohm, "The Biotechnology Revolution," 118.

[112] Institute of Medicine and National Research Council, *Globalization, Biosecurity, and the Future of the Life Sciences*, 161.

[113] Institute of Medicine and National Research Council, 162.

[114] Nixdorff, Brauburger, and Hahlbohm, "The Biotechnology Revolution," 117.

[115] Nixdorff, Brauburger, and Hahlbohm, 117.

[116] Ouagrham-Gormley, *Barriers to Bioweapons*; Vogel, *Phantom Menace or Looming Danger? A New Framework for Assessing Bioweapons Threats*.

work in threat assessments, Vogel developed the biosocial frame as an alternative to the biotech revolution frame (see Fig. 3.1).

This section first examines the biosocial frame, its key elements, the empirical basis of its claims, and its application to assessing bioweapons threats from both state and nonstate actors. Next, the section considers the knowledge and skill involved in applying genetic engineering to agents and more closely examines how knowledge is created and transferred. Afterwards, the chapter reviews the general biological weapons knowledge and skill that is needed to develop biological weapons and how foreign technical assistance can advance, delay, or obstruct work.

As introduced earlier in the chapter, there are different framings that analysts apply to assessments of biological weapons threats. Information that falls within the framing is deemed reliable and included for analysis in the assessment, while information that falls outside the framing is deemed irrelevant and "marginalized, ignored, or suppressed."[117] The literature suggests that persistent shortcomings in United States assessments can be attributed to issues with using traditional biotechnology frames, which are materials focused and do not consider social factors involved in biological weapons work.[118] The biosocial frame, which considers the social context of biological weapons work, is an alternative to traditional biotechnology frames.

Key elements of the biosocial frame are illustrated alongside the biotech revolution frame in Fig. 3.1. The biosocial frame focuses on the following: "(1) the important role of know-how and other uncodified knowledge; (2) the local and contingent character of biological development and use; (3) the importance of understanding past biological work and end products; and (4) the multiple, negotiated biotechnological trajectories that are possible."[119]

First, the biosocial frame emphasizes the importance of uncodified, also referred to as tacit or non-codified, knowledge in carrying out scientific work. As explained later in this chapter during the discussion of the social obstacles to modification, tacit knowledge is described as "skills, know-how, and sensory cues that are vital to the successful use of a technology yet cannot be reduced to writing and must be acquired through hand-on practice and experience."[120] Second, the biosocial frame accounts for the local laboratory knowledge as well as the specialized expertise that goes into scientific work. Knowledge that can be codified, or written down, and transferred to others in scientific publications makes scientific work seem like it is "clean and

---

[117] Vogel, *Phantom Menace or Looming Danger? A New Framework for Assessing Bioweapons Threats*, 13, 17.

[118] Vogel, 4–5.

[119] Vogel, 60.

[120] Jonathan B. Tucker, "Review of the Literature on Dual Use," in *Innovation, Dual Use, and Security: Managing the Risks of Emerging Biological and Chemical Technologies*, ed. Jonathan B. Tucker (Cambridge, Massachusetts: The MIT Press, 2012), 23.

orderly"—when in practice, scientists may have to extensively troubleshoot and work through other contingencies.[121]

Third, unlike the biotech revolution frame that treats technological developments as "unpredictable and disruptive," the biosocial frame cites historical patterns of technological advancements and views these developments as "incremental, evolutionary advances."[122] As opposed to the belief that technology develops autonomously, free from outside forces, the biosocial frame looks at how people shape the development of technology. An important aspect of the biosocial frame is the interpretive flexibility of a technology, which means that social groups can think of and use a technology in different ways. For instance, one group may choose to use a technology to develop bioweapons, while another group may opt to use the technology for religious purposes.[123] Fourth, the biosocial frame "does not assume that developments in biotechnology will inevitability follow a globalized and diffused pattern but recognizes that local and social components shape (and frustrate) standardization, routinization, and diffusion of biotechnologies."[124] Thus, permitting the trajectories of the technological and bioweapons threat to vary and not necessarily follow a linear or exponential course.

Through case studies that reexamine controversies over synthetic genomics and the development of bioweapons in a Soviet facility, Vogel's *Phantom Menace or Looming Danger? A New Framework for Assessing Bioweapons Threats* demonstrates how the biosocial frame provides a more comprehensive understanding of bioweapons threat than the biotech revolution frame. These case studies reveal how the biosocial frame better captures "the contingency and complexity present in the development, use, and transfer of bioweapons technologies" and "provides a more fruitful understanding of bioweapons threats from state or nonstate actors."[125]

As reviewed in more detail below, there are many social obstacles to developing biological weapons that should not be ignored from misuse assessments. While a genetic engineering technology may be technically capable of achieving an end product, it does not guarantee the actor has the knowledge and skills to work through laboratory procedures and issues that come up along the way. Thus, to avoid repeating the same shortcomings as past assessments, this study applies the biosocial frame to assess the risk of misuse of genome editing technologies. Ultimately, the biosocial frame is a better description of the current situation as compared to the biotech revolution because the biosocial frame better accounts for the contingencies and complexities involved in the process of biological weapons development instead of focusing on the end product.

---

[121] Vogel, *Phantom Menace or Looming Danger? A New Framework for Assessing Bioweapons Threats*, 61–63.

[122] Vogel, 66.

[123] Vogel, 66–67.

[124] Vogel, 69.

[125] Vogel, 14.

Knowledge and Skill

A bioweaponeer should have the knowledge and skill necessary to apply genetic engineering to agents. To genetically engineer an agent, bioweaponeers should know what part of the genetic code to modify and in what way to modify it.[126] Reynolds M. Salerno and colleagues advise that bioweaponeers who wish to enhance an agent's infectivity or pathogenicity, or both, need "highly specialized expertise that combines knowledge of a particular pathogen or toxin with knowledge of genetics and DNA processes."[127] Sonia Ben Ouagrham-Gormley explains how "specialized knowledge is not easy to acquire, use, or transfer because much of it is tacit, local, and collective in nature."[128] Give the "local and personal character of bioweapons knowledge, specialized skills, and scientific know-how," it can be difficult to transfer between people and locations.[129] While some knowledge may be conveyed through codified, or explicit, means such as laboratory protocols, other expertise may uncodified, or tacit, and transferred through direct means such as a master-apprentice relationship.[130]

The remainder of this section reviews the differences between the types of knowledge (explicit and tacit) and forms of tacit knowledge. Next, this section considers how knowledge is transferred as well as concerns over video protocols that may erode tacit knowledge requirements. Finally, this section notes that bioweaponeers who apply genetic engineering techniques to create biological weapons also need to have general bioweapons knowledge and skill. Traditionally, bioweaponeers encounter challenges during the acquisition, production, and dissemination of agents.[131] Thus, the final section reviews the knowledge and skills needed for the acquisition, production, and dissemination of agents.

Knowledge Creation

The creation of knowledge is a two-step process. In the first step, data are converted into information. Data are described as "series of observations, measurements, or

---

[126] Salerno et al., "A BW Risk Assessment: Historical and Technical Perspectives," 43.

[127] Salerno et al., 43.

[128] Ouagrham-Gormley, *Barriers to Bioweapons*, 17.

[129] Sonia Ben Ouagrham-Gormley and Kathleen M. Vogel, "The Social Context Shaping Bioweapons (Non)Proliferation," *Biosecurity and Bioterrorism: Biodefense Strategy, Practice, and Science* 8, no. 1 (2010): 9, https://doi.org/10.1089/bsp.2009.0054.

[130] H. M. Collins, "Tacit Knowledge, Trust and the Q of Sapphire," *Social Studies of Science* 31, no. 1 (February 2001): 71–83; Ouagrham-Gormley, *Barriers to Bioweapons*, 17–29; James Revill and Catherine Jefferson, "Tacit Knowledge and the Biological Weapons Regime," *Science and Public Policy* 41, no. 5 (October 1, 2014): 599–603, https://doi.org/10.1093/scipol/sct090; Vogel, *Phantom Menace or Looming Danger? A New Framework for Assessing Bioweapons Threats*, 61–63; Ouagrham-Gormley and Vogel, "The Social Context Shaping Bioweapons (Non)Proliferation," 13.

[131] Salerno et al., "A BW Risk Assessment: Historical and Technical Perspectives," 40–48; Graham S. Pearson, "The Essentials of Biological Threat Assessment," in *Biological Warfare: Modern Offense and Defense*, ed. Raymond A. Zilinskas (Boulder, CO: Lynne Rienner Publishers, 2000), 62–66; Tucker, "The Future of Biological Warfare," 67–72.

facts in the form of numbers, words, sounds and/or images."[132] This raw material (i.e., data) gives way to information. Information is described as "data that have been arranged into a meaningful pattern."[133] For instance, data from a survey could be processed and arranged into charts and graphs that generate informative patterns. In the second step, information is organized, analyzed, converted into knowledge, which is described as "the application and productive use of information."[134] The conversion of information into knowledge depends on a human dimension since scientists draw from prior explicit and tacit knowledge, experience, and information from many sources to understand and assign meaning to the information.[135]

There are two main types of knowledge: explicit and tacit. Explicit, or codified, knowledge, is described as knowledge that is "written down in the form of a recipe or laboratory protocol, and transferred from one individual to another by impersonal means, such as publication in a scientific journal."[136] On the other hand, tacit, or non-codified, knowledge is described as "skills, know-how, and sensory cues that are vital to the successful use of a technology yet cannot be reduced to writing and must be acquired through hand-on practice and experience."[137]

The several forms of tacit knowledge described in the literature are broadly categorized these as weak, somatic, and communal. The first category of weak tacit knowledge refers to knowledge that could be explicit depending on the condition but is uncodified due to the "inability, unwillingness or practicality" to make it codified.[138] The sub-forms of weak tacit knowledge—including concealed knowledge,[139] mismatched saliences,[140] unrecognized knowledge[141] and ostensive

[132] Joanne Roberts, "From Know-How to Show-How? Questioning the Role of Information and Communication Technologies in Knowledge Transfer," *Technology Analysis & Strategic Management* 12, no. 4 (December 1, 2000): 430, https://doi.org/10.1080/713698499.

[133] Roberts, 430.

[134] Roberts, 430; National Research Council, *Life Sciences and Related Fields: Trends Relevant to the Biological Weapons Convention* (Washington, DC: The National Academies Press, 2011), 96, https://doi.org/10.17226/13130.

[135] National Research Council, *Life Sciences and Related Fields*, 96–97.

[136] Tucker, "Review of the Literature on Dual Use," 23.

[137] Tucker, 23.

[138] Revill and Jefferson, "Tacit Knowledge and the Biological Weapons Regime," 599.

[139] Concealed knowledge is knowledge that is intentionally or unintentionally left out, which is illustrated by a scientist who intentionally leaves out tricks of the trade or unintentionally leaves out knowledge from publications due to space limitations.

[140] Mismatched salience occurs when communications are mismatched such as the case when scientist A does not realize scientist B should be directed to do things a particular way, and scientist B does not know what to ask scientist A so that knowledge is transferred.

[141] Unrecognized knowledge occurs when scientists are unaware that they are performing procedures in a certain way; other scientists may unknowingly acquire the habit when working with scientists who know how to perform the procedure—both parties may be unaware that any knowledge was even transferred.

knowledge[142]—were first defined by H. M. Collins and then referred to by several scholars in discussions of biological weapons.[143] Concealed knowledge and mismatched saliences are specific sub-forms of weak tacit knowledge that could be made explicit, while unrecognized knowledge and ostensive knowledge are forms of weak tacit knowledge that could not, or are not, made explicit.[144]

The second category of somatic, or somatic-limited, tacit knowledge describes "things that our bodies can do which we cannot articulate, transfer and replicate as knowledge without the recipient learning by doing."[145] For instance, we can be lectured on or read about how to ride a bike; however, much of the ability to do so comes with actually getting on a bike, practicing, and training our bodies on how to balance.[146] Unlike weak and somatic tacit knowledge, the third category of communal tacit knowledge refers to knowledge that results from the collaborative effort of individuals from different disciplines who work together.[147] One interpretation of communal tacit knowledge is that it occurs when experts from different backgrounds are brought together and work to achieve a common goal attainable through the "sum of their parts, but, nonetheless, can still be separated out into their respective individual parts."[148] The other interpretation of communal tacit knowledge is that it is knowledge created communally and formed by the continued interaction of persons whose experience varies. Importantly, these persons do not exist individually, instead they are integrated within a group.[149]

Knowledge Transfer

Scientists strive to convert knowledge into explicit forms that can be shared among individuals and communities through reservoirs such as scientific articles and experimental protocols. Although knowledge may be codified, it may not be possible to capture all its tacit element, so the transfer of only codified knowledge may not result in the complete transfer of knowledge. Thus, tacit knowledge may need to be

---

142 Ostensive knowledge is knowledge that can be codified, such as through words and diagrams; however, this form of knowledge is too complex to be fully conveyed explicitly and requires some sort of demonstration or pointing to understand.

143 Collins, "Tacit Knowledge, Trust and the Q of Sapphire," 72; Ouagrham-Gormley, *Barriers to Bioweapons*, 19–20; Kathleen Vogel, "Bioweapons Proliferation: Where Science Studies and Public Policy Collide," *Social Studies of Science* 36, no. 5 (2006): 682; Revill and Jefferson, "Tacit Knowledge and the Biological Weapons Regime," 599–600.

144 Ouagrham-Gormley, *Barriers to Bioweapons*, 19–20.

145 Revill and Jefferson, "Tacit Knowledge and the Biological Weapons Regime," 600–601.

146 Collins, "Tacit Knowledge, Trust and the Q of Sapphire," 82; Ouagrham-Gormley, *Barriers to Bioweapons*, 21; Revill and Jefferson, "Tacit Knowledge and the Biological Weapons Regime," 601.

147 Ouagrham-Gormley, *Barriers to Bioweapons*, 22; Revill and Jefferson, "Tacit Knowledge and the Biological Weapons Regime," 602.

148 Revill and Jefferson, "Tacit Knowledge and the Biological Weapons Regime," 602–3.

149 Revill and Jefferson, 603.

transferred along with explicit knowledge.[150] Tacit knowledge may be acquired through "learning by example" from experts and other advisors or "learning by doing" by extensive trial-and-error.[151]

Ben Ouagrham-Gormley emphasizes the "personal, communal, and local" character of knowledge.[152] She advises how knowledge "is created within a specific environment, by a specific set of individuals, and with a specific infrastructure," so it is not easy to transfer this knowledge to other places.[153] Similarly, Joanne Roberts emphasizes the "co-location and co-presence" needed to transfer tacit knowledge.[154] For instance, when adapting a technology to a new site, at least one of the original contributors may need to be present in the new location to demonstrate how to use the technology.[155] Although the transfer of knowledge can be difficult and actors, such as terrorists, may encounter challenges when using "published scientific method to create a biological weapon," Vogel acknowledges: "This does not mean that terrorists would never be able to develop such know-how or alternative synthetic methods, merely that there are very specific factors and timescales required to obtain biological know-how that vary according to the biological material, technique, and practice."[156]

General Biological Weapons Knowledge and Skill

Prior to concluding this section, it is important to point out that bioweaponeers who apply genetic engineering techniques to create biological weapons should also have the general knowledge and skill to make these weapons. General knowledge of biology is not the same thing as the specific knowledge needed to make biological weapons. Ben-Ouagrham Gormley advises how the misconception "that any biology-related knowledge is applicable to bioweapons development and that bioweapons expertise is easily acquired and used" is at the core of faulty bioweapons threat assessments.[157] In addition to knowledge and skill needed for genetic engineering, a bioweaponeer also should have the knowledge and skill to acquire, produce, and disseminate biological weapons.[158]

First, bioweaponeers need to acquire a virulent strain of the agent. Though seemingly straightforward, this task can be regarded as complicated. According to Reynold Salerno and colleagues, there are many differences in "pathogenicity, lethality, transmission rates, environmental susceptibility, and other factors" in

[150] National Research Council, *Life Sciences and Related Fields*, 97; Ouagrham-Gormley, *Barriers to Bioweapons*, 29; Roberts, "From Know-How to Show-How?," 433.

[151] Ouagrham-Gormley, *Barriers to Bioweapons*, 20–21; Vogel, "Bioweapons Proliferation," 670.

[152] Ouagrham-Gormley, *Barriers to Bioweapons*, 22.

[153] Ouagrham-Gormley, 22.

[154] Roberts, "From Know-How to Show-How?," 439.

[155] Ouagrham-Gormley, *Barriers to Bioweapons*, 22–23.

[156] Vogel, *Phantom Menace or Looming Danger? A New Framework for Assessing Bioweapons Threats*, 9.

[157] Ouagrham-Gormley, *Barriers to Bioweapons*, 3.

[158] Salerno et al., "A BW Risk Assessment: Historical and Technical Perspectives," 40–46.

immunologically similar agents, and technical skill is needed to decipher between virulent and avirulent strains.[159] They suggest the skills to identify and isolate a viable, virulent agent from nature would likely "not be available to an untrained individual nor would it be easily mastered by reading a textbook on the subject."[160] In addition to acquiring an agent from nature, they also suggest an agent could be obtained from microbial culture collections or bioscience laboratories; however, legislation and other guidelines make acquisition through these sources difficult.[161]

Second, an agent needs to be produced in a suitable form and sufficient quantity.[162] The bioweaponeer needs to understand the ways to produce the agent, which typically involves preparing the inoculum and nutrient media; cultivating the microorganisms; and concentrating, purifying, preparing, and packaging the product. The details of this process, however, vary depending on several factors such as the agent, scale of production, and type of formulation (dry or liquid).[163] During this process, contamination or genetic mutations can occur, so bioweaponeers should know how to identify and mitigate these issues. In addition to producing the agent in a suitable form, a bioweaponeer needs to know the quantity needed to achieve the objective of its use. The calculation should account for the amount of agent required to injure or kill the victim, which varies depending on the agent and its route of infection. The calculation should also account for as the amount that will die-off due to environmental stressors and the amount of agent that will reach the target.[164] Issues producing the agent are greatly dependent on the amount needed—large-scale production is considered "extremely challenging" because agents may lose one or more qualities, such as infectivity, in the process of scale-up.[165]

Third, delivery is especially difficult and seldom accomplished. While past offensive biological warfare programs in the United States and former Soviet Union developed effective delivery systems, other programs such as the Iraq program only achieved rudimentary delivery methods.[166] The dissemination of "a liquid suspension or freeze-dried powder" of agents is regarded as the "most difficult aspect of weaponization."[167] While aerosol dissemination of agents is recognized as an effective mode of delivery, it can be challenging, especially for hydrated agents, to

---

[159] Salerno et al., 40.

[160] Salerno et al., 41.

[161] Salerno et al., 41.

[162] Salerno et al., 42.

[163] Ken Alibek, "Threat Analysis I: Bacterial Agents." (Lecture series, George Mason University, Fairfax, VA, Fall 2003); Fox, Donna, *Scientific and Technical Aspects of a Large-Scale Biological Weapons Program: Lessons Learned from the Former Soviet Union's Anti-Personnel BW Program, 1975–1990* (PhD in Biodefense at George Mason University, ProQuest Dissertations Publishing, 2006), 120.

[164] Salerno et al., "A BW Risk Assessment: Historical and Technical Perspectives," 42.

[165] National Academies of Sciences, Engineering, and Medicine, *Biodefense in the Age of Synthetic Biology*, 108.

[166] van Aken and Hammond, "Genetic Engineering and Biological Weapons," S58.

[167] Tucker, "The Future of Biological Warfare," 67–68.

generate particles that are the correct diameter (1–5 microns).[168] Genetic engineering could be applied to improve aerosols or develop other distribution methods.[169]

Foreign Technical Assistance

Foreign technical assistance is a type of exogenous variable, or rather factor that originates outside of a program, that influence the "program's stability, continuity, and integration."[170] While foreign technical assistance can aid the advancement of biological weapons programs, this type of assistance can also delay or obstruct progress if the type and timing of assistance does not align with the recipient's program ability to absorb and integrate the help. The organization and management of the recipient's program will also influence how well the foreign technical assistance is integrated.[171]

3.1.4.3 Organization and Management

Organization and management factors are types of endogenous variables, or rather factors that originate within a program, that directly influence the creation, transfer, use, and sustainment of knowledge inside a biological weapons program.[172] Structural organizations designed to support close interactions between people and mobility of those people within the organization facilitate the transfer of tacit knowledge and skill. Additionally, integrative mechanisms established to coordinate independent tasks and stages of complex programs support the successful development of bioweapons.[173]

While organizational factors will assist knowledge transfer, managerial factors will support recipient acceptance and use of the knowledge. Managers who create a trustworthy environment and maintain a shared, technical language among subordinates improve the likelihood of acceptance and use of knowledge transferred from fellow workers. Managers who reward openness and information sharing will support knowledge institutionalization, whereas managers who punish undesirable actions may make it harder to identify and solve problems. Lastly, managers who keep communication open between those who make political decisions and those who implement programs help to secure political and financial backing for bioweapons programs.[174]

[168]Tucker, 69.
[169]Block, "Living Nightmares: Biological Threats Enabled by Molecular Biology," 46–47.
[170]Ouagrham-Gormley, *Barriers to Bioweapons*, 53.
[171]Ouagrham-Gormley, 13, 56–63.
[172]Ouagrham-Gormley, 12, 37, 62.
[173]Ouagrham-Gormley, 38–45.
[174]Ouagrham-Gormley, 45–52.

3.2 Misuse of Genome Editing Technologies

While the literature reviewed so far covers older and more general concerns over the misuse potential of genetic engineering, this section uncovers more specific concerns over the potential misuse of genome editing. First, this section traces the rising concerns over the potential of ZFN, TALEN, and CRISPR to be misused to develop biological weapons. Next, it lays out the fears that people have over the ways that genome editing could be misused: to alter microbes, make gene drives, and augment humans. An assessment of the risk of misuse of genome editing technologies to achieve these applications is conducted in Chaps. 4 and 5.

3.2.1 Monitoring for Dual-Use Technologies

Concerns over the potential misuse of genome editing with programmable nuclease technologies were initially raised over ZFN followed by TALEN and CRISPR. A review of documentation from official meetings of the BWC reveals that worries about ZFN misuse were identified at least as early as 2011. ZFN was identified as a technology that could "enable greater control in manipulating biological systems and networks."[175] Even so, ZFN were considered hard to design and produce—"a task left to specialist contractors."[176] Nevertheless, key advancements made to ZFN in 2011 streamlined the process, increased DNA cleavage activity, and enhanced modular assembly. The updated 2012 BWC background document suggested that these advancements "could open the door for the much wider use of this technology."[177] In spite of these advancements, the document advised that the patent monopoly held over zinc-fingers, by a company called Sangamo, may still impede knowledge of ZFN acquired via its commercial application and affect research endeavors.[178]

In 2013, worries over ZFN as an enabling technology were reinforced and new concerns were initiated over TALEN and CRISPR technologies. The updated 2013 BWC background document identified recent advances in ZFN and developments in

[175] *Seventh Review Conference of the States Parties to the Convention on the Prohibition of the Development, Production and Stockpiling of Bacteriological (Biological) and Toxin Weapons and on Their Destruction*, 6.

[176] "Advances in Enabling Technologies: Background Information Document Submitted by the Implementation Support Unit," in *Meeting of the States Parties to the Convention on the Prohibition of the Development, Production and Stockpiling of Bacteriological (Biological) and Toxin Weapons and on Their Destruction* (Geneva, 2012), 8, https://documents-dds-ny.un.org/doc/UNDOC/GEN/G12/611/26/PDF/G1261126.pdf.

[177] "Advances in Enabling Technologies: Background Information Document Submitted by the Implementation Support Unit," 8–9.

[178] "Advances in Enabling Technologies: Background Information Document Submitted by the Implementation Support Unit," 9.

TALEN that elevated concerns over these technologies. Nevertheless, the document assessed CRISPR as "more specific and cheaper" and "more efficient" than ZFN and TALEN, respectively.[179] Worries over CRISPR quickly overshadowed those over ZFN and TALEN, and, by 2014, concerns focused on CRISPR.[180] In particular, concerns emerged in the literature over the potential misuse of engineered, CRISPR-based gene drives.[181] Gene drives increase the odds that offspring will inherit the gene to almost 100%—in effect *driving* genes through a population much faster than normal.[182]

While there was some discussion, as mentioned above, of the potential risk of misuse of genome-editing technologies pre-2016, the debate amplified following the report from Director Clapper in 2016 (see Chap. 1).[183] Although the report did not specify ways the technologies could be misused, Piers Millett suggested the following possibilities: "altering pathogens, application of gene drives, influencing future human generations, new types of neurological weapons, and enhancement of military capabilities ("super soldiers")."[184] The 2018 NASEM report on synthetic biology, which was limited to activities that directly impacted human health, identified several concerns regarding synthetic biology-enabled capabilities, such as making existing bacteria and viruses more dangerous.[185] CRISPR, as one of the many technologies that enable synthetic biology, has the potential to be misused to achieve these worrisome capabilities. Additionally, the 2018 *Editing Biosecurity* report concisely summarized the biological threats that are enabled by genome editing as "modifying pathogens to be more dangerous," "hijacking the microbiome," "weaponizing gene drives," and "weaponizing gene therapy."[186]

---

[179] *Meeting of the States Parties to the Convention on the Prohibition of the Development, Production and Stockpiling of Bacteriological (Biological) and Toxin Weapons and on Their Destruction*, 7–8.

[180] "Advances in Science and Technology Related to the Convention," in *Meeting of the States Parties to the Convention on the Prohibition of the Development, Production and Stockpiling of Bacteriological (Biological) and Toxin Weapons and on Their Destruction*, 2014, 11, https://unoda-web.s3-accelerate.amazonaws.com/wp-content/uploads/assets/media/1742CC0986E6A683C1257D49004030BC/file/BWC_MSP_2014_MX_INF.3%2BE.pdf.

[181] David Gurwitz, "Gene Drives Raise Dual-Use Concerns," *Science* 345, no. 6200 (August 29, 2014): 1010, https://doi.org/10.1126/science.345.6200.1010-b.

[182] Kevin M. Esvelt et al., "Concerning RNA-Guided Gene Drives for the Alteration of Wild Populations," *Elife* 3, no. e03104 (2014): 3, https://doi.org/10.7554/eLife.03401.

[183] Clapper, 9.

[184] Robin Fears, *Assessing the Security Implications of Genome Editing Technology: Report of an International Workshop* (Herrenhausen, Germany: InterAcademy Partnership, 2017), 11.

[185] National Academies of Sciences, Engineering, and Medicine, *Biodefense in the Age of Synthetic Biology*, 5, 10.

[186] Kirkpatrick et al., "Editing Biosecurity: Needs and Strategies for Governing Genome Editing," 50.

3.2.2 *Applications of Misuse*

Based on the reports reviewed in the previous section on monitoring for dual-use technologies, the potential applications of misuse for genome editing are concisely summarized as follows: altering microbes, weaponizing gene drives, and augmenting humans. Altering microbes includes both traditional applications for making existing pathogens more dangerous as well as new applications for modifying the human microbiome. Augmenting humans includes applications for engineering super soldiers as well as the general population—both of which could be achieved through gene therapy. It is noted that, for neuroweapons, genome editing could be applied in a traditional sense to make existing pathogens more dangerous *or* in a new way to augment humans. Neuroweapons, are defined as "weapons that specifically target the brain or the central nervous system in order to affect the targeted person's mental state, mental capacity and ultimately the person's behaviour in a specific and predictable way."[187] The magnitude of potential harm resulting from the misuse of genome editing technologies to achieve these applications is assessed in Chap. 5.

3.3 Conclusion

Dating back to the 1970s, fears over the misuse of genetic engineering centered around how dual-use technologies could make agents more dangerous. While there are still worries that newer genome editing technologies, namely CRISPR, could make agents more dangerous, there are also concerns over the potential to develop new types of weapons. These new applications of misuse include modifying the human microbiome, weaponizing gene drives, and augmenting super soldiers and the general population. The newer debate over the misuse of genome editing fits into the longstanding debate over the threat of genetic engineering.

There are several types of obstacles that users could run into when genetically engineering agents. In addition to technical obstacles, there are also social obstacles, as well as organization and management obstacles, to developing biological weapons. Early assessments of the threat of genetic engineering embodied elements of the biotech revolution frame. The biotech revolution frame is technically focused with a heavy emphasis on the end product. A limitation in emphasizing the end product, however, is that it marginalizes the process that it took to arrive at the results. While a genetic engineering technology may technically be able to successfully make a modification, it does not mean the actor has the knowledge and skills to work through the procedures and issues that may arise while working with the technology in the laboratory.

[187] Armin Krishnan, *Military Neuroscience and the Coming Age of Neurowarfare* (Taylor & Francis, 2017), 11.

The biosocial frame was examined in this chapter an alternative to the biotech revolution frame. A strength of the biosocial frame is that it places an emphasis on the knowledge and skills needed to use technologies, which could be a limiting factor in the development of biological weapons even if users have all the materials. At the same time, one of the focuses of the biosocial frame is that it considers tacit knowledge requirements to use technologies and the ability to transfer this type of knowledge. Given that tacit knowledge is uncodified, it is inherently more difficult to measure in an assessment than explicit knowledge. Background on the social obstacles to modification, especially tacit knowledge transfer, is applied later in this book to guide the collection and analysis of qualitative data on the types of knowledge and skills needed to use CRISPR (see Chap. 4, Parameter 2).

Unfortunately, scholars and experts have offered limited insight into the feasibility and plausibility over the potential misuse of genome editing technologies to make pathogens more dangerous and to develop new types of biological weapons. The risk of misuse assessment of genome editing technologies is conducted in Chaps. 4 and 5 to help to fill in this gap in knowledge. A unique feature of this risk of misuse assessment is that it is conducted within a biosocial frame. The biosocial frame provides a better description of the current situation because it takes into consideration the contingencies and complexities involved in the process of biological weapons development instead of narrowly focusing on the end product.

References

1. Administration of George W. Bush. "Homeland Security Presidential Directive/HSPD–18—Medical Countermeasures Against Weapons of Mass Destruction." GPO, January 31, 2007.
2. "Advances in Biotechnology and Genetic Engineering: Implications for the Development of New Biological Warfare Agents." Washington, DC: Department of Defense, June 1996.
3. "Advances in Enabling Technologies: Background Information Document Submitted by the Implementation Support Unit." In *Meeting of the States Parties to the Convention on the Prohibition of the Development, Production and Stockpiling of Bacteriological (Biological) and Toxin Weapons and on Their Destruction*. Geneva, 2012. https://documents-dds-ny.un.org/doc/UNDOC/GEN/G12/611/26/PDF/G1261126.pdf.
4. "Advances in Science and Technology Related to the Convention." In *Meeting of the States Parties to the Convention on the Prohibition of the Development, Production and Stockpiling of Bacteriological (Biological) and Toxin Weapons and on Their Destruction*, 2014. https://unoda-web.s3-accelerate.amazonaws.com/wp-content/uploads/assets/media/1742CC0986E6A683C1257D49004030BC/file/BWC_MSP_2014_MX_INF.3%2BE.pdf.
5. Aken, Jan van, and Edward Hammond. "Genetic Engineering and Biological Weapons." *EMBO Reports* 4, no. Suppl 1 (2003): S57–60. https://doi.org/10.1038/sj.embor.embor860.
6. Alibek, Ken. "Threat Analysis I: Bacterial Agents." Lecture series, George Mason University, Fairfax, VA, Fall 2003.
7. Army Medical Research Institute for Infectious Diseases. *Usamriid's Medical Management of Biological Casualties Handbook*. Edited by Mark R. Withers. 8th ed. Government Printing Office, 2014.
8. Beerntsen, Brenda T., Anthony A. James, and Bruce M. Christensen. "Genetics of Mosquito Vector Competence." *Microbiology and Molecular Biology Reviews* 64, no. 1 (March 2000): 115–37.

9. Block, Steven M. "Living Nightmares: Biological Threats Enabled by Molecular Biology." In *The New Terror: Facing the Threat of Biological and Chemical Weapons*, edited by Sidney D. Drell, Abraham D. Sofaer, and George D. Wilson. Stanford, Calif: Hoover Institution Press, 1999.
10. Center for Biologics Evaluation and Research. "Questions about Vaccines – Smallpox." WebContent. Accessed July 8, 2018. https://www.fda.gov/BiologicsBloodVaccines/Vaccines/QuestionsaboutVaccines/ucm070429.htm.
11. Clapper. *Worldwide Threat Assessment of the U.S. Intelligence Community: Statement for the Record before the Senate Armed Services Committee*, 114th Con., 2nd Sess., February 9, 2016 (statement of James R. Clapper, Director of National Intelligence).
12. Collins, H. M. "Tacit Knowledge, Trust and the Q of Sapphire." *Social Studies of Science* 31, no. 1 (February 2001): 71–85.
13. Commission on the Prevention of WMD Proliferation and Terrorism. "World at Risk: The Report of the Commission on the Prevention of WMD Proliferation and Terrorism." New York: Vintage Books, 2008.
14. Coyle, Philip E. "Biological Warfare and Bioterrorism: How Do They Differ from Other WMD Threats?" In *Global Biosecurity: Threats and Responses*, edited by Peter Katona, John P. Sullivan, and Michael D. Intriligator. Routledge, 2010.
15. Dennis, C. "The Bugs of War." *Nature* 411, no. 6835 (May 17, 2001): 232–35. https://doi.org/10.1038/35077161.
16. Esvelt, Kevin M., Andrea L. Smidler, Flaminia Catteruccia, and George M. Church. "Concerning RNA-Guided Gene Drives for the Alteration of Wild Populations." *Elife* 3, no. e03104 (2014). https://doi.org/10.7554/eLife.03401.
17. FDA. "FDA Approves First Live, Non-Replicating Vaccine to Prevent Smallpox and Monkey-pox," September 24, 2019. http://www.fda.gov/news-events/press-announcements/fda-approves-first-live-non-replicating-vaccine-prevent-smallpox-and-monkeypox.
18. Fears, Robin. *Assessing the Security Implications of Genome Editing Technology: Report of an International Workshop*. Herrenhausen, Germany: InterAcademy Partnership, 2017.
19. Fishman, Jonathan M., Katherine Wiles, and Kathryn J. Wood. "Chapter 8: The Acquired Immune System Response to Biomaterials, Including Both Naturally Occurring and Synthetic Biomaterials." In *Host Response to Biomaterials*. Elsevier Inc., 2015.
20. Fortuna, Claudia, Maria Elena Remoli, Marco Di Luca, Francesco Severini, Luciano Toma, Eleonora Benedetti, Paola Bucci, et al. "Experimental Studies on Comparison of the Vector Competence of Four Italian Culex Pipiens Populations for West Nile Virus." *Parasites & Vectors* 8 (September 17, 2015): 463. https://doi.org/10.1186/s13071-015-1067-z.
21. Fox, Donna. *Scientific and Technical Aspects of a Large-Scale Biological Weapons Program: Lessons Learned from the Former Soviet Union's Anti-Personnel BW Program, 1975–1990*. PhD in Biodefense at George Mason University, ProQuest Dissertations Publishing, 2006.
22. Gilsdorf, Janet R., and Raymond A. Zilinskas. "New Considerations in Infectious Disease Outbreaks: The Threat of Genetically Modified Microbes." *Clinical Infectious Diseases: An Official Publication of the Infectious Diseases Society of America* 40, no. 8 (April 15, 2005): 1160–65. https://doi.org/10.1086/428843.
23. Gurwitz, David. "Gene Drives Raise Dual-Use Concerns." *Science* 345, no. 6200 (August 29, 2014): 1010–1010. https://doi.org/10.1126/science.345.6200.1010-b.
24. Institute of Medicine and National Research Council. *Globalization, Biosecurity, and the Future of the Life Sciences*. Washington, DC: National Academies Press, 2006. https://doi.org/10.17226/11567.
25. Jorgensen, Ellen. *What You Need to Know about CRISPR*. Banff, Canada: TEDSummit, 2016.
26. Kadlec, Robert P., and Alan P. Zelicoff. "Implications of the Biotechnology Revolution for Weapons Development and Arms Control." In *Biological Warfare: Modern Offense and Defense*, edited by Raymond A. Zilinskas. Boulder, CO: Lynne Rienner Publishers, 2000.

27. Kirkpatrick, Jesse, Gregory D. Koblentz, Megan J. Palmer, Edward Perello, David A. Relman, and Sarah W. Denton. "Editing Biosecurity: Needs and Strategies for Governing Genome Editing," December 2018.
28. Koblentz, Gregory D. *Living Weapons: Biological Warfare and International Security*. Ithaca: Cornell University Press, 2009.
29. Krishnan, Armin. *Military Neuroscience and the Coming Age of Neurowarfare*. Taylor & Francis, 2017.
30. Lederberg, Joshua. "Views: Biological Warfare: A Global Threat: Will Biomedical Knowledge Be Dedicated to Peaceful Uses by International Agreement or Do We Face a New Weapons Spiral?" *American Scientist* 59, no. 2 (1971): 195–97.
31. Leitenberg, Milton, and Raymond A. Zilinskas. *The Soviet Biological Weapons Program: A History*. Cambridge, Massachusetts: Harvard University Press, 2012.
32. Lucey, Charles. "Summary of Conference Papers." In *Emerging Technologies: Recommendations for Counter-Terrorism*. Institute for Security Technology Studies Dartmouth College, 2001.
33. *Meeting of the States Parties to the Convention on the Prohibition of the Development, Production and Stockpiling of Bacteriological (Biological) and Toxin Weapons and on Their Destruction*. U.N. Doc. BWC/MSP/2013/MX/INF.1/Rev.1. Advances in Science and Technology Related to the Convention. Geneva, Switzerland, 2013.
34. Meselson, Matthew. "The Problem of Biological Weapons." *Bulletin of the American Academy of Arts and Sciences* 52, no. 5 (1999): 46–58. https://doi.org/10.2307/3824391.
35. National Academies of Sciences, Engineering, and Medicine. *A Proposed Framework for Identifying Potential Biodefense Vulnerabilities Posed by Synthetic Biology: Interim Report*. The National Academies Press, 2017. https://doi.org/10.17226/24832.
36. National Academies of Sciences. *Biodefense in the Age of Synthetic Biology*. Washington, DC: The National Academies Press, 2018. https://doi.org/10.17226/24890.
37. National Research Council. *Life Sciences and Related Fields: Trends Relevant to the Biological Weapons Convention*. Washington, DC: The National Academies Press, 2011. https://doi.org/10.17226/13130.
38. National Research Council (US) Committee on Research Standards and Practices to Prevent the Destructive Application of Biotechnology. *Biotechnology Research in an Age of Terrorism*. Washington, DC: National Academies Press, 2004.
39. Nixdorff, Kathryn, Jens Brauburger, and Dorte Hahlbohm. "The Biotechnology Revolution: The Science and Applications." In *Verification of the Biological and Toxin Weapons Convention*, 32:77–124. NATO ASI Series, 1: Disarmament Technologies, 2000. https://doi.org/10.1007/978-94-017-3643-5_3.
40. Ouagrham-Gormley, Sonia Ben. *Barriers to Bioweapons: The Challenges of Expertise and Organization for Weapons Development*. Ithaca: Cornell University Press, 2014.
41. Ouagrham-Gormley, Sonia Ben, and Kathleen M. Vogel. "The Social Context Shaping Bioweapons (Non)Proliferation." *Biosecurity and Bioterrorism: Biodefense Strategy, Practice, and Science* 8, no. 1 (2010): 9–24. https://doi.org/10.1089/bsp.2009.0054.
42. Pearson, Graham S. "The Essentials of Biological Threat Assessment." In *Biological Warfare: Modern Offense and Defense*, edited by Raymond A. Zilinskas. Boulder, CO: Lynne Rienner Publishers, 2000.
43. Petro, James B., Theodore R. Plasse, and Jack A. McNulty. "Biotechnology: Impact on Biological Warfare and Biodefense." *Biosecurity and Bioterrorism: Biodefense Strategy, Practice, and Science* 1, no. 3 (September 1, 2003): 161–68. https://doi.org/10.1089/153871303769201815.
44. ScienceDaily. "Reference Terms: Pathogen." Accessed August 17, 2017. https://www.sciencedaily.com/terms/pathogen.htm.
45. Revill, James, and Catherine Jefferson. "Tacit Knowledge and the Biological Weapons Regime." *Science and Public Policy* 41, no. 5 (October 1, 2014): 597–610. https://doi.org/10.1093/scipol/sct090.

46. Roberts, Joanne. "From Know-How to Show-How? Questioning the Role of Information and Communication Technologies in Knowledge Transfer." *Technology Analysis & Strategic Management* 12, no. 4 (December 1, 2000): 429–43. https://doi.org/10.1080/713698499.
47. Rouvroit, Axel Lambert de, and Erik D. Heegaard. "Total Costs Associated with Replicating and Non-Replicating Smallpox Vaccines." *Global Security: Health, Science and Policy* 1, no. 1 (January 1, 2016): 3–9. https://doi.org/10.1080/23793406.2016.1171944.
48. Salerno, Reynolds M., Jennifer Gaudioso, Rebecca L. Frerichs, and Daniel Estes. "A BW Risk Assessment: Historical and Technical Perspectives." *The Nonproliferation Review* 11, no. 3 (2004): 25–55. https://doi.org/10.1080/10736700408436977.
49. *Seventh Review Conference of the States Parties to the Convention on the Prohibition of the Development, Production and Stockpiling of Bacteriological (Biological) and Toxin Weapons and on Their Destruction*. U.N. Doc. BWC/CONF.VII INF.3. New Scientific and Technological Developments Relevant to the Convention. Geneva, Switzerland, 2011.
50. The United States Commission on National Security/21st Century. *New World Coming: American Security in the 21st Century, Supporting Research and Analysis, The Phase I Report on the Emerging Global Security Environment for the First Quarter of the 21st Century*, 1999.
51. The U.S. National Response Team. "NRT Quick Reference Guide: Bacillus Anthracis (Causes the Disease Anthrax)," September 2011.
52. Thomas, Stephen R., and Joseph S. Elkinton. "Pathogenicity and Virulence." *Journal of Invertebrate Pathology* 85, no. 3 (March 1, 2004): 146–51. https://doi.org/10.1016/j.jip.2004.01.006.
53. Tucker, Jonathan. "The Future of Biological Warfare." In *The Proliferation of Advanced Weaponry: Technology, Motivations, and Responses*, edited by Eric H. Arnett and W. Thomas Wander. Washington, DC: American Association for the Advancement of Science, 1992.
54. Tucker, Jonathan B. "Review of the Literature on Dual Use." In *Innovation, Dual Use, and Security: Managing the Risks of Emerging Biological and Chemical Technologies*, edited by Jonathan B. Tucker, 19–44. Cambridge, Massachusetts: The MIT Press, 2012.
55. Vogel, Kathleen. "Bioweapons Proliferation: Where Science Studies and Public Policy Collide." *Social Studies of Science* 36, no. 5 (2006): 659–90.
56. Vogel, Kathleen M. *Phantom Menace or Looming Danger? A New Framework for Assessing Bioweapons Threats*. Baltimore: Johns Hopkins University Press, 2013.
57. Voosen, Paul. "Synthetic Biology Comes Down to Earth." *The Chronicle of Higher Education*, March 4, 2013. http://www.chronicle.com/article/Synthetic-Biology-Comes-Down/137587.
58. Zilinskas, Raymond A., ed. *Biological Warfare: Modern Offense and Defense*. 1 edition. Boulder, CO: Lynne Rienner Pub, 2000.
59. ———. "Possible Terrorist Use of Modern Biotechnology Techniques," 1–14. Istituto Diplomatico "Mario Toscano" Villa Madama, Rome, Italy, 2000.
60. ———. *The Soviet Biological Weapons Program and Its Legacy in Today's Russia*. CSWMD Occasional Paper 11. Washington, DC: National Defense University Press, 2016.

Chapter 4
Risk of Misuse Assessment: Part I

In 2017, scientists and security experts were alarmed when they recognized that rapid advances in genome editing were outpacing the ability of domestic and international security communities to discuss and develop strategies to address potential security concerns. In October of that year, the InterAcademy Partnership, NASEM, European Academies Science Advisory Council, and German National Academy of Sciences Leopoldina assembled a major international workshop in Hanover, Germany for experts to discuss the status quo of genome editing research, capabilities and limitations of the technologies, and subsequent concerns of the security community. One of the key questions that emerged from the workshop was whether advances in genome editing technologies would enable a greater range of people to use, or even misuse, these technologies. Traditional research communities (i.e., academic and private institutions), who for the longest time were the only actors able to access and use genome editing technologies, were still a concern. Participants of the workshop, however, now drew additional attention to the threat from actors who resided outside of this realm.[1]

In this chapter, I tackle this question and others through an in-depth examination of the accessibility and ease of misuse of genome editing technologies. Accessibility and ease of misuse are both specified within Tucker's *Decision Framework for Governance of Dual-Use Technologies* as parameters that should be used to assess risk of misuse. The first parameter—accessibility—is determined through the ease of acquiring hardware and software, cost, and dependence on other technologies. The second parameter—ease of misuse—is determined through the skill and tacit knowledge needed to use the technologies, ways to acquire skill and knowledge, and transferability of skill and knowledge. It is noted that the other two parameters that

[1] "International Workshop: Assessing the Security Implications of Genome Editing Technology, October 11–13, 2017, Hanover, Germany," The National Academies of Sciences, Engineering, and Medicine, accessed December 27, 2021, http://nas-sites.org/dels/events/ge-security/

K. Paris, *Genome Editing and Biological Weapons*,
https://doi.org/10.1007/978-3-031-21820-0_4

Tucker uses to assess risk of misuse (i.e., magnitude of potential harm and imminence of potential misuse) are examined in Chap. 5.

Not only does this chapter address accessibility, but it also determines the ease of misuse of genome editing technologies through a unique lens because I consider the threat that falls outside of traditional research communities. Here, I include analysis of first-hand data that I collected when using CRISPR at Genspace, the world's first DIYbio community laboratory, in Brooklyn, New York. DIYbio laboratories are considered non-traditional research institutions. Genspace formed in 2009 when a group of scientists and other hobbyists, artists, and entrepreneurs who all shared a deep passion and curiosity for biotechnology came together and pioneered the idea for a new way to practice science in their community. They dreamed of a community-level biotechnology laboratory where anyone, including those without formal laboratory training, could come and have access to work.

The founders of Genspace recognized the drawback of traditional research communities that are intended for trained practitioners and off limits to the everyday person. Additionally, they realized that even some highly trained and skilled practitioners wanted to have a space, outside of a traditional research community, to pursue other research interests. Although Genspace was the first of its kind, it was widely successful and has paved the way for over 100 other community laboratories. Even so, skeptics worried that non-traditional spaces could be ground zero for catastrophic laboratory accidents, could be used to create doomsday pathogens, or could be places where reckless actors create and then self-inject their own therapeutics. These narratives are damaging, and scholars are working to disprove them. One of the greatest advocates for the biosafety and biosecurity of DIYbio laboratories is the DIYbio community itself. The DIYbio laboratories have a shared culture of responsibility as reflected in their self-created code of ethics.[2]

In this chapter, I provide an inside look into what it is like to use CRISPR in a DIYbio laboratory. Not only does my account provide more insight into the practical application of CRISPR, but it also reveals more about the safety and security measures that are put in place in DIYbio laboratories to prevent mishap and misuse. Scholars and security experts have long speculated about the potential for the misuse of genome editing by non-traditional research communities, including the DIYbio community; however, these concerns have not been formally studied. I use the experience gained while working with CRIPSR within the Genspace DIYbio laboratory later in Chap. 7 to inform the possible strategies that domestic and international governments and organizations could use to govern genome editing technologies.

Chapter 4 is organized into two main sections. The first section covers the accessibility of genome editing technologies, and the second section focuses on the ease of misuse of genome editing technologies. Both sections begin by describing the methodology used to determine each parameter. Data for this assessment was

---

[2]Todd Kuiken, "Biosafety in a World Without Walls," accessed December 27, 2021 https://vimeo.com/277161865

mainly obtained from databases and first-hand laboratory experience. Parameter 1 (accessibility) was largely operationalized through indicators that were quantitatively assessed. Alternatively, parameter 2 (ease of misuse) was mostly operationalized through indicators that measured the ability and knowledge of individuals, which were assessed using qualitative methods.

4.1 Parameter 1: Accessibility

4.1.1 Methodology

Indicators used to analyze parameter 1 (accessibility) included ease of acquiring hardware (in this case, the reagents) and software, cost, and dependence on other technologies. To determine the ease of acquiring reagents, searches were conducted to identify providers of ZFN, TALEN, and CRISPR that advertise on the Science Exchange. Science Exchange is a leading marketplace for R&D services with data from over 2500 providers around the world. Additional providers were identified in a review article in which researchers affiliated with Harvard and other institutions reviewed companies that use genome editing technologies.[3] Information on pricing and availability was obtained from Science Exchange as well as providers' websites and other online publications for 87 providers (see Appendix A).

Given that Addgene is not only a principal provider of genome editing tools to academic and nonprofit institutions but also the company that was used to purchase the CRISPR-Cas9 plasmid for the Genspace study, it was selected for additional analysis. Addgene is a repository that functions as a middleman and handles the storage and transfer of plasmids to other scientists—this relieves the burden from scientists who would have to coordinate directly with each other to share plasmids. Scientists are encouraged and incentivized to send plasmids, especially those that are featured in published papers, to Addgene.[4] Additional data on ZFN, TALEN, and CRISPR plasmids and kits was requested and obtained from Addgene. Addgene data was analyzed to determine the extent that genome editing material have been deposited by and shipped to laboratories around the world.

After companies that sell reagents were analyzed, the availability of software that could be used to assist the design of genome editing components was determined. Since many software design tools—notably those used for CRISPR—continue to emerge each year, dated publications did not capture the full extent of software

---

[3]Katelyn Brinegar et al., "The Commercialization of Genome-Editing Technologies," *Critical Reviews in Biotechnology* 37 (January 18, 2017): 1–12, https://doi.org/10.1080/07388551.2016.1271768

[4]"Join the Community of Addgene Depositors," Addgene, accessed November 26, 2018, https://www.addgene.org/deposit/; "Addgene's Collection," Addgene, accessed February 1, 2019, https://www.addgene.org/browse/. "Addgene: Mission," accessed July 28, 2017, http://www.addgene.org/mission/

available to users. OmicX, an online platform that assists users in searching for bioinformatics tools, was selected as a data source for this study because the platform is continually updated with new information. Thus, omicX produced the most comprehensive list of genome editing software. Data for 7 ZFN design tools, 18 TALEN design tools, and 90 CRISPR design tools was obtained from omicX (see Appendix B). Data was analyzed to determine the relative availability and restrictions of using ZFN, TALEN, and CRISPR software.

Genome editing technologies depend on technologies to deliver genome editing cargo to the target and technologies to validate experimental results. For delivery technologies, the Genspace study was assessed to determine how the CRISPR plasmid was delivered to cells. Data from the Science Exchange on providers of genome editing services was also assessed to determine the availability of services that help users deliver cargo to the target. For validation technologies, the availability of service providers of Sanger sequencing—the gold standard method of validation—was determined via Science Exchange. A search for "Sanger sequencing" in experiments returned data for 44 service providers from around the world.[5]

4.1.2 Ease of Acquiring Hardware and Software

4.1.2.1 Hardware

Most providers of genome editing technologies offer CRISPR products and services, while few providers offer ZFN or TALEN products and services (Fig. 4.1). Of the 87 providers surveyed, all offer CRISPR products and services, and only about 8% and 17% offer ZFN and TALEN products and services, respectively. Many providers for each technology are from the commercial sector (ZFN: 85.7%, TALEN: 73.3%, CRISPR: 78.2%), and the rest of the providers are from the nonprofit (nonacademic) sector (ZFN: 14.3%, TALEN: 13.3%, CRISPR: 4.6%) and the academic sector (TALEN: 13.3% and CRISPR: 17.2%). Providers are located throughout the world including Austria, Belgium, Canada, China, Denmark, Germany, India, South Korea, Spain, United Kingdom, and United States. For a list of providers, see Appendix A.

While Sigma-Aldrich (now Millipore Sigma) remains the principal commercial provider of ZFN, others are phasing out the tool. For instance, Labomics, a biotechnology company based in Belgium and provider of ZFN, TALEN, and CRISPR tools, advises how the efficiency and specificity of TALEN and CRISPR surpasses that of ZFN. Thus, Labomics only provides ZFN upon request.[6] Addgene, a

[5] "Search Results for 'Sanger Sequencing,'" Science Exchange, accessed April 22, 2019, https://www.scienceexchange.com/marketplace/experimental-services?q=sanger%20sequencing&page=1&servicesFacets=%5B%22experiment%3ASanger%20Sequencing%22%5D

[6] "ZFNs—CRISPR-RGEN, TALEN & ZFN - Products," Labomics, accessed January 18, 2019, http://labomics.com/products/rgen-talen-zfn/zfns.html

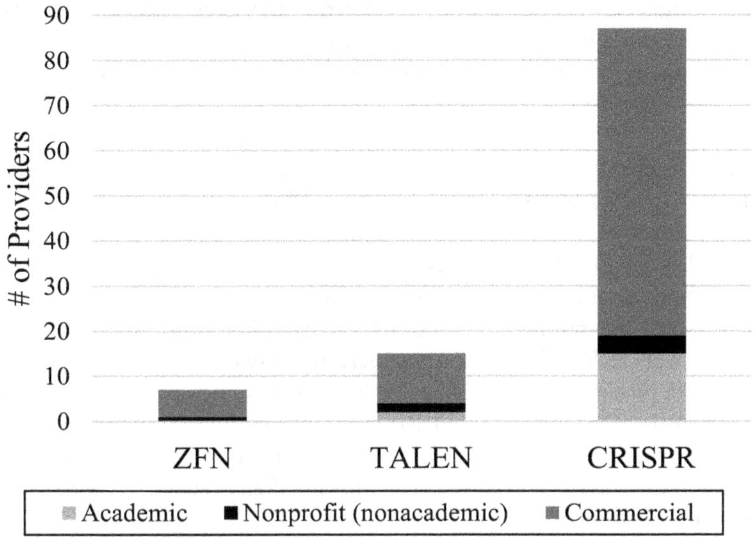

Fig. 4.1 Providers of ZFN, TALEN, and CRISPR products and services (Data for the figure were obtained from the Science Exchange on April 18, 2019)

nonprofit plasmid repository, still receives requests for ZFN; however, the number of ZFN units requested from Addgene by academic and nonprofit organizations around the world has been steadily decreasing since 2011. A little over 100 ZFN units were requested from Addgene in 2018.[7] Genome editing units ordered from and deposited at Addgene are assessed in more detail later in this section.

Similar to ZFN, TALEN products and services are available from a limited number of providers. Commercial companies that were leading providers of TALEN include Cellectis and Life Technologies (now Thermo Fisher Scientific).[8] While Thermo Fisher Scientific still provides TALEN products and services, Cellectis uses TALEN in-house to develop immunotherapies but no longer provides this tool. In short, Cellectis had trouble selling TALEN due to the emergence of CRISPR.[9] Like ZFN, the nonprofit company Addgene is also a provider of TALEN. Even so, the number of TALEN units requested from Addgene has been declining

[7] Data courtesy of Addgene, January 31, 2019.

[8] "Cellectis Nabs TAL Effector IP from Minnesota and Iowa State Universities," GEN—Genetic Engineering and Biotechnology News, January 20, 2011, accessed January 19, 2019, https://www.genengnews.com/topics/omics/cellectis-nabs-tal-effector-ip-from-minnesota-and-iowa-state-universities/; "TALEN Design to Validation Services - US," Thermo Fisher Scientific, accessed January 19, 2019, https://www.thermofisher.com/us/en/home/life-science/genome-editing/genome-engineering-services/talen-design-validation-services.html

[9] "The Most Important Battle in Gene Editing: CRISPR versus TALEN," Labiotech.eu, March 13, 2018, accessed November 9, 2018, https://labiotech.eu/features/crispr-talen-gene-editing/

since 2013. Only 460 TALEN units were requested from Addgene in 2018.[10] In addition to commercial and nonprofit (nonacademic) providers, a few academic providers—namely those at the University of California, San Francisco and University of Utah—offer TALEN constructs.[11]

In contrast to ZFN and TALEN, there are more providers of CRISPR across commercial, nonprofit (nonacademic), and academic sectors. Even the prominent commercial providers of ZFN and TALEN—Millipore Sigma and Thermo Fisher Scientific, respectively—heavily market CRISPR products and services on company websites.[12] Like ZFN and TALEN, the nonprofit company Addgene provides CRISPR. While the number of requests for ZFN and TALEN from Addgene have been decreasing, the number of requests for CRISPR from Addgene have been increasing—over 31,500 CRISPR units were requested in 2018.[13] While there are many CRISPR providers from the academic sector, some advertised restrictions—such as the Animal Models Shared Resource at Georgetown University Medical Center that limits its services to academic and nonprofit institutions as well as the ES Cell Targeting Core at the University of California, San Francisco (UCSF) that limits its services to UCSF researchers.[14]

Addgene

Overall, Addgene has received over ten-fold more orders for units of CRISPR than for units of ZFN and TALEN, combined. Due to the configuration of Addgene reports, a unit here is counted as either a plasmid or kit. Kits are described as "a collection of reagents or devices that can perform a specific biological experiment."[15] Kit items, regardless of size, were counted as one item. For instance, even though the ZFN kit from the Wolfe-Lawson laboratory contains 73 plasmids,

[10]Data courtesy of Addgene, January 31, 2019.

[11]"ES Cell Targeting Core, UCSF - Science Exchange," accessed June 4, 2019, https://www.scienceexchange.com/labs/es-cell-targeting-core; "Mutation Generation and Detection Facility - Science Exchange," Science Exchange, accessed June 3, 2019, https://www.scienceexchange.com/labs/mutation-generation-and-detection-facility

[12]"CRISPR," Sigma-Aldrich, accessed June 9, 2019, https://www.sigmaaldrich.com/life-science/functional-genomics-and-rnai/crispr-cas9-gene-editing.html; "CRISPR Products and Services - US," Thermo Fisher Scientific, accessed June 9, 2019, https://www.thermofisher.com/us/en/home/life-science/genome-editing/geneart-crispr.html

[13]Data courtesy of Addgene, January 31, 2019.

[14]"Animal Models Shared Resource (AMSR) - Zebrafish at Georgetown University Medical Center," Science Exchange, accessed June 9, 2019, https://www.scienceexchange.com/labs/animal-models-shared-resource-amsr-zebrafish; "ES Cell Targeting Core, UCSF - Science Exchange."

[15]"Kits," Creative Biogene, accessed June 2, 2019, https://www.creative-biogene.com/products/kits.html

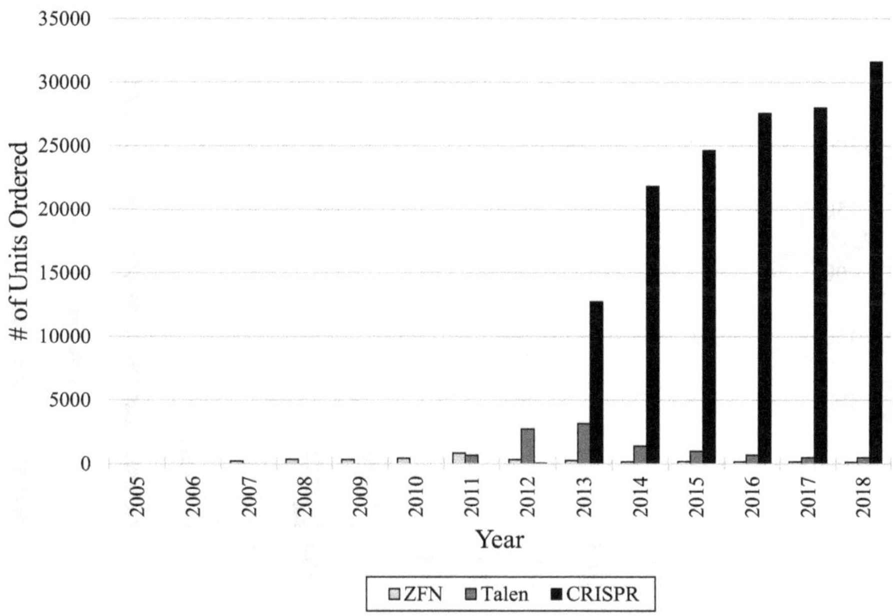

Fig. 4.2 ZFN, TALEN, and CRISPR units ordered per year from Addgene (2005–2018). The graph depicts the total number of ZFN, TALEN, and CRISPR plasmid orders shipped from Addgene. This includes plasmids as well as kits (which are made up of plasmids). Note that kits, regardless of size, are counted as one item (Data courtesy of Addgene, January 31, 2019)

it was counted as one item (i.e., one unit) when ordered.[16] As Fig. 4.2 depicts, in total, over 146,000 CRISPR units have been ordered from Addgene and shipped to laboratories, which is more than ten-times the number of ZFN and TALEN units, combined, that have been ordered and shipped.

As Fig. 4.2 shows, the number of units of ZFN ordered increased from about 200 units in 2007 to over 800 units in 2011. While the number of ZFN units ordered began to decline the year after Addgene offered TALEN units, the number of TALEN units ordered sharply increased. The number of TALEN units ordered peaked in 2013 with just over 3000 units; however, that same year, the number of CRISPR units surpassed the number of TALEN units ordered from Addgene. Since Addgene began to distribute CRISPR units in 2012, the number of CRISPR units ordered each year was more than the previous year. Although this trend slowed some in 2017, the number of CRISPR units ordered picked back up again in 2018. The overall increasing trend suggests that, in the future, orders for CRISPR will continue to increase and that many more academic and nonprofit organizations are using, and will continue to use, CRISPR in lieu of ZFN and TALEN.

[16]"Addgene: Wolfe-Lawson ZFN Modular Assembly Kit," accessed February 6, 2019, https://www.addgene.org/kits/wolfe-lawson-zfn/#kit-contents

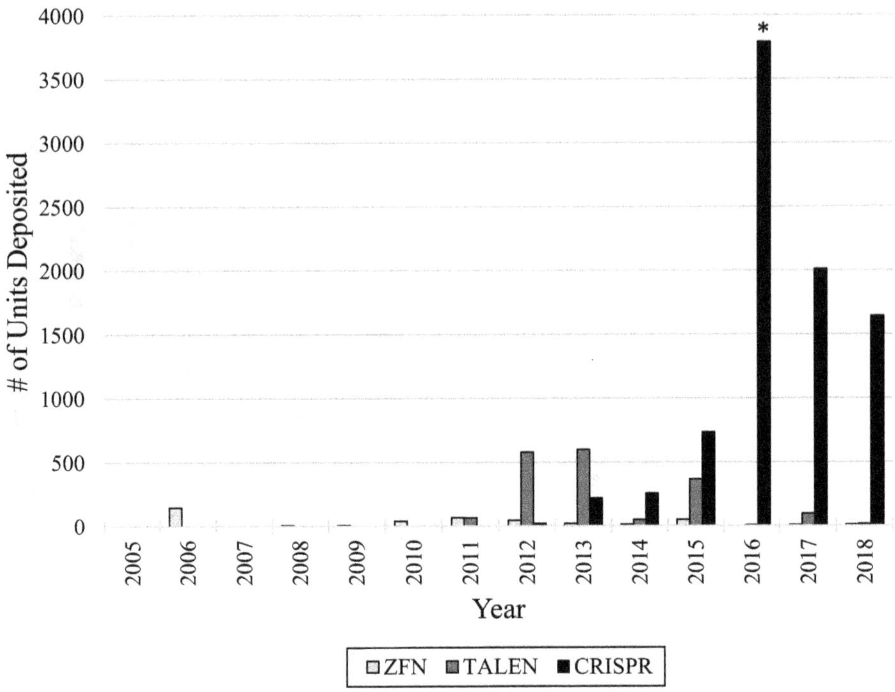

Fig. 4.3 ZFN, TALEN, and CRISPR units deposited per year to Addgene (2005–2018). The graph depicts the total number of ZFN, TALEN, and CRISPR plasmid orders deposited to Addgene. This includes plasmids as well as kits (which are made up of plasmids). Note that kits, regardless of size, are counted as one item. This reflects only plasmids that were deposited, passed Addgene's quality control process, and were released for other laboratories to order. Asterisk, Addgene received an unusually large sgRNA deposit of 2696 plasmids in 2016 (Data courtesy of Addgene, January 31, 2019)

In addition to ordering plasmids and kits from Addgene, laboratories also deposit plasmids and kits to Addgene. Figure 4.3 depicts the number of ZFN, TALEN, and CRISPR units deposited to Addgene. Note that this reflects only plasmids that were deposited, passed Addgene's quality control process, and were released for other laboratories to order. In general, the number of ZFN units deposited at Addgene is low—less than 100 units per year—and remains low. While the number of TALEN units did take off some in 2012, no more than 600 TALEN units have been deposited to Addgene in a given year. In 2015, the number of CRISPR units deposited to Addgene surpassed those units deposited for ZFN and TALEN. By 2016, the number of CRISPR units deposited soared to almost 3800 units due to an unusually large sgRNA deposit of nearly 2700 plasmids. Nevertheless, the number of CRISPR units deposited in 2017 and 2018 remained high—over 1500 units each year. Overall, these trends indicate that laboratories are working with and depositing units for CRISPR—not really for ZFN or TALEN.

While Addgene offers affordable plasmids for ZFN, TALEN, and CRISPR, it is important to point out that there are restrictions that limit distribution of materials. First, Addgene provides material to academic and nonprofit organizations—only a limited portion of the Addgene library can be distributed to for-profit organizations. Second, to purchase material through Addgene, users register for an account, which is verified by Addgene, and then place the order online.[17] Third, a Material Transfer Agreement (MTA) is needed for each order.

The MTA is an agreement that is often needed by the depositing and requesting institutions when material is shared. MTAs offer information on intellectual property rights, legal liability protection, and use of the material. Addgene uses the Uniform Biological Materials Transfer Agreement since it is a commonly accepted MTA in the United States and abroad. The recipient scientist is instructed to complete the Recipient Scientist Acknowledgement to acknowledge the terms of the MTA. The MTA needs to be signed by an authorized representative, usually in the technology transfer or legal office, at the recipient's institution and returned to Addgene.[18]

There are also restrictions when shipping orders to other countries. Addgene follows export control laws, and, as of November 2018, does not ship orders to the following countries: "Cuba, Iran, North Korea, Sudan, Syria, Somalia, Myanmar/ Burma and Rwanda."[19] Recipients in Australia need the Australian Quarantine and Inspection Service permit, an import permit, for orders. Additionally, recipients in New Zealand need the Permit to Import Restricted Biological Products of Animal, an import permit, for orders.[20] Furthermore, Addgene works with recipients in Kazakhstan and Russia to help orders through customs.[21]

Figures 4.4, 4.5, and 4.6 depict the global distribution of orders for ZFN, TALEN, and CRISPR units, respectively, that are shipped from Addgene to laboratories. Addgene has received orders and shipped units of ZFN to 42 countries, TALEN to 55 countries, and CRISPR to 78 countries. Compared to ZFN, orders for TALEN spread to southern North America, southern South America, Middle East, and Southern Asia. Orders for CRISPR, compared to ZFN and TALEN, spread to Central America, northern and western South America, Eastern Europe, Western and Eastern Africa, and Asia Pacific. This analysis indicates that orders for CRISPR are more widespread than ZFN and TALEN, which suggests that academic and nonprofit laboratories in far more countries are working with CRISPR compared to ZFN and TALEN.

---

[17] "Create an Account," Addgene, accessed November 26, 2018, https://www.addgene.org/users/register?next=/

[18] "Addgene: How to Order," accessed June 12, 2019, https://www.addgene.org/ordering/; "Technology Transfer Information," Addgene, accessed November 26, 2018, https://www.addgene.org/techtransfer/

[19] "Global Plasmid Distribution: Shipping and Customs," Addgene, accessed November 26, 2018, https://www.addgene.org/ordering/international/

[20] "Permits," Addgene, accessed November 26, 2018, https://www.addgene.org/ordering/permits/#new-zealand

[21] "Global Plasmid Distribution: Shipping and Customs."

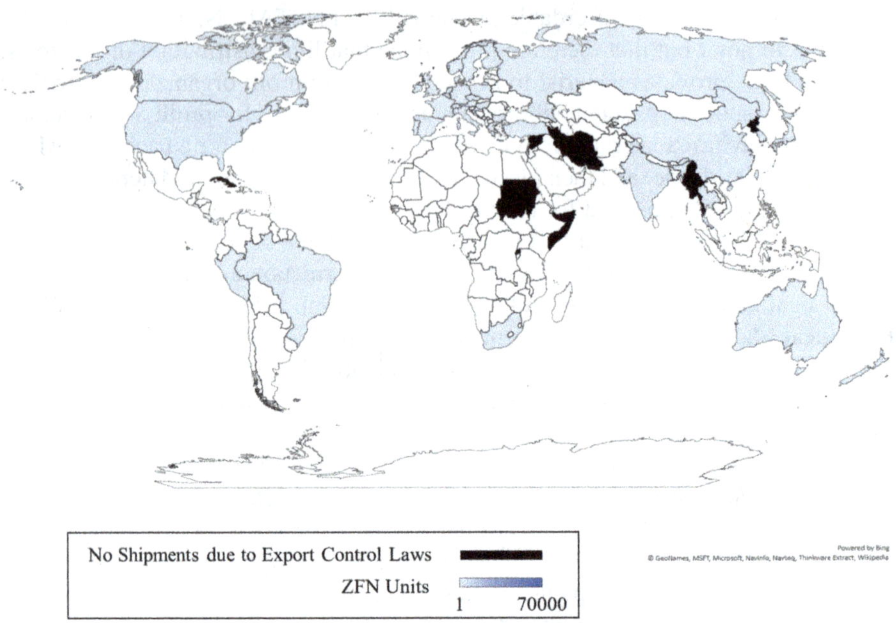

Fig. 4.4 Global distribution of orders for ZFN units from Addgene. Addgene began to receive orders for ZFN units in 2007. The map covers countries where ZFN units were ordered from 2007 through 2018. Black indicates countries where Addgene does not ship (Data courtesy of Addgene, January 31, 2019. The map was generated in Microsoft Excel, powered by Bing, © GeoNames, MSFT, Microsoft, NavInfo, Navteq, Thinkware Extract, Wikipedia)

4.1.2.2 Software

There are a variety of computational technologies available to help researchers improve the efficiency and specificity of genome editing experiments. These computational technologies range from general technologies developed to assist in the overall conceptualization and planning of experiments to those developed more specifically for genome editing tools to assist in the design of genome editing components and validation of experimental results. Often, a combination of computational technologies can be used during an experiment. This is exemplified by the Genspace study in which three computational tools were used: the *Saccharomyces* Genome Database (SGD), CRISPRdirect, and SnapGene. First, the sequence of the *ade2* gene was obtained via the SGD. Next, CRISPRdirect, a web server that supports the rational design of gRNA, was used to assist the design of gRNA. Based on the search results, a gRNA was selected and then visualized using

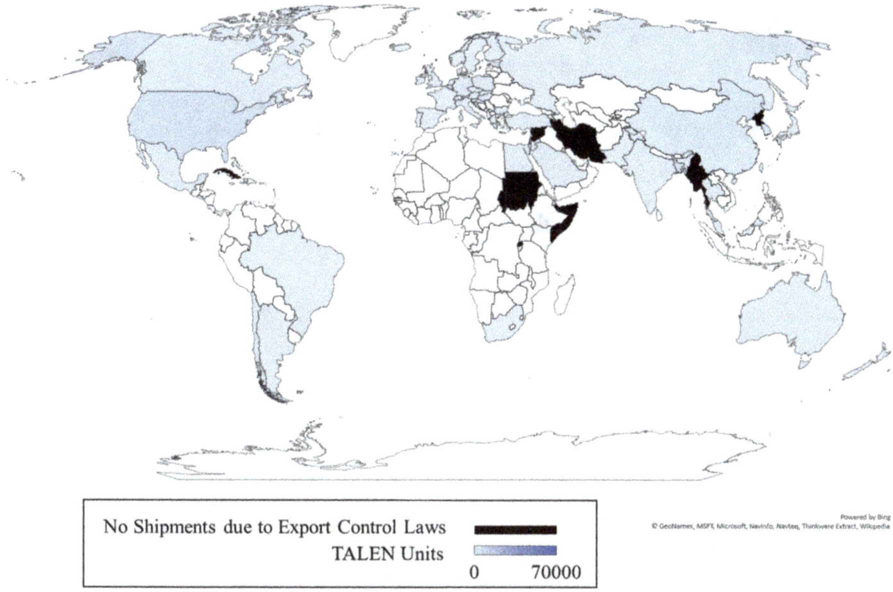

Fig. 4.5 Global distribution of orders for TALEN units from Addgene. Addgene began to receive orders for TALEN units in 2011. The map covers countries where TALEN units were ordered from 2011 through 2018. Black indicates countries where Addgene does not ship (Data courtesy of Addgene, January 31, 2019. The map was generated in Microsoft Excel, powered by Bing, © GeoNames, MSFT, Microsoft, NavInfo, Navteq, Thinkware Extract, Wikipedia)

SnapGene, a molecular biology software tool that supports the planning, visualization, and documentation of procedures.[22]

Out of the computational technologies described above, only one—CRISPRdirect—was created specifically to support CRISPR experiments. In addition to CRISPRdirect, there are many other software design tools developed to assist CRISPR as well as ZFN and TALEN experiments. Figure 4.7 depicts the number of new genome editing software design tools that were created each year for ZFN, TALEN, and CRISPR.

Overall, there are 3.6-times more design tools for CRISPR than for ZFN and TALEN, combined. The lines on Fig. 4.7 depict the cumulative number of technologies over time. The number of total CRISPR design tools continues to increase linearly, while the number of ZFN and TALEN design tools remains stagnant. Most of these tools have no restrictions for their use–only a few CRISPR software tools have license purchase requirements or are only available to academic or non-commercial users (Fig. 4.8). The availability of CRISPR software is concerning as these tools help to reduce the knowledge needed to design gRNA as discussed in

[22] Ben Glick and Aline Glick, "Identifying Sequence Elements with SnapGene's Feature Database," Addgene, accessed February 11, 2018, http://blog.addgene.org/identifying-sequence-elements-with-snapgenes-feature-database

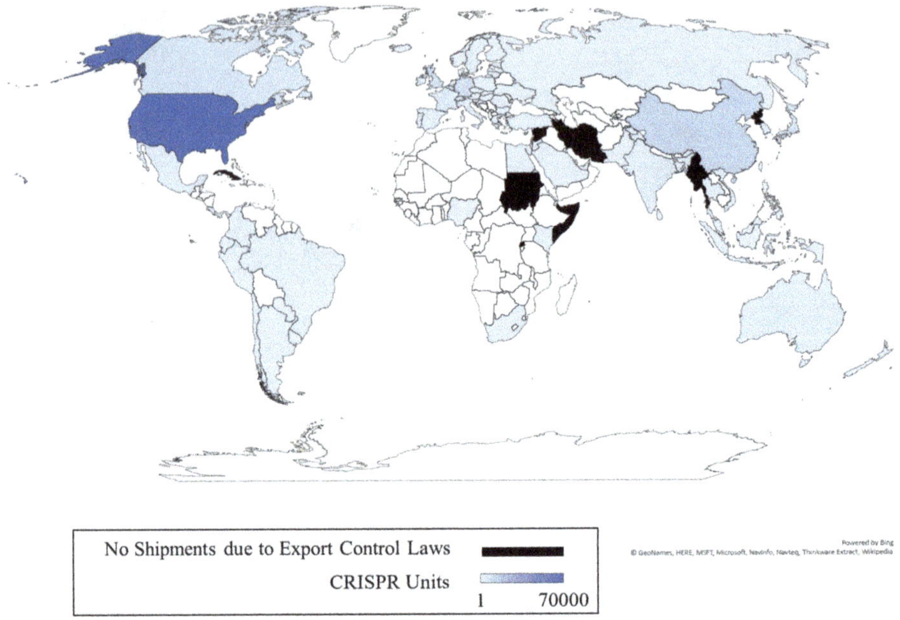

Fig. 4.6 Global distribution of orders for CRISPR units from Addgene. Addgene began to receive orders for CRISPR units in 2012. The map covers countries where CRISPR units were ordered from 2012 through 2018. Black indicates countries where Addgene does not ship (Data courtesy of Addgene, January 31, 2019. The map was generated in Microsoft Excel, powered by Bing, © GeoNames, MSFT, Microsoft, NavInfo, Navteq, Thinkware Extract, Wikipedia)

Parameter 2. Table 4.1 provides a summary for the assessment of the ease of acquiring software, as well as hardware, for genome editing.

4.1.3 Cost

In general, CRISPR is less expensive than ZFN and TALEN, and many companies are opting to use CRISPR to cut research costs. Novartis, a Switzerland-based pharmaceutical company, collaborates with Intellia Therapeutics, a Massachusetts-based biotechnology company, to use CRISPR-Cas9 to develop medical technologies, favors using CRISPR over TALEN given the affordability of CRISPR. CRISPR is 200-times cheaper to use than TALEN. It only costs Novartis approximately \$20 per gene to use CRISPR, whereas it costs the company approximately \$4000 per gene and many more months of work to use TALEN.[23] In another

[23] "CRISPR Genome Editing Fuels Cancer Drug Discovery," Novartis, accessed October 28, 2018, https://www.novartis.com/stories/from-our-labs/crispr-genome-editing-fuels-cancer-drug-discovery

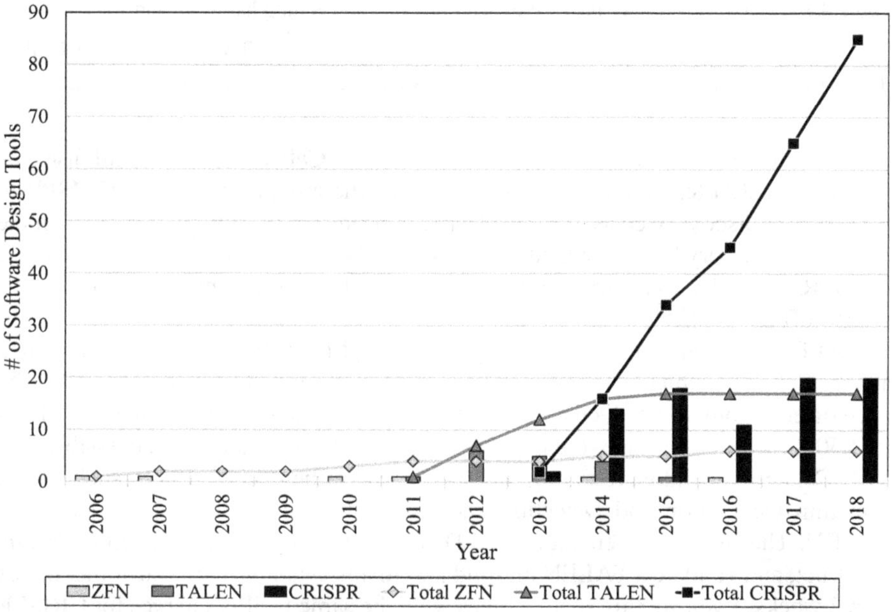

Fig. 4.7 Software design tools for genome editing (2006–2018). The bars depict the number of ZFN, TALEN, and CRISPR software design tools that emerged each year. The lines represent the total number of ZFN, TALEN, and CRISPR software design tools that accumulated over time (Data was obtained using omicX on December 29, 2018)

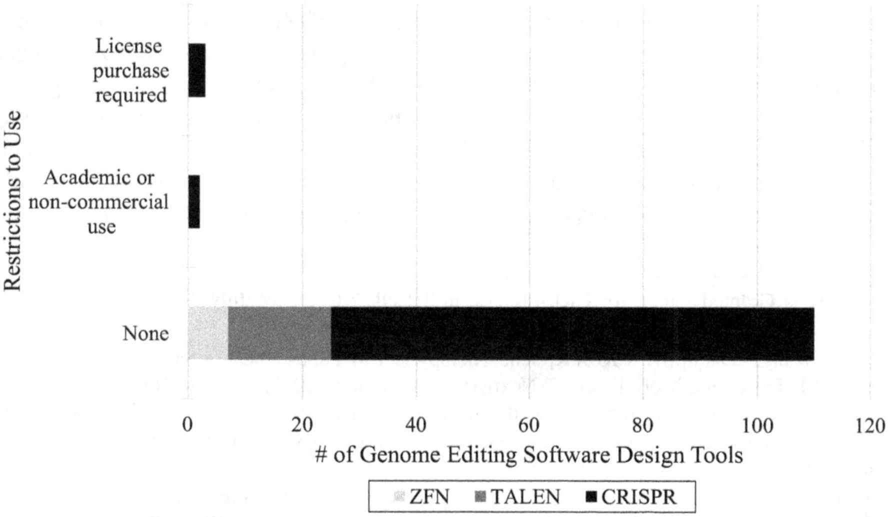

Fig. 4.8 Restrictions to use genome editing software design tools (Data was obtained using omicX on December 29, 2018)

Table 4.1 Overview score for the ease of acquiring genome editing hardware and software

	ZFN	TALEN	CRISPR
Ease of acquiring hardware and software	Moderate	Moderate	Easy

instance, the French biopharmaceutical company Cellectis, which still uses its flagship TALEN technology to develop immunotherapy products, uses CRISPR to cut research discovery costs.[24] James Haber, a professor of biology who was on the Scientific Advisory Board for Cellectis, reported that it can cost as little as $30 to use CRISPR since the RNA fragment can be ordered and other material can be purchased off the shelf.[25]

TALEN has had trouble competing with CRISPR in the market. As mentioned above, Cellectis, which was eager early in 2011 to manufacture affordable TALEN, once offered nonvalidated, custom TALEN for $3360 and design validated TALEN for $5000.[26] However, due to the emergence of CRISPR as a highly accessible new tool, Cellectis abandoned selling TALEN.[27] Nevertheless, providers such as the Mutation Generation and Detection Facility at the University of Utah, still offer TALEN. The Mutation Generation and Detection Facility charges $1125 to design, construct, and validate a TALEN plasmid pair that is available to the customer in less than 3 weeks. Even so, this is three times what the same facility charges for CRISPR ($375 per construct) and takes three times as long.[28]

While ZFN is the most expensive of the technologies, it has significantly come down in price since Sigma-Aldrich first began to sell zinc-finger products over a decade ago for upwards of $25,000 for custom products. Two important events—promotion of the open sharing of zinc-finger materials and emergence of TALEN as another genome editing tool—marked the decline of ZFN costs. By 2008, open-access protocols on zinc-finger protein engineering were available and zinc-finger materials were available for $5000.[29] Around 2010 to 2011, as TALEN emerged in the market, Sigma-Aldrich lowered the price and offered less expensive, off-the-self, validated ZFNs for knockout that cost $4000 to $5000 as well as custom, validated ZFNs for knockout or targeted integration that cost $6000 to $7000. Sigma-Aldrich's flagship ZFN-related products are termed CompZr (CompoZr ZFNs). Predesigned

[24]"Cellectis Granted Patent for CRISPR Use in T-Cells," Cellectis, July 24, 2017, http://www.cellectis.com/en/press/cellectis-granted-patent-for-crispr-use-in-t-cells

[25]Brian Wang, "Disruptive CRISPR Gene Therapy Is 150 Times Cheaper than Zinc Fingers and CRISPR Is Faster and More Precise," NextBigFuture, June 9, 2015, accessed November 7, 2018, https://www.nextbigfuture.com/2015/06/disruptive-crispr-gene-therapy-is-150.html; Cellectis, "2012 Annual Report," 8, accessed January 22, 2019, https://www.cellectis.com/uploads/files/ra-eng-papbd.pdf

[26]Laura DeFrancesco, "Move over ZFNs," *Nature Biotechnology* 29 (August 5, 2011): 684, https://doi.org/10.1038/nbt.1935; Jeffrey M. Perkel, "Genome Editing with CRISPRs, TALENs and ZFNs," Biocompare, August 27, 2013, http://www.biocompare.com/Editorial-Articles/144186-Genome-Editing-with-CRISPRs-TALENs-and-ZFNs/

[27]"The Most Important Battle in Gene Editing."

[28]"Mutation Generation and Detection Facility - Science Exchange."

[29]Pearson, "Protein Engineering," 161–64.

Table 4.2 Addgene kit prices. Kit prices exclude shipping costs ("Addgene: Plasmid Kits"; "Addgene: How to Order.")

	Kit Pricing	
	Minimum (US)	Maximum (US)
ZFN	$350.00	$650.00
TALEN	$325.00	$1,400.00
CRISPR	$275.00	$400.00

Table 4.3 Overview score for the cost of genome editing

	ZFN	TALEN	CRISPR
Cost	High	Moderate	Low

CompZr ZFN is validated and takes 3–11 weeks to reach customers. Custom CompZr ZFN is designed, assembled, validated, and sent to customers in 10–11 weeks.[30]

An affordable solution for all technologies is to purchase ZFN, TALEN, and CRISPR plasmids through Addgene, the nonprofit repository that provides plasmids to academic and other nonprofit organizations (Table 4.2). The price for all plasmids begins at $65 per plasmid and drops to as low as $45 per plasmid for orders of 21 or more plasmids.[31] The starting price of kits is also comparable for all technologies. The price of ZFN, TALEN, and CRISPR kits start at $350, $325, and $275 respectively.[32] A summary of the cost of ZFN, TALEN, and CRISPR is presented in Table 4.3.

4.1.4 Dependence on Other Technologies

4.1.4.1 Delivery Technologies

For genome editing technologies to work, they need to be delivered to the target. However, this can be challenging as delivery is limited by the method used to introduce the genome editing cargo.[33] Delivery of CRISPR to the cell is considered

---

[30] Joe Frangipane, PhD, Field Applications Specialist, Sigma-Aldrich Corporation, "Targeted Genome Editing in the Eukaryotic Systems Using Zinc Finger Nucleases" (Rutgers ORA Research Technologies Expo, Rutgers University, 2013), accessed August 1, 2017, https://www.youtube.com/watch?v=yDmB_NMGKWo

[31] "Addgene: How to Order."

[32] "Addgene: Plasmid Kits," accessed June 12, 2019, https://www.addgene.org/kits/#ge

[33] Qurrat Ul Ain, Jee Young Chung, and Yong-Hee Kim, "Current and Future Delivery Systems for Engineered Nucleases: ZFN, TALEN and RGEN," *Journal of Controlled Release: Official Journal of the Controlled Release Society* 205 (May 10, 2015): 124, https://doi.org/10.1016/j.jconrel.2014.12.036; Gaj, Gersbach, and Barbas, "ZFN, TALEN and CRISPR/Cas-Based Methods for Genome Engineering," 402.

the "greatest obstacle" for its use.[34] While there are a variety of delivery methods (i.e., delivery vehicles)—including physical delivery, viral vectors, and non-viral vectors—there are struggles using each system, which are highlighted below.

Microinjection (the gold standard for CRISPR delivery) and electroporation, which are popular physical delivery methods, are respectively regarded as "a technically challenging and laborious process" and "toxic and restricted to certain cell types"[35] Even so, viral vectors can be limited by the size of the delivery cargo (as is the case for Adeno-associated viruses, abbreviated AAV), cause strong inflammatory responses (which can happen with Adenovirus), and limited to users who are experts with "significant experimental experience."[36] Although non-viral vectors provide virus-free options for delivery, the commonly used lipid nanoparticles and liposome delivery methods suffer from issues caused by the degradation of cargo.[37]

Despite these problems, providers offer kits and services that make it easier for users to deliver cargo. In the case of the Genspace study, CRISPR was easily delivered to yeast cells using a transformation kit purchased from Zymo Research (see Table 4.4. Genspace Lab CRISPR Experimental Workflow, Step 4: Delivery). Furthermore, as the assessment of Parameter 2: Ease of Misuse reveals, there are many more providers offering viral packaging services for CRISPR compared to ZFN and TALEN (see Fig. 4.11).

4.1.4.2 Validation Technologies

Validation technologies are essential to determine whether genome editing was successful. As the Genspace study workflow demonstrates (Table 4.4), validation is needed throughout an experiment to ensure genome editing components (step 3), and any additional repair template (extra step), are correctly assembled prior to delivery into cells. Furthermore, a final validation of results is needed to determine whether the modification was achieved (step 5). Genspace sent transformed yeast colonies to GENEWIZ for Sanger sequencing to verify whether CRISPR successfully knocked out *ade2* and whether *gfp* was integrated during the repair.

Sanger sequencing is the gold standard for validating experimental results. Many providers across academic, commercial, and non-profit (non-academic) sectors around the world offer inexpensive and fast Sanger sequencing services. Over 40 providers advertise as Sanger sequencing providers through Science Exchange, a popular R&D services marketplace. Most of these service providers for Sanger

---

[34] Christopher A. Lino et al., "Delivering CRISPR: A Review of the Challenges and Approaches," *Drug Delivery* 25, no. 1 (May 25, 2018): 1252, https://doi.org/10.1080/10717544.2018.1474964

[35] Lino et al., 1243; Gaj, Gersbach, and Barbas, "ZFN, TALEN and CRISPR/Cas-Based Methods for Genome Engineering," 402.

[36] Lino et al., "Delivering CRISPR," 1242.

[37] Lino et al., 1242–47.

Table 4.4 Genspace lab CRISPR experimental workflow. The workflow used during the Genspace CRISPR lab followed the general format described in Chap. 2 of this assessment: choose target, design gRNA, build gRNA, deliver CRISPR components (i.e., Cas effector and gRNA), and validate. Given the purpose of this lab was not only to disrupt the *ade2* gene in S. cerevisiae (yeast) but also to insert the gene that codes for green fluorescent protein (GFP), it was necessary to conduct an additional step to prepare a repair template that contained the *gfp* gene. CRISPR cleaved the *ade2* gene in yeast, which yeast then repaired via HDR using the repair template. During the repair, yeast integrated the *gfp* gene from the template into the yeast's own genome (Data obtained from Genspace NYC, "Genome Editing with CRISPR-Cas9 Class" (Brooklyn, NY, November 2017))

Workflow step	Task	Kit or mix used	Computational tool used	Outsourced to a private company	Critical component for the workflow	Validation required
1. Choose target	• *ade2* gene (*Saccharomyces* Genome Database)		X			
2. Design gRNA	• *in silico* design (CRISPRdirect)		X			
	• Planning & visualization (SnapGene)		X			
3. Build gRNA	• Order gRNA (Integrated DNA Technologies, Inc.)			X		
	• Ligate target sequence to digested pML107 • Blunt/TA Ligase Master Mix (NEB M0367)	X				
	• Heat shock transformation of *E. coli* with targeted pML107 to finish building					
	• Validate integrated insert via colony PCR & check size via gel electrophoresis (224 bp) • One Taq® 2X Master Mix (NEB M0482)	X				X
	• Amplify in liquid culture					
	• Isolate targeted pML107 & quantify via Qubit fluorometer • Monarch® Plasmid Miniprep Kit (NEB T1010)	X			X	
Extra Step: Build repair template	• Add homology arms to *gfp* gene via PCR • Q5® High Fidelity 2X Master Mix (NEB M0492)	X				
	• Validate repair template is correctly assembled using gel electrophoresis (~1500 bp)					X

(continued)

Table 4.4 (continued)

Workflow step	Task	Kit or mix used	Computational tool used	Outsourced to a private company	Critical component for the workflow	Validation required
	• Purify GFP repair template via PCR & quantify via Qubit fluorometer • Monarch® PCR & DNA Cleanup Kit (NEB T1030)	X			X	X
4. Deliver	• Transform yeast with targeted pmL107 & GFP repair template • Frozen-EZ Yeast Transformation II Kit™ (Zymo T2001)	X				
5. Validate	• Visually inspect					X
	• Isolate DNA from yeast transformants					
	• Prep for sequencing: PCR the *ade2* locus • Q5® High Fidelity 2x Master Mix (NEB M0492)	X				
	• Genetic validation via Sanger sequencing of *ade2* (GENEWIZ)			X		X

Table 4.5 Overview score for the dependence of genome editing on other technologies

	ZFN	TALEN	CRISPR
Dependence on other technologies	All rely on delivery and validation technologies		

sequencing are in the United States (73%). The rest of the providers are in Australia (2%), Canada (9%), Germany (5%), Japan (2%), Spain (2%), Switzerland (5%), and United Kingdom (2%). As advertised on Science Exchange, the starting price for Sanger sequencing ranges from \$2.00 to \$10.00 per sample—the average starting price is \$5.02 per sample. Companies, such as the Swiss-based biotechnology company Fasteris SA, even advertise that they can provide next day results for Sanger sequencing.[38] Table 4.5 summarizes the assessment of the dependence of genome editing on other technologiesand demonstrates how all the genome editing technologies rely on delivery and validation technologies.

4.2 Parameter 2: Ease of Misuse

4.2.1 Methodology

Indicators used to analyze parameter 2 (ease of misuse) included skill and tacit knowledge needed to use the genome editing technologies, ways to acquire skill and knowledge, and transferability of skill and knowledge. The main source of data used to assess skill and tacit knowledge requirements were first-person observations made during a CRISPR class taken at Genspace, the first DIY biology community laboratory in the world, located in Brooklyn, New York. The class consisted of 4-h class sessions held four times over the course of 2 weeks in November 2017. During the class, CRISPR was used to knockout the *ade2* gene in *Saccharomyces cerevisiae* (yeast). In other words, CRISPR cleaved the *ade2* gene in yeast, which prevented yeast from expressing this gene. A repair template was then provided to yeast cells to repair the cleaved DNA while simultaneously inserting the *gfp* gene.

The instructor led ten student participants through protocols that supported the CRISPR workflow. Students ranged from those with no laboratory experience or formal education in college-level biology to those with 20 or more years of laboratory experience and advanced biology degrees. The instructor holds an undergraduate degree in biology and worked for several years as a pharmaceutical researcher prior to coming to Genspace in 2014. Another lab instructor who was on hand to help students during the class holds a Ph.D. that focused on medical mycology and infectious disease, has taught at a university and community college, and has been with Genspace since 2012. The Institutional Review Board (IRB) at George Mason

[38] Data is provided for 44 academic, commercial, and non-profit (non-academic) service providers. "Search Results for 'Sanger Sequencing.'"

University was consulted before beginning the class at Genspace. Given that general observations were made about the students in the class and no personally identifiable information on the students was used, the IRB determined that permission was not needed from the IRB for this study.

To determine the skill and knowledge needed to use the technologies, specifically CRISPR, the following sources of data were used: notes taken while performing the CRISPR lab at Genspace; a lab handout from the Genspace instructor that contained 100 pages of background and written protocols; and written protocols, supplemental information, and other instructional videos found on manufacturers' websites. A detailed workflow of the procedures carried out during the Genspace lab was prepared to support the systematic examination of data (Table 4.4).

Class notes were analyzed to determine the skill and knowledge used to perform the steps of the workflow. Additionally, to determine skills needed to design components (i.e., step 2), data on the skills (basic, medium, or advanced) required to use software design tools, as listed in Appendix B, were obtained from omicX. Protocols from the Genspace lab handout were compared to those from manufacturers to determine how the lab was modified to fit the class. Moreover, class notes were compared to the Genspace lab handout to determine the impromptu changes made to the protocol during the lab by the instructor, based on the instructor's experience and expertise, to maximize the likelihood of successful results.

While an actor may not have the skill or knowledge needed to use the genome editing technologies, it does not mean that the actor could not acquire these requirements. There are two main strategies for users to fulfill the skill and knowledge requirements needed to use genome editing technologies: contract out the work to private companies or acquire the expertise yourself through in-person or online resources. First, companies listed in Appendix A were reviewed to determine the types of services and online resources offered by companies. Additional data on the availability of online videos to learn skill and knowledge needed to use genome editing technologies was obtained from the *Journal of Visualized Experiments* (*JoVE*).

JoVE is a scientific, peer reviewed, video journal headquartered in Cambridge, Massachusetts that features visual demonstrations of written protocols.[39] Between 2017 and 2019, the number of *JoVE* video articles doubled to nearly 10,000.[40] Given that *JoVE* is a popular and rapidly growing source used to learn skills and knowledge needed to perform protocols, the journal was selected as a key data source.

Searches were conducted in *JoVE* for video articles on ZFN, TALEN, and CRISPR. Each search result was reviewed, and only video articles in which the search term was in the protocol section of the article were selected for inclusion in

---

[39] "JoVE | Peer Reviewed Scientific Video Journal - Methods and Protocols," accessed February 18, 2019, https://www-jove-com.mutex.gmu.edu

[40] JoVE contained over 5500 video articles in March 2017 and over 9600 video articles in February 2019. "About JoVE (2017)," Journal of Visualized Experiments, accessed March 20, 2017, https://www.jove.com/about; "About JoVE (2019)," Journal of Visualized Experiments, accessed February 19, 2019, https://www.jove.com/about

this assessment.[41] Data on the number of ZFN, TALEN, and CRISPR *JoVE* video article protocols was obtained for 47 articles published between 2012 and 2018. Additionally, data on the number of views of these articles was also gathered to determine the extent that ZFN, TALEN, and CRISPR video article protocols available through *JoVE* are viewed. A list of video articles included in this study is provided in Appendix C.

Next, searchers were conducted in Google for ZFN, TALEN, and CRISPR classes, workshops, and laboratories to determine the availability of in-person forums (Appendix D). Additional information on DIY biology laboratories was obtained through the Genspace study and communications with DIY laboratories. To determine the transferability of skill and knowledge from these forums, data was obtained from three key sources: (1) undergraduate biology classes at DeSales University and Clemson University that watched *JoVE* videos on basic lab skills and methods as a pre-lab requirement, (2) an undergraduate biology class at the University of New Mexico that used CRISPR to modify *Drosophila* genes, and (3) the DIY biology class at the Genspace lab in Brooklyn, New York that used CRISPR to modify yeast.

First, data on the effectiveness of *JoVE* videos on performance was obtained from an independent, controlled study that was carried out by a company called TERC, an acronym short for the Technical Education Research Centers that was coined when the company formed in 1965. TERC is a nonprofit research organization located in Cambridge, Massachusetts.[42] The study was published in January 2018. Two undergraduate biology courses—one course with four sections from DeSales University (94 students) and one course with 14 sections from Clemson University (252 students)—participated in the study. The purpose of the study was to determine the effectiveness of a pre-lab requirement to watch *JoVE* videos on student performance and learning experiences.[43]

For the JoVE study, students were either part of the treatment group, known as the video group, or the control group. The control group was assigned a lab handout to read before class, while the video group was assigned a video to watch before and after reading the lab handout—the video group watched the video twice before the class.[44] The video group watched videos on basic lab skills (light microscopy and spectrophotometry) and basic cellular and molecular biology methods (protein separation with SDS-PAGE and plasmid purification).[45]

---

[41] In many instances, search terms were only listed in the title of reference material. Thus, irrelevant search results such as these were excluded from assessment.

[42] "About Us," TERC, accessed September 4, 2019, https://www.terc.edu/display/About/About+Us

[43] Karen Mutch-Jones and Yueming Jia, *Effectiveness of JoVE Videos in Improving STEM Students' Performance: A Study in Undergraduate Biology Labs* (Cambridge, MA: STEM Education Evaluation Center at TERC, 2018), 1–5.

[44] Mutch-Jones and Jia, 2.

[45] JoVE Science Education Database, *General Laboratory Techniques, Introduction to Light Microscopy* (Cambridge, MA: JoVE, 2019); JoVE Science Education Database, *General Laboratory Techniques, Introduction to the Spectrophotometer* (Cambridge, MA: JoVE, 2019); JoVE

Students performance was measured through one pre-lab assessment and one post-lab assessment. The assessments asked students questions about lab techniques and concepts. The study on *JoVE* videos found that students who watched videos on spectrophotometry, protein separation with SDS-PAGE, and protein purification did significantly better on pre- and post-lab assessments than the control group. Thus, data on the raw scores for pre- and post-lab assessments for the video and control groups were obtained to determine how much higher the video group scored on pre- and post-lab assessments compared to the control group. Additionally, data was obtained to determine how much the videos contributed to the video group's higher score—these values were controlled for the following factors that could influence higher scores: "student confidence and general knowledge of basic biology concepts."[46]

Second, data on the success of using CRISPR at the undergraduate level were obtained from a journal article that published the results of a university study. During the Spring of 2015, six undergraduate students at the University of New Mexico completed a 15-week semester laboratory class that used CRISPR-Cas9 to modify *Drosophilia.* The class met for 2 h and 50 min, 2 days each week. Each student targeted one *Drosophila* gene, which did not have any mutations that currently existed, and used CRISPR to delete that gene.[47] Lastly, data on the results of the DIY biology class at the Genspace lab, as described above, that were obtained while taking the class were analyzed to determine how successful students were in achieving the knockouts of the *ade2* gene and insertion of the *gfp* gene.

4.2.2 Skill and Tacit Knowledge Needed to Use Technology

While CRISPR uses an RNA-based DNA binding domain that is easier to engineer, ZFN and TALEN use protein-based DNA binding domains that are more difficult to engineer. Feng Zhang, a molecular biologist and core member at the Broad Institute of MIT and Harvard, describes the challenges of engineering ZFN and TALEN DNA-binding domains during one of the National Institutes of Health Director's Lecture Series: "the complexity of reengineering each one of these DNA-binding

Science Education Database, *Basic Methods in Cellular and Molecular Biology, Separating Protein with SDS-PAGE* (Cambridge, MA: JoVE, 2019); JoVE Science Education Database, *Basic Methods in Cellular and Molecular Biology, Plasmid Purification* (Cambridge, MA: JoVE, 2019).

[46] Mutch-Jones and Jia, *Effectiveness of JoVE Videos in Improving STEM Students' Performance: A Study in Undergraduate Biology Labs*, 1, 5–16.

[47] Vanesa Adame et al., "An Undergraduate Laboratory Class Using CRISPR/Cas9 Technology to Mutate Drosophila Genes," *Biochemistry and Molecular Biology Education* 44, no. 3 (May 6, 2016): 263–75, https://doi.org/10.1002/bmb.20950

domains can be pretty challenging."[48] For ZFN, "it can take quite of bit of directed evolution and engineering effort, and, even then, it's not guaranteed."[49] For TALEN, "the modules are very repetitive, and, if you were to use standard PCR based assembly, it's quite cumbersome and challenging to assemble."[50]

Although there are open-source protocols available for the engineering of ZFN, they do not always work, are not simple, and still require skill to use. The modular assembly method became publicly available to researchers in 2006.[51] Unfortunately, the method suffered from high failure rates (at least 76%), and users were warned to "expect this method to fail."[52] Another method for engineering ZFN, known as Oligomerized Pool Engineering (OPEN), was published in 2009.[53] The protocol, which is about 20 pages and 117 steps, is described as "hardly workaday"—taking approximately 6 months to 1 year.[54] Its widespread use was reportedly limited by "the labor and expertise required to screen combinatorial libraries."[55] Context-dependent assembly (CoDA), which was published in 2011, is described as a simpler version of OPEN because it eliminates "some of the more laborious screening and selection steps"—taking only 1 month.[56]

Even so, if the assembly of ZFN does not work and needs to be troubleshooted, it is reported that "only a few laboratories have the kind of expertise that can address this type of problem."[57] Although there are affordable options available for laboratories to prepare ZFN themselves, it is reported that "commercially available ZFNs generally work better."[58] Sigma-Aldrich (now, Millipore Sigma), the commercial provider of ZFN, explained that it takes full-time employees 6–10 weeks to validate custom orders of ZFN.[59] A group of Australian researchers who published a *JoVE*

[48]Feng Zhang, "From Microbial Immunity to Genome Editing," accessed June 26, 2019, https://www.youtube.com/watch?v=ihmt9MkQY34&list=PLPuptUEHTSjDGWQ16svQQuOXf78g4z4Bn&index=5&t=0s

[49]Zhang.

[50]Zhang.

[51]David A. Wright et al., "Standardized Reagents and Protocols for Engineering Zinc Finger Nucleases by Modular Assembly," *Nature Protocols; London* 1, no. 3 (August 2006): 1637–52, http://dx.doi.org.mutex.gmu.edu/10.1038/nprot.2006.259

[52]Cherie L. Ramirez et al., "Unexpected Failure Rates for Modular Assembly of Engineered Zinc Fingers," *Nature Methods* 5, no. 5 (May 2008): 374–75, https://doi.org/10.1038/nmeth0508-374

[53]Morgan L. Maeder et al., "Oligomerized Pool Engineering (OPEN): An 'open-Source' Protocol for Making Customized Zinc-Finger Arrays," *Nature Protocols* 4, no. 10 (2009): 1471–1501, https://doi.org/10.1038/nprot.2009.98

[54]Pearson, "Protein Engineering," 164; DeFrancesco, "Move over ZFNs," 681.

[55]Jeffry D. Sander et al., "Selection-Free Zinc-Finger-Nuclease Engineering by Context-Dependent Assembly (CoDA)," *Nature Methods* 8, no. 1 (January 2011): 67, https://doi.org/10.1038/nmeth.1542

[56]DeFrancesco, "Move over ZFNs," 681–82; Sander et al., "Selection-Free Zinc-Finger-Nuclease Engineering by Context-Dependent Assembly (CoDA)."

[57]DeFrancesco, "Move over ZFNs," 683.

[58]Kim and Kim, "A Guide to Genome Engineering with Programmable Nucleases," 327.

[59]DeFrancesco, "Move over ZFNs," 682–83.

video article on using ZFN to target human embryonic stem cells advised that, since it is tough to design ZFN in-house from scratch, the researchers ordered custom ZFN from a commercial provider.[60]

In addition to explaining the difficulties of designing ZFN, the Australian researchers recommended swapping TALEN plasmids for ZFN mRNA because TALENs would cut research costs and could be designed in-house.[61] Dan Voytas, an inventor of TALEN, reportedly said that that it would take any user who has molecular biology skill a week to engineer TAL effectors.[62] George Church, a genetics professor at Harvard and MIT, called attention to researchers who were able to make a collection of TALENs that could target "almost all the protein coding genes" in the human genome—over 18,000 genes—and suggested that "it can't have been that hard to do, even though it wasn't CRISPR."[63] Even so, the aforementioned references are experts in genome editing with many years of laboratory experience. In contrast, a DIY biologist member of Genspace tried to use TALEN several years prior to November 2017; however, the member experienced several difficulties and abandoned the technology. Now, Genspace uses and teaches classes dedicated to CRISPR.[64]

While CRISPR is regarded as easy to use, applications of its use suggest that, at a minimum, basic laboratory skills and knowledge of biology are needed to use the tool. Jon Cohen, a staff writer for *Science*, tested the belief that anyone can use CRISPR. Using CRISPR, Cohen targeted *CD32*, a gene that is part of the immune response to dengue virus, for knockout. Cohen hypothesized the knockout would lead to a better understanding of how Zika virus reproduces in individuals who have acquired immunity from dengue virus.[65] Cohen performed the experiment in a laboratory under the close supervision and hands-on guidance from Roland Wagner, a postdoctoral researcher who uses CRISPR. In one of the first steps, Cohen attempted to pipette water; however, the technique used to pipette had changed since he was last in a laboratory as an undergraduate student over 30 years ago.[66] Cohen, who went too fast and created bubbles in the water, appeared unaware of his

[60] Brigham J. Hartley et al., "Zinc-Finger Nuclease Enhanced Gene Targeting in Human Embryonic Stem Cells," *JoVE (Journal of Visualized Experiments)*, no. 90 (August 23, 2014): e51764, https://doi.org/10.3791/51764

[61] Hartley et al.

[62] DeFrancesco, "Move over ZFNs," 682.

[63] George Church, "The Future of the Genomic Editing Revolution" (UC Berkeley, January 26, 2017), accessed June 28, 2019, https://www.youtube.com/watch?v=Wswvf8Nrubg; Yongsub Kim et al., "A Library of TAL Effector Nucleases Spanning the Human Genome," *Nature Biotechnology* 31, no. 3 (March 2013): 251–58, https://doi.org/10.1038/nbt.2517

[64] Genspace NYC, "Genspace Class, November 2017."

[65] Jon Cohen, "One of Our Reporters Tried to Do CRISPR. He Failed Miserably," *Science*, November 3, 2016, https://doi.org/10.1126/science.aal0335

[66] Cohen.

error until Wagner corrected the technique. Wagner said Cohen's inability to perform this routine skill set the experiment "on a bad path."[67]

Ultimately, Cohen failed in a crucial step early on to prepare the CRISPR plasmid for cloning, which left him unable to successfully complete the experiment. Wagner said his experiments did not always work out when he first began, and he admitted there are many "setbacks in science."[68] Wagner added, "I think there is a certain skill level . . . if you take a random person from the street and say knock out *CD32* . . . there is no way this person would do it. Not in a hundred years."[69] Cohen concluded, "I've already learned that any idiot cannot do CRISPR: It takes, at least, basic laboratory skills."[70]

The *CD32* knockout experiment described above supports the notion that a certain set of laboratory skills and knowledge are needed to use CRISPR. Yet, it is unclear what skills and knowledge are needed throughout an experiment. While the Genspace study was a basic experiment that utilized a commonly studied organism (yeast) and genes (*ade2 and gfp*), it is assessed here to provide a baseline of the types of skill and knowledge needed to use CRISPR. There is an inherent understanding that, depending on the application of misuse, additional skill and knowledge would be required to create a biological weapon. Table 4.6 provides a summary of skills and knowledge requirements for using CRISPR as obtained via the Genspace study.

4.2.2.1 Step 1: Choose Target

To begin the genome editing workflow, scientists need to know what genomic sequence to target for modification. In the case of the Genspace study, yeast (*S. cerevisiae*), which is a well-studied model organism, was selected for modification: its eukaryotic genome, which is approximately 13 million base pairs coding 6000 proteins, was the first to be sequenced.[71] The *ade2* gene in yeast was selected as the target sequence. This sequence was easily and freely accessible to the public online through the *Saccharomyces* Genome Database (SGD).[72] SGD provides users with the reference genomic chromosomal sequence as well as information on genes, proteins, and other encoded characteristics for yeast.[73] A search for "*ade2*" returned one result for the reference genome sequence from the S288C laboratory strain of

[67] Jon Cohen, "One of Our Reporters Tried to Do CRISPR. He Failed Miserably [Video]," *Science*, November 3, 2016, https://doi.org/10.1126/science.aal0335

[68] Cohen, "One of Our Reporters Tried to Do CRISPR. He Failed Miserably."

[69] Cohen, "One of Our Reporters Tried to Do CRISPR. He Failed Miserably [Video]."

[70] Cohen, "One of Our Reporters Tried to Do CRISPR. He Failed Miserably."

[71] Jonathan Pevsner, *Bioinformatics and Functional Genomics*, 2nd ed. (New Jersey: John Wiley & Sons, Inc., 2009), 15, 466.

[72] "ADE2 / YOR128C Overview," Saccharomyces Genome Database, accessed February 10, 2018, https://www.yeastgenome.org/locus/S000005654

[73] "SGD Help," accessed March 2, 2019, https://sites.google.com/view/yeastgenome-help/about

Table 4.6 Summary of skills and knowledge requirements for using CRISPR (Data for the table were obtained during the Genspace NYC, "Genspace Class, November 2017")

Workflow step	Task	Skills and knowledge
Step 1: Choose target	Obtaining the *ade2* gene target sequence	• Applied bioinformatic skill (use of SGD). • Basic understanding of biology (understanding genomic sequences).
Step 2: Design components	Design of gRNA	• Specialized understanding of genome editing—selection of good gRNA designs (consulting literature, previous experience, and trial and error). • Reduction of knowledge requirements due to use of design software (CRISPRdirect): applied bioinformatic skill to use design software.
Step 3: Build components	Obtaining gRNA	• Skill and knowledge reduction by outsourcing to a private company.
	Ligating gRNA insert to the plasmid backbone	• Basic molecular cloning skill. • Knowledge to adapt ligation protocol to Genspace lab (previous experience, consulting manufacturer's website). • Reduction of knowledge requirements due to use of master mixes and kit: basic lab skills to use (micropipetting); knowledge to adapt manufacturer protocol to Genspace lab (previous experience, trial and error, impromptu protocol changes made during the lab based on experience and expertise).
Extra Step: Build repair template	Preparing the repair template by adding homology arms to the *gfp* gene	• Basic molecular cloning skill. • Reduction of knowledge requirements due to use of master mix and kit: basic lab skills to use (micropipetting), and knowledge to troubleshoot protocol (discussion among experts).
Step 4: Deliver	Transform yeast with targeted plasmid (from step 3) and GFP repair template (from extra step)	• Reduction of knowledge requirements due to use of kit: basic lab skills to use (micropipetting); impromptu protocol changes made during the lab based on experience and expertise.
Step 5: Validate	Visual inspection of yeast colonies	• Basic lab skill to visually identify; knowledge of marker genes for selection and screening.
	Genetic validation via Sanger sequencing of *ade2*	• Skill and knowledge reduction by outsourcing to a private company.

(continued)

Table 4.6 (continued)

Workflow step	Task	Skills and knowledge
		• Specialized knowledge of how to interpret Sanger sequencing results.

S. cerevisiae. Users could easily download the DNA or protein sequence for the gene.[74]

Besides SGD, there are other genome browsers can be used to obtain sequence information for yeast. In particular, the National Center for Biotechnology Information (NCBI), which is part of the United States National Institutes of Health, hosts "approximately 40 online literature and molecular biology databases," which includes databases that store genome sequence data.[75] GenBank at the NCBI is highly regarded as the "database of most known nucleotide and protein sequences."[76] As of February 2019, GenBank contained over 300 billion bases of DNA from over 200 million sequences—the amount of bases doubles about every 18 months.[77] GenBank has sequences from 420,000 species.[78]

While there are databases available online to obtain genome sequence data for many species, this step can still present its challenges regardless of the genome editing technology (ZFN, TALEN, or CRISPR) that is used. A common pitfall is conducting vague searches that return too much information.[79] For instance, a search for the term "*ade-2*" on NCBI's Nucleotide, a database that contains sequences from many sources such as GenBank, returned almost 150 entries. A revised search for "*ade-2 S. cerevisiae*" in Nucleotide, however, conveniently provided a link at the top of the search results page for the reference sequence for DNA and protein sequences.[80]

Users need to understand molecular and genetic biology as well as bioinformatics to use the web browsers. NIH defines bioinformatics as, "The organization and analysis of biological and related information, usually involving the use of computers to develop databases, retrieval mechanisms, and data analysis tools, especially in the fields of molecular biology, structural biology, and genetics."[81] It can be

---

[74] "ADE2 / YOR128C Overview," 2.

[75] *The NCBI Handbook*, 2nd ed. (National Center for Biotechnology Information (US), 2013).

[76] Pevsner, *Bioinformatics and Functional Genomics*, 14.

[77] "GenBank and WGS Statistics," accessed March 2, 2019, https://www.ncbi.nlm.nih.gov/genbank/statistics/

[78] Eric W Sayers et al., "GenBank," *Nucleic Acids Research* 47, no. Database issue (January 8, 2019): D94, 96, https://doi.org/10.1093/nar/gky989

[79] Pevsner, *Bioinformatics and Functional Genomics*, 42.

[80] Search conducted on NCBI's Nucleotide on March 2, 2019. "Nucleotide - NCBI," accessed March 2, 2019, https://www.ncbi.nlm.nih.gov/nucleotide/

[81] U.S. National Library of Medicine, NIH, "Collection Development Manual," accessed March 2, 2019, https://www.nlm.nih.gov/tsd/acquisitions/cdm/subjects12.html

overwhelming for users with little or no experience to navigate the abundance of web resources and know what database and what form of sequence data (e.g., DNA or protein) to use.[82] There are books, webinars, and tutorials, especially those from NCBI, as well as classes available to learn how to use the tools and conduct better searchers that generate more appropriate search results.[83] By using these resources and practicing, users who understand biology should be able to improve their bioinformatics skills over time.

4.2.2.2 Step 2: Design Components

The design of components (step 2) takes specialized knowledge of the genome editing technology. As described in Chap. 2, the design process is more difficult for ZFN than for TALEN and CRISPR. Although the nucleic acid binding domains of ZFN and TALEN interact with DNA through a protein-DNA interaction, recall that each zinc-finger recognizes three nucleotides and each TALE repeat only recognizes one nucleotide. Design of TALEN is simplified by the one-to-one interaction between each TALE repeat domain with a single nucleotide in the target-binding site.[84] Similarly, there is a one-to-one interaction between nucleotides of the target sequence with each nucleotide of the CRISPR binding domain (i.e., the gRNA). To clarify, recall that CRISPR interacts with the target sequence via RNA and not protein.

As described earlier in Chap. 4, there are software design tools that aid in the design of components (i.e., zinc finger, TALE, and gRNA) that interact with the target sequence (see Parameter 1: Accessibility, Ease of acquiring hardware and software, Software). Figure 4.9 depicts the amount of skill—basic, medium, or advanced—needed to use these design software tools. Skill rankings were obtained through omicX.[85] A great percentage of ZFN and TALEN software design tools (83–86%) can be used with basic skills—few require advanced skills (14–17%). In contrast to ZFN and TALEN, a much greater percentage of CRISPR software tools operate at an advanced or medium level (41%). With so many CRISPR design tools available to use, a misperception might form that they are all easy to use, which is not the case. While there are many more software design tools for CRISPR than there are for ZFN and TALEN, this analysis of skill exposes an important trend in that a little less than half of the CRISPR design tools need medium-to-advanced skill to use.

[82] Pevsner, *Bioinformatics and Functional Genomics*, 42.

[83] "Learn - NCBI," accessed March 6, 2019, https://www.ncbi.nlm.nih.gov/home/learn/

[84] J. Keith Joung and Jeffry D. Sander, "TALENs: A Widely Applicable Technology for Targeted Genome Editing," *Nature Reviews. Molecular Cell Biology* 14, no. 1 (January 2013): 49, https://doi.org/10.1038/nrm3486

[85] OmicX does not specify how skill was determined. An analysis of the data suggests that the level of skill was based on the interface: web user interfaces require basic skill, graphical user interfaces require medium skill, and command line interfaces require advanced skill.

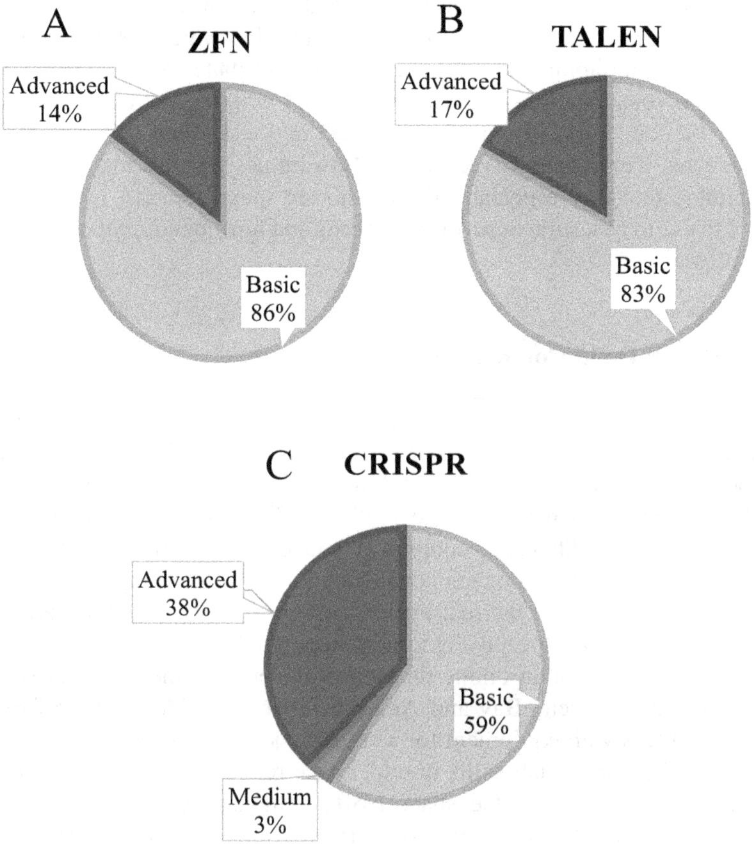

Fig. 4.9 Skill to use genome editing software design tools. The figures depict the skill (basic, medium, or advanced) used to operate (**a**) ZFN, (**b**) TALEN, and (**c**) CRISPR software design tools (As ranked by omicX. Data was obtained using omicX on December 29, 2018)

Tools can help to a certain extent (e.g., scoring and ranking potential gRNA), but it is up to the user to make the final decision. During the Genspace study, CRISPRdirect, was used to support the design of gRNA to target *ade2*. The following information was entered in CRISPRdirect: the sequence of *ade2* (obtained from SGD), PAM sequence requirement (5′-NGG-3′), and organism (yeast). CRISPRdirect generated a list of 140 potential gRNAs to bind to target sites; however, it did not rank gRNAs. Instead, CRISPRdirect listed gRNAs in the order in which the gRNAs were arranged on the gene. CRISPRdirect provided information on sequence information including G:C content; melting temperature (a function of G:C content); constructions of five or more consecutive thymines, which were automatically grayed out because these were not good sequences to use unless transcribing gRNA *in vitro*; and restriction sites.

While it takes basic skill to use CRISPRdirect, it was ultimately up to the discretion of the Genspace lab instructor to determine which gRNA would likely work. The instructor applied knowledge of known attributes of good gRNA designs as well as previous experience and trial and error working with CRISPR to select gRNA for the lab. Typically, users will select and test a few different gRNAs to determine what works best. Given all the software design tools available, it can be overwhelming to users, especially those who are inexperienced, to know exactly what tool to use to best suite experimental needs and how to interpret output from the design software.

4.2.2.3 Step 3: Build Components

For step 3, and the remaining steps, CRISPR users needed to at least have basic lab skills to be able to perform molecular cloning. Molecular cloning is described as, "the process by which recombinant DNA molecules are produced and transformed into a host organism, where they are replicated."[86] Molecular cloning is a basic laboratory method.[87] Standard equipment found in modern molecular biology laboratories—such as the autoclave, centrifuge, microwave, fridge, freezer, micropipette and tips, incubator, thermal cycler, and vortex—was used to carry out the Genspace protocol. No special equipment was used. This equipment is widely available and easy to obtain online through providers like the Carolina Biological Supply Company or even eBay and Amazon. There are videos online for how to stock DIY bio labs with equipment for as little as $1000 to $3000.[88]

Users with little-to-no lab skills or experience relied on the instructor and other proficient participants in the Genspace lab to demonstrate skills and guide users through the steps of the protocol. While there were ways to reduce knowledge requirements (e.g., order components from private companies and use kits), inexperienced and unskilled users typically did not know how to modify protocol to fit the needs of the lab, make impromptu changes to the protocol to adjust to the contingencies of the lab, or troubleshoot failed steps. The basic lab skills, knowledge requirements, and troubleshooting expertise involved in carrying out the lab are addressed in the remainder of this section.

From the start of step 3, knowledge requirements were reduced by ordering the gRNA from a private company. To do this, the Genspace instructor ordered two strands of custom-made DNA—a top and bottom strand that were annealed together to form double-stranded DNA—which contained the gRNA target sequence from

[86] New England Biolabs, "Molecular Cloning Technical Guide," 2019, 2, accessed October 20, 2019, https://www.neb.com/-/media/nebus/files/brochures/cloning_tech_guide.pdf?la=en&rev=5e4ee766c39f49e08fe1a378c4cbd2e0

[87] "Molecular Cloning: Basics and Applications | Protocol," JoVE, accessed October 20, 2019, https://www-jove-com.mutex.gmu.edu/science-education/5074/molecular-cloning

[88] "How to Stock a Biology/Genetics Lab - YouTube," accessed October 20, 2019, https://www.youtube.com/watch?v=zqQ4x9rmkKg

Integrated DNA Technologies (IDT). IDT is a company that "develops, manufactures, and markets nucleic acid products" and is recognized as a leader in custom oligonucleotide synthesis.[89] Prior to ligating the target sequence to the plasmid backbone (pML107), the plasmid was digested with restriction enzymes, *Swa*I and *Bcl*I, which were purchased from New England BioLabs.[90] The digestion cut pML107 to allow for the insertion of the target sequence.

pML107 is a programmable vector used to deliver CRISPR into yeast cells. In additional to containing restriction enzyme sites that facilitate the cloning of gRNA, pML107 contains the gene and regulatory sequences that drive expression of Cas9. pML107 (Addgene plasmid # 67639) was deposited to Addgene by the John Wyrick lab, and the protocol for how to retarget the plasmid was published by Laughery and colleagues.[91] Genspace purchased pML107 through Addgene. The Genspace instructor reviewed the Laughery protocol, and, based on previous experience and trial and error, figured out how to modify the protocol to suit the needs of the Genspace lab.

The instructor made changes in the Laughery protocol to enhance the efficiency of ligating the gRNA insert to the plasmid backbone. As mentioned above, pML107 was digested with the restriction enzymes *Bcl*I and *Swa*I, which resulted in an overhang end and blunt end, respectively. The Laughery protocol recommended using a T4 DNA ligase, which works well for overhang ends but not as well for blunt ends. Based on prior experience working with ligases and knowing that there are dozens of ligases on the market, the Genspace instructor consulted the NEB manufacturer website and searched the ligase glossary for a ligase that worked well with both blunt and overhang ends. The ligase selected for the Genspace lab, Blunt/TA Ligase Master Mix purchased from New England BioLabs (NEB M0367), increased ligation efficiency about ten-fold and saved time because the ligation no longer needed to sit overnight as required in the original protocol.[92]

Use of the Blunt/TA Ligase Master Mix, as well as all the other mixes and kits, required participants to be skilled in micropipetting. The instructor as well as more skilled and experienced lab participants demonstrated the technique and offered advice (e.g., how to load the sample, deliver the volume, and switch pipette tips between reagents) to users with less skill and experience. The instructor repeatedly demonstrated and guided participants on good micropipetting technique and other skills used throughout the lab. In general, those who were more skilled and

---

[89] "About Us," Integrated DNA Technologies, accessed February 12, 2018, https://www.idtdna.com/pages/about

[90] The Genspace lab instructor performed the cut protocol prior to the lab because this step took 12 h.

[91] John Wyrick Lab, "PML107 (Plasmid #67639)," Addgene, 107, accessed November 18, 2017, https://www.addgene.org/67639/; Marian F. Laughery et al., "New Vectors for Simple and Streamlined CRISPR–Cas9 Genome Editing in Saccharomyces Cerevisiae," *Yeast* 32, no. 12 (December 1, 2015): 711–20, https://doi.org/10.1002/yea.3098

[92] "Blunt/TA Ligase Master Mix," New England BioLabs, accessed February 12, 2018, https://www.neb.com/products/m0367-blunt-ta-ligase-master-mix#Product%20Information

experienced took the lead and helped those who were less skilled and experienced to accomplish tasks.

To support the reproducibility of the lab, the Genspace lab protocol included the manufacturer and product number of custom reagents. The Genspace lab protocol aligned closely with those of the manufacturers; however, slight adjustments were sometimes made by Genspace. In the case of NEB M0367, the Genspace lab protocol used 100-fold molar excess of the insert, while the manufacturer protocol used three-fold molar excess of the insert.[93] Given previous experience, trial and error, and knowledge that some of the plasmid backbones close up on themselves, Genspace decided to make this adjustment in the protocol to maximize the likelihood of ligation of the target sequence insert into the plasmid backbone.[94]

Furthermore, an impromptu additional centrifuge step was added at the end of the protocol. This step was not written in the Genspace or manufacturer's protocols, but the instructor decided to add it while performing the lab because it was good lab practice to spin the small volume of reagents down to the bottom of the tube. Again, during this lab, the instructor demonstrated and walked participants through the basic lab skill of how to use and program a centrifuge. Other skilled participants also helped less skilled participants to learn the basics of how to balance tubes in the centrifuge (Fig. 4.10).

After the target sequence was ligated to pML107, *E. coli* was used to finish building the plasmid. Chemically competent[95] *E. coli* was prepared in-house by Genspace and then transformed with pML107 using heat shock. This involved the incubation of transformed *E. coli* on ice for 5 min, heat shock for 1 min, and recovery on ice for another 5 min. The heat shock protocol used by Genspace for this lab became more abbreviated over time. Having repeated this lab several times, the instructor learned through trial and error where it was possible to abbreviate the protocol and how to optimize the procedures to make them tolerate of the abbreviations. In this case, the instructor found that the ligation is very efficient—especially because of the change that the lab instructor made in the protocol to flood every

[93] "Ligation Protocol for Cloning with Blunt/TA Ligase Master Mix (M0367) | NEB," accessed March 8, 2019, https://www.neb.com/protocols/2012/08/27/transfer-master-mix-to-ice-prior-to-reaction-set-up-mix-tube-by-finger-flicking-before-use-combin

[94] It was possible to have such a high molar excess of insert because inserts were ordered from Integrated DNA Technologies without phosphate ends. Thus, two inserts could not ligate to each other and could only ligate to the plasmid backbone. *E. coli* was used to finish building the plasmid since *E. coli* has more robust DNA repair mechanisms and could phosphorylate the ends of the insert to allow ligation to the plasmid. This results in approximately 50% of plasmids that have the insert.

[95] Chemically competent cells are prepared in an environment abundant in calcium. The positive charges of the calcium help to overcome the electrostatic repulsion between the plasmid and *E. coli* cell wall—the bacterial cell wall weakens. The heat shock, or abrupt rise in temperature, causes pores to form in *E. coli* and permits the entry of the plasmid into the cell. For more on this process see JoVE Science Education Database, *Basic Methods in Cellular and Molecular Biology*, Bacterial Transformation: The Heat Shock Method (Cambridge, MA, 2018).

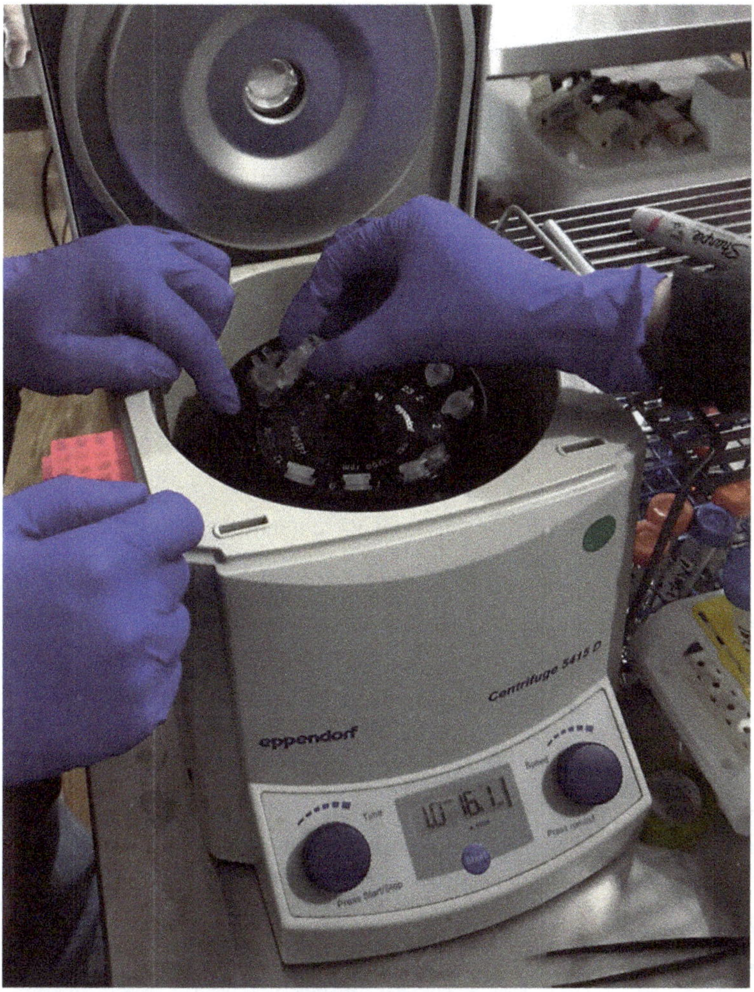

Fig. 4.10 Guidance on basic lab skills: centrifugation. Here, a more experienced and skilled participant guides a less experienced and skilled participant of the Genspace study on how to load the centrifuge (Centrifuge used in the Genspace laboratory, November 6, 2017. Katherine Paris)

1 copy of the plasmid backbone with 100 copies of the gRNA insert. Thus, not as much time was needed for recovery between steps of the heat shock protocol.

Following the heat shock, transformed *E. coli* were incubated overnight at 37 °C on LB agar plates that contained the antibiotic ampicillin. Since pML107 contains the gene that codes for ampicillin resistance (*ampR*), only *E. coli* that took up pML107 survived the plating process. All 10 participants of the Genspace lab had *E. coli* colonies. Just because *E. coli* took up the plasmid, however, did not necessarily mean the plasmid contained the insert (i.e., the target sequence). The ligation was only expected to have about a 50% efficiency, so about half of the *E. coli* colonies were expected to have the targeting insert and the other half were not expected to have the insert.

Thus, colony PCR was run using OneTaq® 2X Master Mix (NEB M0482) to screen for pML107 with the correct integration of the insert. Primers for the PCR were synthesized by IDT. PCR products were loaded onto a 2% gel for electrophoresis: correctly assembled PCR products were 224 bp. Again, it was expected that half the PCR would fail—not due to user error—because the ligation efficiency was only 50%.[96] Out of the ten student participants, four achieved the ligation while six did not achieve the ligation. Thus, the success rate of the ligation efficiency of students was 40%.

E. coli with correctly assembled plasmids were amplified in liquid culture overnight. Students who did not have correctly assembled plasmids used *E. coli* from students who did have the correctly assembled plasmid, so, moving forward, all students were working with the correctly assembled plasmid. Next, the Monarch® Plasmid Miniprep Kit (NEB T1010) was used to lyse *E. coli* and isolate the plasmids from the cells. The plasmid concentrations were quantified using the Qubit fluorometer to ensure all concentrations were high enough to transform yeast. Throughout the Genspace study, the lab instructor demonstrated several of the skills including how to use the Qubit fluorometer.

4.2.2.4 Extra Step: Build Repair Template

In addition to preparing the vector for CRISPR-Cas9 during the Genspace lab, a template to repair cleaved DNA was also synthesized. The repair template contained a gene that coded for GFP. Under ultraviolet light, GFP fluoresces green. GFP served as a reporter of expression to validate the integration of the repair template since yeast cells that successfully integrated GFP glowed green (see step 5). The lab skills to build the repair template were much the same as those needed to build the CRISPR-Cas9 plasmid with the gRNA insert in the previous step (i.e., step 3). Again, master mixes and kits were used, which reduced tacit knowledge requirements, and other impromptu changes were made to the protocol during the lab.

For the repair template to interface with DNA at the correct location, homology arms that were 60 bp each were attached to the repair template via PCR using the Q5® High Fidelity 2X Master Mix purchased from New England BioLabs (NEB M0492).[97] The Genspace instructor had communicated with researchers at high profile labs while troubleshooting the Genspace protocol to determine the best length

[96] Some of the plasmid backbones were expected to close on themselves, and some of the gRNA inserts may not have integrated in the plasmid backbone in the correct direction.

[97] "Q5® High-Fidelity 2X Master Mix," New England BioLabs, accessed February 16, 2018, https://www.neb.com/products/m0492-q5-high-fidelity-2x-master-mix#Product%20Information

for the homology arms. Ultimately, the Genspace instructor combined approaches from multiple experts and decided to use 60 bp homology arms.

Homology arms were built onto the ends of the repair template by using customized primers purchased from IDT. To ensure PCR products were correctly assembled, repair template PCRs were loaded onto a 1% gel for electrophoresis: correctly assembled PCR products were approximately 1500 bp. Four out of ten student participants, a little less than half, did not have correctly assembled PCR products and had to use pre-prepared repair template from the Genspace instructor to complete the lab.

Participants who did not successfully prepare the repair template tended to have the least amount of lab skill and experience. Several participants who failed this step took turns using the same micropipette, which a highly skilled and experienced participant discovered was broken during the next protocol. The highly skilled and experienced student had noticed the volume of liquid that was pipetted was much lower than it should have been, and, after looking at the liquid in pipette tip, confirmed the pipette was not properly working. However, unskilled and inexperienced students who had just used the micropipette could not tell the micropipette was broken. Although the micropipette was removed from further use, the malfunctioning equipment likely contributed to the failure of students to achieve the correctly assembled repair template.

The Monarch® PCR & DNA Cleanup Kit, purchased from New England BioLabs (NEB T1030), was used to purify the repair template (DNA) from the PCR.[98] The kit helped to isolate and concentrate the repair template from other impurities in the solution such as proteins, primers, and other nucleotides. Although the Genspace protocol text was the same as the protocol text provided by the manufacturer for NEB T1030, there was some ambiguity in the last step (i.e., step 6) of the manufacturer's protocol. The manufacturer's protocol said to add ≥6 μL of DNA Elution Buffer and suggested that volumes range from 6 to 20 μL.[99] The Genspace protocol text said to add 6 μL of DNA Elution Buffer; however, the Genspace instructor made an impromptu change during the lab by verbally directing participants to instead add 10 μL of DNA Elution Buffer. Through trial and error, the Genspace instructor determined that this volume of DNA Elution Buffer was optimal for maximizing the concentration of the DNA while minimizing the amount of DNA left behind. After PCR cleanup, concentrations of the repair template DNA were measured with the Qubit fluorometer to ensure they were high enough to transform yeast.

[98] "Monarch® PCR & DNA Cleanup Kit (5 Mg) | Purification Kits," New England BioLabs, accessed February 16, 2018, https://www.neb.com/products/t1030-monarch-pcr-dna-cleanup-kit-5-ug#Product%20Information

[99] "Quick Protocol for DNA Cleanup and Concentration Using the Monarch® PCR & DNA Cleanup Kit (5 Mg) (NEB #T1030) | NEB," accessed March 13, 2019, https://www.neb.com/protocols/2015/12/08/quick-protocol-for-monarch-pcr-dna-cleanup-kit-5-g-t1030

4.2.2.5 Step 4: Deliver

Yeast cells were transformed with pML107 and the repair template following the preparation of these components. Basic lab skills including use of the micropipette and centrifuge were used for the Frozen-EZ Yeast Transformation II Kit™ (Zymo T2001), which facilitated the transformation of chemically competent yeast. Again, the Genspace instructor made an impromptu change and added a centrifuge step to the end of the yeast transformation protocol to maximize the plating of yeast colonies. The Genspace instructor consulted with another Genspace instructor, who also participated and helped the class, to determine the best speed and time for centrifugation of yeast cells. Although the second lab instructor had not worked with the Zymo T2001 kit before, the second instructor did have previous experience centrifuging yeast in the lab. Based on prior experience with yeast, the instructors decided to spin down slower and for a longer time to minimize disruption of yeast cells. While the other centrifugations in the Genspace protocol occurred at 16,000 RCF, the centrifugation of yeast cells occurred at 4000 RCF for 3 min.

The transformation kit used for the Genspace study represents one way to deliver genome editing components; however, there are other ways as discussed in Parameter 1. Not all methods of delivery are easy. A recorded interview with Technician Laura Willett captures the delivery challenges experienced by the United States Department of Agriculture Agricultural Research Service team during the injection of CRISPR into moth eggs. According to Technician Willett: "The needle is very long and skinny. It's kind of like trying to inject something with a rubber hose. It [i.e., the needle] is glass, and we break it a little bit to get something sharp and at an angle … once you get that down, hopefully you don't break it … it's a lot of practice, you don't do it right the first ten times."[100]

4.2.2.6 Step 5: Validate

Results of the Genspace study were validated through the visual inspection of yeast cells followed by the analysis of the genetic sequence. Yeast colonies were red if CRISPR cleaved *ade2*, and yeast glowed green under UV light if cells integrated and expressed the GFP repair template. Four out of ten student participants had red yeast colonies, which indicated the CRISPR was successful. Out of the red yeast colonies, 10% fluoresced under UV light, which indicated successful integration of GFP.

Results were genetically validated by sequencing the *ade2* gene. To do this, basic lab skills were applied to isolate DNA from yeast transformants by suspending the yeast in 10% chelex (a purifying material) followed by vortexing, boiling, and

[100] GoodFruitGrower, *The Challenges of Using CRISPR to Modify Codling Moth Genes*, pt. 01: 20 min, accessed March 14, 2017, https://www.youtube.com/watch?v=My911P0meKE

Table 4.7 Overview score for the skill and tacit knowledge needed to use genome editing technologies

	ZFN	TALEN	CRISPR
Skill and tacit knowledge needed to use technology	High	High to moderate	Moderate to low

centrifuging.[101] PCR of the *ade2* locus was performed using Q5® High Fidelity 2X Master Mix (NEB M0492). PCR products were sent to a company called GENEWIZ for Sanger sequencing. Specialized knowledge of how to read Sanger sequencing results was applied by the instructor in order to interpret the results for the students. The assessment of the skill and tacit knowledge needed to use genome editing technologies is summarized in Table 4.7.

4.2.3 Ways to Acquire Skill and Knowledge

4.2.3.1 Services Offered by Providers

Providers of genome editing technologies offer services that allow the user to outsource portions of the user's lab work. Overall, a greater number of providers advertise more services for CRISPR than for TALEN and ZFN (Fig. 4.11). Over five times as many providers offer services to design, construct, and validate genome editing components for CRISPR compared to TALEN and ZFN, combined. Additionally, for users who decide on viral-based delivery, over 20 providers advertise services to make ready-to-use AAV or lentiviral preparations of CRISPR. About the same number of providers also offer services to make CRISPR libraries, which are useful for applications such as determining gene function and discovering drug targets. Bioinformatics services are also available mostly for CRISPR, but also ZFN and TALEN, to help users go from designing genome components to ordering these products from the provider.

In addition to services that help users to obtain genome editing components, there are services that provide users with genetically engineered model organisms and cell lines—many of which are modified using CRISPR. A commonly held understanding of scientists is as follows: "fundamental problems are most easily solved in the simplest and most accessible system in which the problem can be addressed."[102] For

[101] DNA from transformed yeast was extracted based on a technique from Benjamin A. Blount, Maureen R. M. Driessen, and Tom Ellis, "GC Preps: Fast and Easy Extraction of Stable Yeast Genomic DNA," *Scientific Reports* 6, no. 26863 (2016), https://doi.org/10.1038/srep26863

[102] James D. Watson et al., *Molecular Biology of the Gene, Sixth Edition*, 6 edition (San Francisco: Cold Spring Harbor, N.Y: Pearson, 2007), 783.

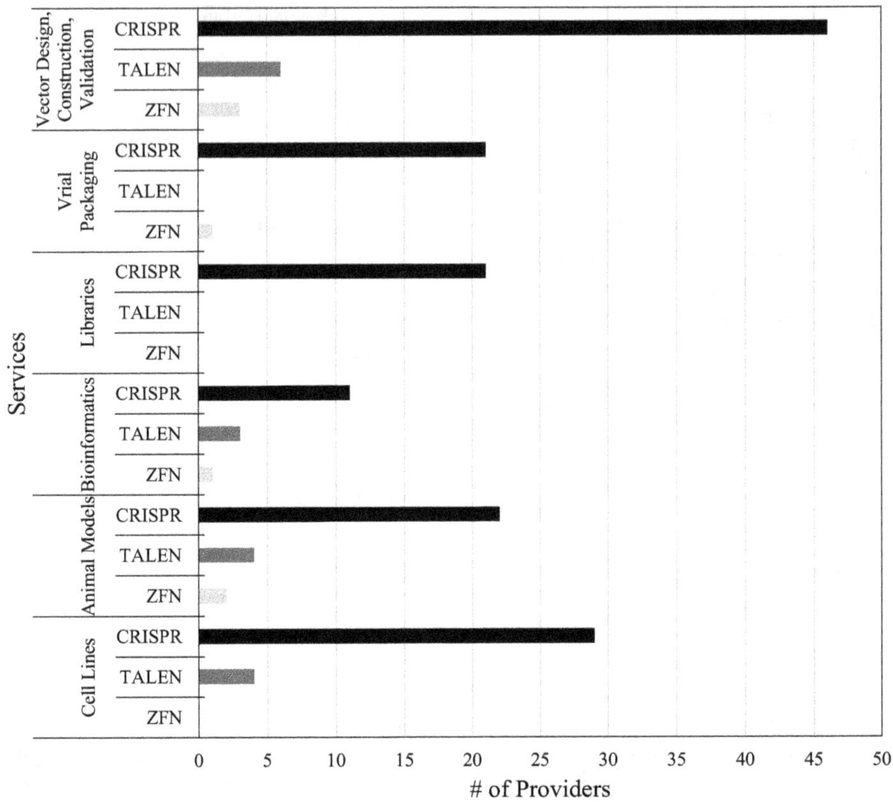

Fig. 4.11 Services offered by providers for ZFN, TALEN, and CRISPR (Data for the figure were obtained from the Science Exchange on April 18, 2019)

instance, the mouse is used extensively to model the development of mammals, understand gene function, and model human diseases.[103] Several providers, such as the New-York-based biotechnology company Taconic Biosciences, use CRISPR to make genetic modifications, such as knock-outs, in mice or other organisms to create custom animal models.[104] Taconic Biosciences prefers to use CRISPR over ZFN and TALEN due to limitations with "complex design and assembly as well as high costs, which limit their routine use in the laboratory."[105]

[103] Strachan and Read, *Human Molecular Genetics, Fourth Edition*, 304–5.

[104] Taconic Biosciences, "CRISPR Gene Editing | CRISPR Knockout Mice," Taconic, accessed October 5, 2018, https://www.taconic.com/genetically-engineered-animal-models/crispr-gene-editing/

[105] Adriano Flora, Jost Seibler, and Ralph Gareus, *Application of CRISPR/Cas to the Generation of Genetically Engineered Mice* (Taconic, 2018), 4.

4.2.3.2 Forums

Online

Companies that sell genome editing products and services also provide helpful online resources such as written and video instructions, FAQs, blogs, and other troubleshooting tips. New England Biolabs (NEB) is a key source that provided products for the Genspace case study such as the Monarch Plasmid Miniprep Kit (NEB #T1010) that was used to purify plasmid DNA (i.e., the CRISPR-Cas9 plasmid). The NEB website contains three videos on the NEB #T1010 kit protocol, tips for using, and how to recycle the material. Additionally, the NEB website offers NEB #T1010 kit product information, FAQs and tech tips, product manual, troubleshooting advice, and safety data sheets.[106]

Other companies, such as GenScript, Millipore-Sigma, and Thermo Fisher Scientific, offer several webinars on how to use CRISPR.[107] There many videos available on YouTube that convey knowledge and skills such as how to use Benchling and CRISPRdirect, which are gRNA software design tools.[108] *JoVE,* another popular and rapidly growing source, offers peer-reviewed video article protocols on ZFN, TALEN, and CRISPR. Video article protocols from *JoVE* are analyzed in more detail below.

JoVE *Video Article Protocols*

Overall, there are many more *JoVE* video articles for CRISPR than ZFN or TALEN. Out of 47 *JoVE* video articles published between 2012 and 2018 on genome editing, there were 4, 8, and 41 video articles that provided protocols on ZFN, TALEN, and CRISPR, respectively—note that some video articles provided protocols on more than one genome editing tool. As Fig. 4.12 depicts, the total number of CRISPR video article protocols is exponentially increasing, while the total number of ZFN and TALEN video article protocols has remained stagnant since about 2016.

[106]"Monarch® Plasmid Miniprep Kit," New England BioLabs, accessed August 30, 2018, https://www.neb.com/products/t1010-monarch-plasmid-miniprep-kit#Product%20Information

[107]"2018 Webinar Series," GenScript, accessed February 26, 2019, https://www.genscript.com/webinars.html?src=leftbar; "CRISPR Design Tools," Millipore Sigma, accessed February 14, 2019, https://www.milliporesigmabioinfo.com/bioinfo_tools/?domain=STANDARD; "Genome Editing Videos & Webinars - US," Thermo Fisher Scientific, accessed February 26, 2019, https://www.thermofisher.com/us/en/home/life-science/genome-editing/genome-editing-learning-center/genome-editing-resource-library/genome-editing-videos-webinars.html

[108]Benchling, *CRISPR GRNA Design and Analysis Using Benchling (Updated)*, n.d., accessed February 26, 2019, https://www.youtube.com/watch?v=jfCx9LxXr08&t=1s; togotv, *CRISPRdirect: Designing CRISPR/Cas Guide RNA Sequence*, n.d., accessed February 26, 2019, https://www.youtube.com/watch?v=OGdBU29v3jY

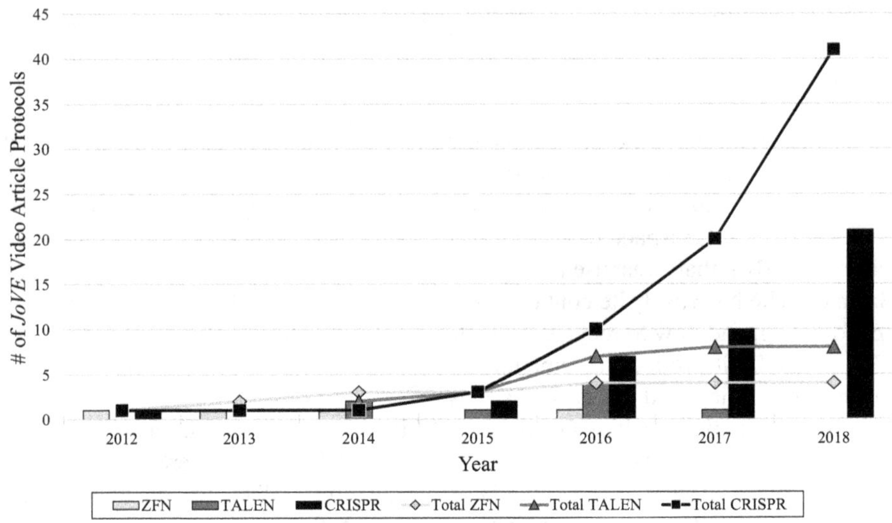

Fig. 4.12 Number of JoVE video articles per year (2012–2018). The figure is based on 4 ZFN, 8 TALEN, and 41 CRISPR video articles on *JoVE*. Solid bars represent the number of new video articles each year. The lines represent the total number of video articles accumulated since 2012 (Data for the figure were obtained from *JoVE* on February 19, 2019)

The total number of views for all *JoVE* video articles is also much higher for CRISPR than ZFN or TALEN. As Fig. 4.13 depicts, the total number of views of CRISPR video articles increases over time, while the total number of views of ZFN and TALEN videos remains about the same. In 2018, there were almost 5-times as many views of CRISPR video article protocols as there were for ZFN and TALEN video articles, combined. While the views of ZFN and TALEN video article protocols on *JoVE* are not increasing, for the most part, they are not declining either. This indicates that a consistent population of users continue to view *JoVE* video article protocols to help use ZFN and TALEN.

In-Person

In general, ZFN and TALEN skill and knowledge tend to be taught in forums where users have higher-level laboratory skills and biological knowledge, and CRISPR skill and knowledge tends to be taught in forums where users have all levels of, including little to no, laboratory skill and biological knowledge. Workshops—where users have higher-level knowledge and skill—offer training on all three technologies, whereas DIY laboratory classes and college classes only offer training on CRISPR (Fig. 4.14). Workshops range from 2 to 5 days, offer a mix of lecture and hands-on exercises, and are geared toward training professionals and PhD-level

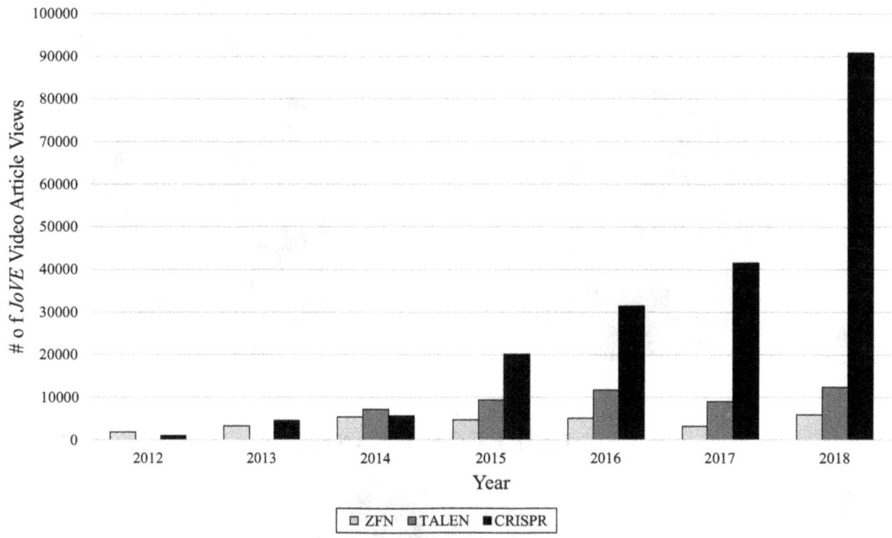

Fig. 4.13 Number of views of JoVE video articles per year (2012–2018). The figure is based on 4 ZFN, 8 TALEN, and 41 CRISPR videos articles on *JoVE*. The bars represent the total views for all the video articles available on *JoVE* that year (Data for the figure were obtained from *JoVE* on February 19, 2019)

researchers who have at least a basic understanding of molecular biology and genetics as well as intermediate laboratory skills. On the other hand, college classes offered so far are typically semester-long classes at the undergraduate level. Similarly, DIY biology classes usually last a few weeks and are attended by a variety of users ranging from those with no experience to professionals and college professors.

Workshops are phasing out ZFN and TALEN in favor of CRISPR. All the workshops currently offered are CRISPR-focused—only one workshop, offered through the National Institutes of Health in Maryland, also covers ZFN and TALEN design and assembly. While the European Molecular Biology Laboratory (EMBL) in Heidelberg, Germany once offered an advanced ZFN workshop in 2011, current classes offered through EMBL are for CRISPR. Additionally, DIY biology laboratories are also sticking with using CRISPR in lieu of ZFN or TALEN. While a DIY biologist member of the Genspace laboratory tried to use TALEN several years prior to November 2017, the member experienced several difficulties assembling TALEN and quickly abandoned the technology.[109] Table 4.8 summarizes the assessment of the ways to acquire genome editing skill and knowledge.

[109] Genspace NYC, "Genspace Class, November 2017."

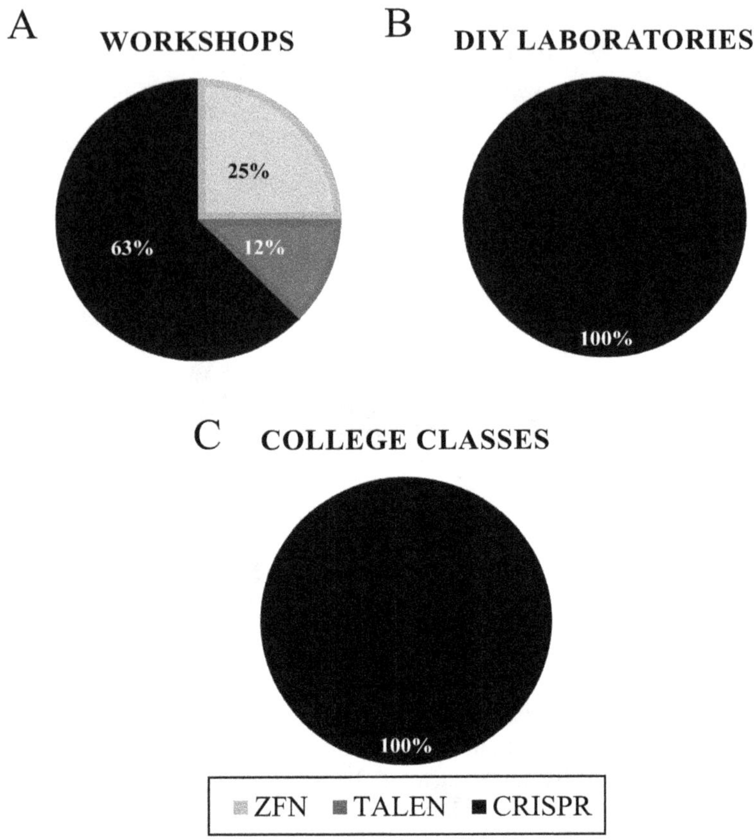

Fig. 4.14 Forums to acquire skill and knowledge to use genome editing technologies. Forums to learn how to use genome editing technologies include (**a**) workshops, (**b**) DIY laboratories, and (**c**) college classes. See Appendix D for more on workshops, DIY laboratories, and college classes

Table 4.8 Overview score for the ways to acquire genome editing skill and knowledge

	ZFN	TALEN	CRISPR
Ways to acquire skill and knowledge	Modest resources available	Modest resources available	Ample resources available

4.2.4 Transferability of Skill and Knowledge

4.2.4.1 Videos

Results from a study on undergraduate students and *JoVE* videos suggest that videos are an effective way to transfer skill and knowledge. Undergraduate students at DeSales University and Clemson University who watched *JoVE* videos on basic lab

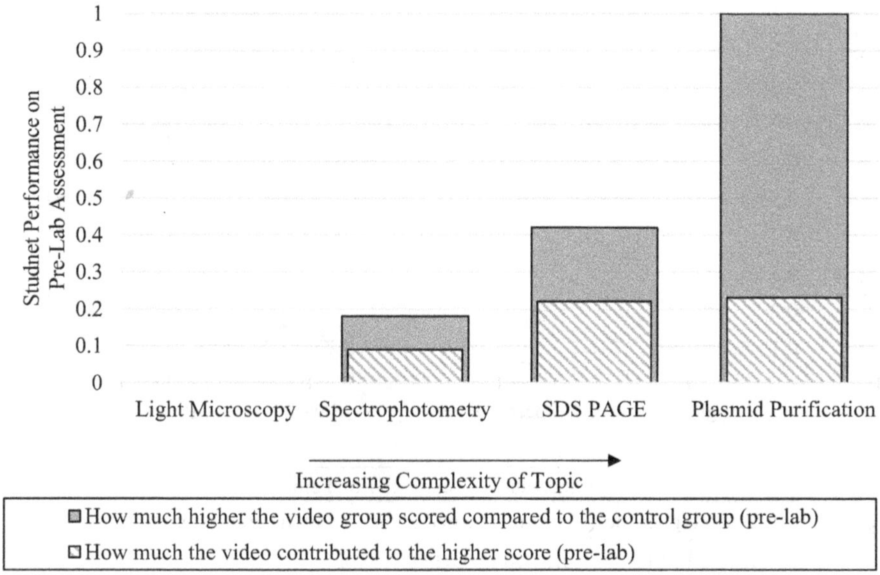

Fig. 4.15 The effectiveness of watching JoVE videos on undergraduate student performance (pre-lab assessment) (Data from the *JoVE* study by Mutch-Jones and Jia)

skills and methods as a pre-lab requirement performed better on pre- and post-lab assessments than students who did not watch the videos. This trend is increasingly more pronounced as topics become more complex. On the pre-lab assessment, the video group scored 18%, 42%, and 100% higher than the non-video group for spectrophotometry, SDS PAGE, and plasmid purification labs, respectively (Fig. 4.15). On the post-lab assessment, the video group scored 6%, 16%, and 76% higher than the non-video group for spectrophotometry, SDS PAGE, and plasmid purification labs, respectively (Fig. 4.16). Note that there was no significant difference in score on pre- and post-lab assessments for the least complex topic, light microscopy.[110]

Videos tended to help students more on more advanced topics. On the pre-lab, watching videos contributed at least 22% and 23% of the higher score for the video group's assessment on SDS PAGE and plasmid purification, respectively, and only 9% of the higher score for spectrophotometry. On the post-lab, watching videos contributed at least 12% and 18% of the higher score for the video group's assessment on SDS PAGE and plasmid purification, respectively. For both assessments, scores were controlled for other factors that also likely contributed to higher scores including "student confidence and general knowledge of basic biology concepts."[111] While this is only one study that is small (346 students), limited to undergraduates,

[110]Data from Mutch-Jones and Jia, *Effectiveness of JoVE Videos in Improving STEM Students' Performance: A Study in Undergraduate Biology Labs.*

[111]Mutch-Jones and Jia, 6.

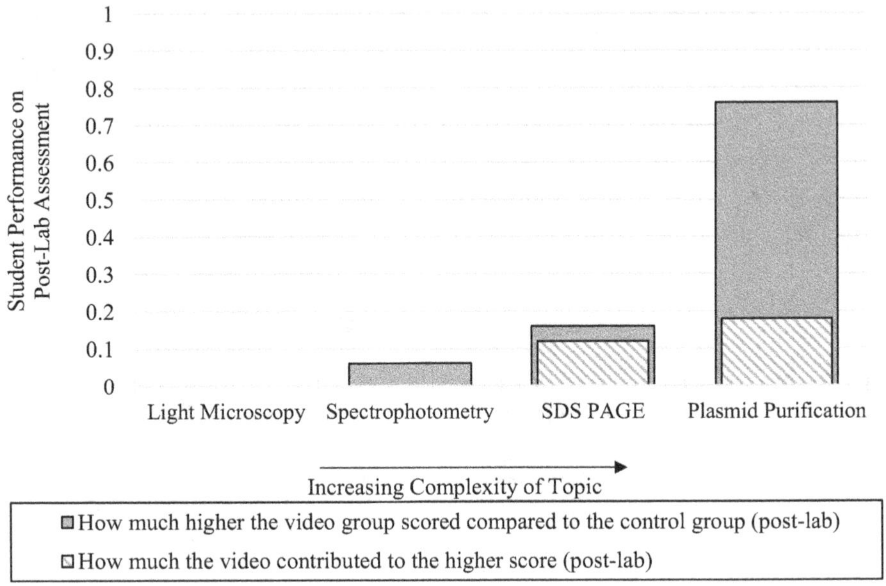

Fig. 4.16 The effectiveness of watching JoVE videos on undergraduate student performance (post-lab assessment) (Data from the *JoVE* study by Mutch-Jones and Jia)

and includes only videos on general laboratory techniques, the techniques—especially plasmid purification—that were demonstrated in the videos are skills that can be used to perform genome editing.

4.2.4.2 Classes

Results from the Genspace DIY bio class and undergraduate class at the University of New Mexico suggest that these classes are moderately effective in transferring CRISPR skill and knowledge. Both forums covered the skills and knowledge needed to navigate the CRISPR workflow from beginning to end. The results of both forums were similar: 40% of students of the Genspace DIY biology class achieved the *ade2* knockout in yeast, and 50% of students in the undergraduate class achieved the gene knockout in Drosophila (Fig. 4.17).[112] In other words, CRISPR worked for 40–50% of students.

Although results from both classes were similar, it is important to point out the limitations and differences between the studies that impact their generalization. First, these studies were both limited by their size—there were only 10 students in the DIY

[112] Of the Genspace DIY biology students who achieved the *ade2* knockout, 10% also achieved integration of the *gfp* gene, which indicated HDR occurred using the repair template that students supplied to the yeast cells. This success rate is in line with the expected success rate.

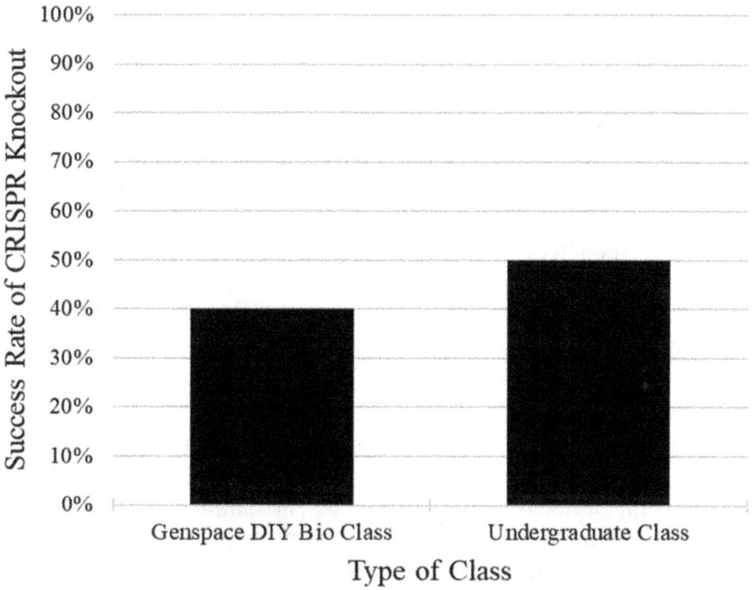

Fig. 4.17 Success rates of CRISPR knockout for the genspace DIY bio class and undergraduate class at the university of new Mexico (Data from the Genspace DIY Bio Lab in November 2017 and Adame et al., "An Undergraduate Laboratory Class Using CRISPR/Cas9 Technology to Mutate Drosophila Genes")

Table 4.9 Overview score for the transferability of genome editing skill and knowledge

	ZFN	TALEN	CRISPR
Transferability of skill and knowledge	Low transfer through videos. In person transfer not assessed.	Low transfer through videos. In person transfer not assessed.	Low transfer through videos. Moderate transfer in person.

biology class and 6 students in the undergraduate class. Second, both classes worked with CRISPR; however, the experimental objectives of the undergraduate class were generally more difficult. All students in the Genspace DIY biology class targeted *ade2*, a well-studied gene, in an organism (yeast), which was transformed using the Frozen-EZ Yeast Transformation II Kit™ (Zymo T2001). On the other hand, each of the six students in the undergraduate class targeted a different gene and injected CRISPR components into fly embryos. As described earlier, injection is a more difficult procedure and takes time and practice to develop this skill. Finally, the Genspace DIY biology lab was carried out in about a quarter of the time that it took for the undergraduate lab. Altogether, the transferability of the skill and knowledge needed to use genome editing technologies is low through videos (Table 4.9). The transferability of the skill and knowledge needed to use CRISPR is moderate in person. Note, the transferability of skill and knowledge needed to use ZFN and TALEN was not assessed.

4.3 Conclusion

Overall, CRISPR is more accessible than ZFN and TALEN. CRISPR products and services are available from many more providers across the academic, nonprofit (nonacademic), and commercial sectors. Analysis of the Addgene data revealed that even a company that is a provider of all three technologies has received over 10-times more overall requests for CRISPR compared to the other two technologies. The ease of acquiring CRISPR software is also high given the abundance of design tools—most of which do not have restrictions for use. While it can cost thousands of dollars to use ZFN or TALEN to edit a gene, it can cost as little as $20 using CRISPR. Still, all technologies need to overcome the challenge of delivering genome editing cargo to target cells. In the case of CRISPR, there are many providers that advertise specific services to assist the viral packaging of CRISPR delivery cargo. Finally, the validation of results for experiments involving any of the technologies can be achieved via Sanger sequencing by providers around the world for an average price of about five dollars per sample.

In general, it would be easier to misuse CRISPR than it would be to misuse ZFN and TALEN. It is more difficult to engineer ZFN and TALEN compared to CRISPR. CRISPR users still need to have a basic understanding of molecular biology and proficiency working in a laboratory. As the Genspace lab demonstrates, skill and knowledge requirements can be reduced by acquiring mixes and kits from commercial sources or by outsourcing portions of the lab work. A review of the providers of genome editing technologies on Science Exchange reveals that many more providers offer services for CRISPR compared to ZFN and TALEN. Furthermore, there are more online (e.g., video article protocols) and in-person sources to acquire skill and knowledge to use CRISPR. Furthermore, a review of assessment scores for undergraduate students in biology classes who watched videos as part of a pre-lab requirement suggest that videos are effective in transferring skill. Even so, only about a half of the students in the Genspace DIY biology class and University of New Mexico undergraduate class were successful in using CRISPR. Part II of the risk of misuse assessment, which investigates the magnitude of potential harm and imminence of potential misuse of genome editing technologies, is carried out in Chap. 5.

References

1. GenScript. "2018 Webinar Series." Accessed February 26, 2019. https://www.genscript.com/webinars.html?src=leftbar.
2. Journal of Visualized Experiments. "About JoVE (2017)." Accessed March 20, 2017. https://www.jove.com/about.
3. Journal of Visualized Experiments. "About JoVE (2019)." Accessed February 19, 2019. https://www.jove.com/about.

4. TERC. “About Us.” Accessed September 4, 2019. https://www.terc.edu/display/About/About+Us.
5. Integrated DNA Technologies. “About Us.” Accessed February 12, 2018. https://www.idtdna.com/pages/about.
6. Adame, Vanesa, Holly Chapapas, Marilyn Cisneros, Carol Deaton, Sophia Deichmann, Chauncey Gadek, TyAnna L. Lovato, Maria B. Chechenova, Paul Guerin, and Richard M. Cripps. “An Undergraduate Laboratory Class Using CRISPR/Cas9 Technology to Mutate Drosophila Genes.” *Biochemistry and Molecular Biology Education* 44, no. 3 (May 6, 2016): 263–75. doi:https://doi.org/10.1002/bmb.20950.
7. “Addgene: How to Order.” Accessed June 12, 2019. https://www.addgene.org/ordering/.
8. “Addgene: Mission.” Accessed July 28, 2017. http://www.addgene.org/mission/.
9. “Addgene: Plasmid Kits.” Accessed June 12, 2019. https://www.addgene.org/kits/#ge.
10. “Addgene: Wolfe-Lawson ZFN Modular Assembly Kit.” Accessed February 6, 2019. https://www.addgene.org/kits/wolfe-lawson-zfn/#kit-contents.
11. Addgene. “Addgene’s Collection.” Accessed February 1, 2019. https://www.addgene.org/browse/.
12. Saccharomyces Genome Database. “ADE2/YOR128C Overview.” Accessed February 10, 2018. https://www.yeastgenome.org/locus/S000005654.
13. Science Exchange. “Animal Models Shared Resource (AMSR) - Zebrafish at Georgetown University Medical Center.” Accessed June 9, 2019. https://www.scienceexchange.com/labs/animal-models-shared-resource-amsr-zebrafish.
14. Ben Glick and Aline Glick. “Identifying Sequence Elements with SnapGene’s Feature Database.” Addgene. Accessed February 11, 2018. http://blog.addgene.org/identifying-sequence-elements-with-snapgenes-feature-database.
15. Benchling. *CRISPR GRNA Design and Analysis Using Benchling (Updated)*, n.d. https://www.youtube.com/watch?v=jfCx9LxXr08&t=1s.
16. Blount, Benjamin A., Maureen R. M. Driessen, and Tom Ellis. “GC Preps: Fast and Easy Extraction of Stable Yeast Genomic DNA.” *Scientific Reports* 6, no. 26863 (2016). doi:https://doi.org/10.1038/srep26863.
17. New England BioLabs. “Blunt/TA Ligase Master Mix.” Accessed February 12, 2018. https://www.neb.com/products/m0367-blunt-ta-ligase-master-mix#Product%20Information.
18. Brinegar, Katelyn, Ali K. Yetisen, Sun Choi, Emily Vallillo, Guillermo U. Ruiz-Esparza, Anand Prabhakar, Ali Khademhosseini, and Seok-Hyun Yun. “The Commercialization of Genome-Editing Technologies.” *Critical Reviews in Biotechnology* 37 (January 18, 2017): 1–12. doi:https://doi.org/10.1080/07388551.2016.1271768.
19. Cellectis. “2012 Annual Report.” Accessed January 22, 2019. https://www.cellectis.com/uploads/files/ra-eng-papbd.pdf.
20. Cellectis. “Cellectis Granted Patent for CRISPR Use in T-Cells,” July 24, 2017. http://www.cellectis.com/en/press/cellectis-granted-patent-for-crispr-use-in-t-cells.
21. GEN - Genetic Engineering and Biotechnology News. “Cellectis Nabs TAL Effector IP from Minnesota and Iowa State Universities,” January 20, 2011. https://www.genengnews.com/topics/omics/cellectis-nabs-tal-effector-ip-from-minnesota-and-iowa-state-universities/.
22. Church, George. “The Future of the Genomic Editing Revolution.” UC Berkeley, January 26, 2017. https://www.youtube.com/watch?v=Wswvf8Nrubg.
23. Cohen, Jon. “One of Our Reporters Tried to Do CRISPR. He Failed Miserably.” *Science*, November 3, 2016. doi:https://doi.org/10.1126/science.aal0335.
24. ———. “One of Our Reporters Tried to Do CRISPR. He Failed Miserably [Video].” *Science*, November 3, 2016. doi:https://doi.org/10.1126/science.aal0335.
25. Addgene. “Create an Account.” Accessed November 26, 2018. https://www.addgene.org/users/register?next=/.
26. Sigma-Aldrich. “CRISPR.” Accessed June 9, 2019. https://www.sigmaaldrich.com/life-science/functional-genomics-and-rnai/crispr-cas9-gene-editing.html.

27. Millipore Sigma. "CRISPR Design Tools." Accessed February 14, 2019. https://www.milliporesigmabioinfo.com/bioinfo_tools/?domain=STANDARD.
28. Novartis. "CRISPR Genome Editing Fuels Cancer Drug Discovery." Accessed October 28, 2018. https://www.novartis.com/stories/from-our-labs/crispr-genome-editing-fuels-cancer-drug-discovery.
29. Thermo Fisher Scientific. "CRISPR Products and Services - US." Accessed June 9, 2019. https://www.thermofisher.com/us/en/home/life-science/genome-editing/geneart-crispr.html.
30. Data provided directly from Addgene on January 31, 2019, January 31, 2019.
31. DeFrancesco, Laura. "Move over ZFNs." *Nature Biotechnology* 29 (August 5, 2011): 681–84. https://doi.org/https://doi.org/10.1038/nbt.1935.
32. "ES Cell Targeting Core, UCSF - Science Exchange." Accessed June 4, 2019. https://www.scienceexchange.com/labs/es-cell-targeting-core.
33. Flora, Adriano, Jost Seibler, and Ralph Gareus. *Application of CRISPR/Cas to the Generation of Genetically Engineered Mice*. Taconic, 2018.
34. Gaj, Thomas, Charles A. Gersbach, and Carlos F. Barbas. "ZFN, TALEN and CRISPR/Cas-Based Methods for Genome Engineering." *Trends in Biotechnology* 31, no. 7 (July 2013): 397–405. https://doi.org/https://doi.org/10.1016/j.tibtech.2013.04.004.
35. "GenBank and WGS Statistics." Accessed March 2, 2019. https://www.ncbi.nlm.nih.gov/genbank/statistics/.
36. Thermo Fisher Scientific. "Genome Editing Videos & Webinars - US." Accessed February 26, 2019. https://www.thermofisher.com/us/en/home/life-science/genome-editing/genome-editing-learning-center/genome-editing-resource-library/genome-editing-videos-webinars.html.
37. Genspace NYC. "Genome Editing with CRISPR-Cas9 Class." Brooklyn, NY, November 2017.
38. Addgene. "Global Plasmid Distribution: Shipping and Customs." Accessed November 26, 2018. https://www.addgene.org/ordering/international/.
39. GoodFruitGrower. *The Challenges of Using CRISPR to Modify Codling Moth Genes*. Accessed March 14, 2017. https://www.youtube.com/watch?v=My9l1P0meKE.
40. Hartley, Brigham J., Stewart A. Fabb, Ben A. L. Finnin, John M. Haynes, and Colin W. Pouton. "Zinc-Finger Nuclease Enhanced Gene Targeting in Human Embryonic Stem Cells." *JoVE (Journal of Visualized Experiments)*, no. 90 (August 23, 2014): e51764. https://doi.org/https://doi.org/10.3791/51764.
41. "How to Stock a Biology/Genetics Lab - YouTube." Accessed October 20, 2019. https://www.youtube.com/watch?v=zqQ4x9rmkKg.
42. The National Academies of Sciences, Engineering, and Medicine. "International Workshop: Assessing the Security Implications of Genome Editing Technology, October 11-13, 2017, Hanover, Germany." Accessed December 27, 2021. http://nas-sites.org/dels/events/ge-security/.
43. Joe Frangipane, PhD, Field Applications Specialist, Sigma-Aldrich Corporation. "Targeted Genome Editing in the Eukaryotic Systems Using Zinc Finger Nucleases." Presented at the Rutgers ORA Research Technologies Expo, Rutgers University, 2013. https://www.youtube.com/watch?v=yDmB_NMGKWo.
44. John Wyrick Lab. "PML107 (Plasmid #67639)." Addgene. Accessed November 18, 2017. https://www.addgene.org/67639/.
45. Addgene. "Join the Community of Addgene Depositors." Accessed November 26, 2018. https://www.addgene.org/deposit/.
46. Joung, J. Keith, and Jeffry D. Sander. "TALENs: A Widely Applicable Technology for Targeted Genome Editing." *Nature Reviews. Molecular Cell Biology* 14, no. 1 (January 2013): 49–55. https://doi.org/https://doi.org/10.1038/nrm3486.
47. "JoVE | Peer Reviewed Scientific Video Journal - Methods and Protocols." Accessed February 18, 2019. https://www-jove-com.mutex.gmu.edu.
48. JoVE Science Education Database. *Basic Methods in Cellular and Molecular Biology*. Bacterial Transformation: The Heat Shock Method. Cambridge, MA, 2018.

49. ———. *Basic Methods in Cellular and Molecular Biology, Plasmid Purification*. Cambridge, MA: JoVE, 2019.
50. ———. *Basic Methods in Cellular and Molecular Biology, Separating Protein with SDS-PAGE*. Cambridge, MA: JoVE, 2019.
51. ———. *General Laboratory Techniques, Introduction to Light Microscopy*. Cambridge, MA: JoVE, 2019.
52. ———. *General Laboratory Techniques, Introduction to the Spectrophotometer*. Cambridge, MA: JoVE, 2019.
53. Kim, Hyongbum, and Jin-Soo Kim. "A Guide to Genome Engineering with Programmable Nucleases." *Nature Reviews Genetics* 15, no. 5 (May 2014): 321–34. https://doi.org/https://doi.org/10.1038/nrg3686.
54. Kim, Yongsub, Jiyeon Kweon, Annie Kim, Jae Kyung Chon, Ji Yeon Yoo, Hye Joo Kim, Sojung Kim, et al. "A Library of TAL Effector Nucleases Spanning the Human Genome." *Nature Biotechnology* 31, no. 3 (March 2013): 251–58. https://doi.org/https://doi.org/10.1038/nbt.2517.
55. Creative Biogene. "Kits." Accessed June 2, 2019. https://www.creative-biogene.com/products/kits.html.
56. Laughery, Marian F., Tierra Hunter, Alexander Brown, James Hoopes, Travis Ostbye, Taven Shumaker, and John J. Wyrick. "New Vectors for Simple and Streamlined CRISPR–Cas9 Genome Editing in Saccharomyces Cerevisiae." *Yeast* 32, no. 12 (December 1, 2015): 711–20. https://doi.org/https://doi.org/10.1002/yea.3098.
57. "Learn - NCBI." Accessed March 6, 2019. https://www.ncbi.nlm.nih.gov/home/learn/.
58. "Ligation Protocol for Cloning with Blunt/TA Ligase Master Mix (M0367) | NEB." Accessed March 8, 2019. https://www.neb.com/protocols/2012/08/27/transfer-master-mix-to-ice-prior-to-reaction-set-up-mix-tube-by-finger-flicking-before-use-combin.
59. Lino, Christopher A., Jason C. Harper, James P. Carney, and Jerilyn A. Timlin. "Delivering CRISPR: A Review of the Challenges and Approaches." *Drug Delivery* 25, no. 1 (May 25, 2018): 1234–57. https://doi.org/https://doi.org/10.1080/10717544.2018.1474964.
60. Maeder, Morgan L., Stacey Thibodeau-Beganny, Jeffry D. Sander, Daniel F. Voytas, and J. Keith Joung. "Oligomerized Pool Engineering (OPEN): An 'open-Source' Protocol for Making Customized Zinc-Finger Arrays." *Nature Protocols* 4, no. 10 (2009): 1471–1501. https://doi.org/https://doi.org/10.1038/nprot.2009.98.
61. JoVE. "Molecular Cloning: Basics and Applications | Protocol." Accessed October 20, 2019. https://www-jove-com.mutex.gmu.edu/science-education/5074/molecular-cloning.
62. New England BioLabs. "Monarch® PCR & DNA Cleanup Kit (5 Mg) | Purification Kits." Accessed February 16, 2018. https://www.neb.com/products/t1030-monarch-pcr-dna-cleanup-kit-5-ug#Product%20Information.
63. New England BioLabs. "Monarch® Plasmid Miniprep Kit." Accessed August 30, 2018. https://www.neb.com/products/t1010-monarch-plasmid-miniprep-kit#Product%20Information.
64. Science Exchange. "Mutation Generation and Detection Facility - Science Exchange." Accessed June 3, 2019. https://www.scienceexchange.com/labs/mutation-generation-and-detection-facility.
65. Mutch-Jones, Karen, and Yueming Jia. *Effectiveness of JoVE Videos in Improving STEM Students' Performance: A Study in Undergraduate Biology Labs*. Cambridge, MA: STEM Education Evaluation Center at TERC, 2018.
66. New England Biolabs. "Molecular Cloning Technical Guide," 2019. https://www.neb.com/-/media/nebus/files/brochures/cloning_tech_guide.pdf?la=en&rev=5e4ee766c39f49e08fe1a378c4cbd2e0.
67. "Nucleotide - NCBI." Accessed March 2, 2019. https://www.ncbi.nlm.nih.gov/nucleotide/.
68. Pearson, Helen. "Protein Engineering: The Fate of Fingers." *Nature* 455, no. 7210 (September 11, 2008): 160–64. https://doi.org/https://doi.org/10.1038/455160a.

69. Perkel, Jeffrey M. "Genome Editing with CRISPRs, TALENs and ZFNs." Biocompare, August 27, 2013. http://www.biocompare.com/Editorial-Articles/144186-Genome-Editing-with-CRISPRs-TALENs-and-ZFNs/.
70. Addgene. "Permits." Accessed November 26, 2018. https://www.addgene.org/ordering/permits/#new-zealand.
71. Pevsner, Jonathan. *Bioinformatics and Functional Genomics*. 2nd ed. New Jersey: John Wiley & Sons, Inc., 2009.
72. New England BioLabs. "Q5® High-Fidelity 2X Master Mix." Accessed February 16, 2018. https://www.neb.com/products/m0492-q5-high-fidelity-2x-master-mix#Product%20Information.
73. "Quick Protocol for DNA Cleanup and Concentration Using the Monarch® PCR & DNA Cleanup Kit (5 Mg) (NEB #T1030) | NEB." Accessed March 13, 2019. https://www.neb.com/protocols/2015/12/08/quick-protocol-for-monarch-pcr-dna-cleanup-kit-5-g-t1030.
74. Ramirez, Cherie L., Jonathan E. Foley, David A. Wright, Felix Müller-Lerch, Shamim H. Rahman, Tatjana I. Cornu, Ronnie J. Winfrey, et al. "Unexpected Failure Rates for Modular Assembly of Engineered Zinc Fingers." *Nature Methods* 5, no. 5 (May 2008): 374–75. https://doi.org/https://doi.org/10.1038/nmeth0508-374.
75. Sander, Jeffry D., Elizabeth J. Dahlborg, Mathew J. Goodwin, Lindsay Cade, Feng Zhang, Daniel Cifuentes, Shaun J. Curtin, et al. "Selection-Free Zinc-Finger-Nuclease Engineering by Context-Dependent Assembly (CoDA)." *Nature Methods* 8, no. 1 (January 2011): 67–69. https://doi.org/https://doi.org/10.1038/nmeth.1542.
76. Sayers, Eric W, Mark Cavanaugh, Karen Clark, James Ostell, Kim D Pruitt, and Ilene Karsch-Mizrachi. "GenBank." *Nucleic Acids Research* 47, no. Database issue (January 8, 2019): D94–99. https://doi.org/10.1093/nar/gky989.
77. Science Exchange. "Search Results for 'Sanger Sequencing.'" Accessed April 22, 2019. https://www.scienceexchange.com/marketplace/experimental-services?q=sanger%20sequencing&page=1&servicesFacets=%5B%22experiment%3ASanger%20Sequencing%22%5D.
78. "SGD Help." Accessed March 2, 2019. https://sites.google.com/view/yeastgenome-help/about.
79. Strachan, Tom, and Andrew Read. *Human Molecular Genetics, Fourth Edition*. 4th edition. New York: Garland Science, 2010.
80. Taconic Biosciences. "CRISPR Gene Editing | CRISPR Knockout Mice." Taconic. Accessed October 5, 2018. https://www.taconic.com/genetically-engineered-animal-models/crispr-gene-editing/.
81. Thermo Fisher Scientific. "TALEN Design to Validation Services - US." Accessed January 19, 2019. https://www.thermofisher.com/us/en/home/life-science/genome-editing/genome-engineering-services/talen-design-validation-services.html.
82. Addgene. "Technology Transfer Information." Accessed November 26, 2018. https://www.addgene.org/techtransfer/.
83. Labiotech.eu. "The Most Important Battle in Gene Editing: CRISPR versus TALEN," March 13, 2018. https://labiotech.eu/features/crispr-talen-gene-editing/.
84. *The NCBI Handbook*. 2nd ed. National Center for Biotechnology Information (US), 2013.
85. Todd Kuiken. "Biosafety in a World Without Walls." Presented at the Webinar: Genome Editing and Security: Governance of Non-Traditional Communities?, June 25, 2018. https://vimeo.com/277161865.
86. togotv. *CRISPRdirect: Designing CRISPR/Cas Guide RNA Sequence*, n.d. https://www.youtube.com/watch?v=OGdBU29v3jY.
87. Ul Ain, Qurrat, Jee Young Chung, and Yong-Hee Kim. "Current and Future Delivery Systems for Engineered Nucleases: ZFN, TALEN and RGEN." *Journal of Controlled Release: Official Journal of the Controlled Release Society* 205 (May 10, 2015): 120–27. https://doi.org/https://doi.org/10.1016/j.jconrel.2014.12.036.
88. U.S. National Library of Medicine, NIH. "Collection Development Manual." Accessed March 2, 2019. https://www.nlm.nih.gov/tsd/acquisitions/cdm/subjects12.html.

89. Wang, Brian. "Disruptive CRISPR Gene Therapy Is 150 Times Cheaper than Zinc Fingers and CRISPR Is Faster and More Precise." NextBigFuture, June 9, 2015. https://www.nextbigfuture.com/2015/06/disruptive-crispr-gene-therapy-is-150.html.
90. Watson, James D., Tania A. Baker, Stephen P. Bell, Alexander Gann, Michael Levine, and Richard Losick. *Molecular Biology of the Gene, Sixth Edition*. 6 edition. San Francisco : Cold Spring Harbor, N.Y: Pearson, 2007.
91. Wright, David A., Stacey Thibodeau-beganny, Jeffry D. Sander, Ronnie J. Winfrey, Andrew S. Hirsh, Magdalena Eichtinger, Fengli Fu, et al. "Standardized Reagents and Protocols for Engineering Zinc Finger Nucleases by Modular Assembly." *Nature Protocols; London* 1, no. 3 (August 2006): 1637–52. http://dx.doi.org.mutex.gmu.edu/10.1038/nprot.2006.259.
92. Labomics. "ZFNs - CRISPR-RGEN, TALEN & ZFN - Products." Accessed January 18, 2019. http://labomics.com/products/rgen-talen-zfn/zfns.html.
93. Zhang, Feng. "From Microbial Immunity to Genome Editing." Presented at the NIH Intramural Research Program, June 28, 2017. https://www.youtube.com/watch?v=ihmt9MkQY34&list=PLPuptUEHTSjDGWQ16svQQuOXf78g4z4Bn&index=5&t=0s.

Chapter 5
Risk of Misuse Assessment: Part II

On October 12, 2015, a team of Chinese scientists unveiled the world's first genome edited puppies, a pair of extra-muscular beagles. The scientists named the puppies Tiangou, after the Chinese legendary heaven dog, and Hercules, after the Greco-Roman legendary hero of exceptional strength. To achieve this phenotype, they used CRISPR-Cas9 to genetically enhance the muscle mass of beagles by knocking out a gene that encodes a protein called myostatin. Myostatin normally inhibits the growth of muscle cells, so muscles do not become too large.[1] On one hand, this study has legitimate applications because dogs are good genetic and clinical models to use to study human disease. Researchers were hopeful that a similar approach could be applied to engineer dogs with DNA mutations to mimic human diseases like muscular dystrophy. This prediction came to fruition 3 years later when scientists used CRISPR to restore the expression of dystrophin, a protein that strengthens and protects muscle fibers, in dog models of Duchenne muscular dystrophy.[2]

On the other hand, security experts worry that legitimate genome editing research and progress made in developing human therapies to treat diseases could be misused to develop super soldiers. Only months before the 2015 Chinese study, scientists were already beginning to speculate that genome editing could be used to engineer the "perfect baby."[3] It was not a far leap to make to suggest that the research on the genome edited beagles could be used to engineer a real-life superman with super-human strength. A 2017 report, titled *Human Genome Editing: Science, Ethics, and Governance,* from the NASEM reinforced these concerns by suggesting that genome

---

[1] Qingjian Zou et al., "Generation of Gene-Target Dogs Using CRISPR/Cas9 System," *Journal of Molecular Cell Biology* 7, no. 6 (October 12, 2015): 580–83, doi:10.1093/jmcb/mjv061.

[2] Leonela Amoasii et al., "Gene Editing Restores Dystrophin Expression in a Canine Model of Duchenne Muscular Dystrophy," *Science* 362, no. 6410 (October 5, 2018): 86–91, doi:10.1126/science.aau1549.

[3] Antonio Regalado, "Engineering the Perfect Baby," MIT Technology Review, March 5, 2015, accessed January 11, 2022, https://www.technologyreview.com/2015/03/05/249167/engineering-the-perfect-baby/

K. Paris, *Genome Editing and Biological Weapons*,
https://doi.org/10.1007/978-3-031-21820-0_5

editing-based therapy that tries to restore the muscle mass on persons with muscular dystrophy could be misused on persons with normal muscle mass to increase muscle strength to an extreme or even beyond.[4] Additionally, a report prepared in 2017 for the Deputy Assistant Secretary of the Army for Research and Technology (DASA R&T) cited the 2015 study, in which CRISPR was used to double the muscle mass of dogs, and suggested that achievements like this "skirt the border between therapeutic breakthroughs" and "genetically-engineered 'super soldiers'."[5]

World leaders have already shown interest in human augmentation. At the 19th World Festival of Youth and Students held in October 2017, Russian President Vladimir Putin suggested that it could be possible to genetically engineer a super soldier to "fight without fear or compassion, mercy, or pain"—saying that it could be "more terrifying than a nuclear bomb."[6] In 2020, DNI John Ratcliffe revealed, "U.S. intelligence shows that China has even conducted human testing on members of the People's Liberation Army in hope of developing soldiers with biologically enhanced capabilities."[7] In December 2020, French Minister of the Armed Forces Florence Parly acknowledged that France does not want to fall behind other countries that do not have the same ethical scruples when it comes to human augmentation. Nevertheless, Minister Parly was firm in France's commitment to uphold ethical values and not to cross the body barrier by announcing, "we say yes to the Iron Man armor and no to the Spiderman augmentation and genetic mutation."[8]

In the past, several barriers (including lack of knowledge of the relationship between genotype and phenotype, questions of ethics—mostly from Western scientists, and lack of tools that could carry out the modifications) have prevented actors from editing the human genome to make super soldiers. While scientific knowledge and ethical boundaries are still constraints, advancements in genome editing technologies are chipping away at a major barrier that once stood in the way. There are a vast number of modifications that could be made to augment humans. Which augmentations are actors most likely to make to develop super soldiers, and how

---

[4]National Academies of Sciences, Engineering, and Medicine, *Human Genome Editing: Science, Ethics, and Governance* (Washington, DC: The National Academies Press, 2017), 9, 145, https://doi.org/10.17226/24623

[5]"Emerging Science and Technology Trends: 2017-2047" (Office of the Deputy Assistant Secretary of the Army (Research & Technology), November 21, 2017), 13.

[6]Vladimir Putin, "Youth 2030: The Image of the Future Panel Session" (19th World Festival of Youth and Students, Russia, October 2017), accessed October 2, 2019, https://www.youtube.com/watch?v=9v3TNGmbArs

[7]John Ratcliffe, "China Is National Security Threat No. 1," The Wall Street Journal, December 3, 2020, accessed January 25, 2022, https://www.wsj.com/articles/china-is-national-security-threat-no-1-11607019599

[8]"Speech by Florence Parly, Minister of the Armed Forces, Introducing the Round Table 'Ethics and Augmented Soldier' at the Digital Defense Innovation Forum," Ministry of the Armed Forces, December 4, 2020, accessed March 6, 2022, https://www.defense.gouv.fr/salle-de-presse/discours/discours-de-florence-parly-ministre-des-armees-introduisant-la-table-ronde-ethique-et-soldat-augmente-au-digital-forum-innovation-defense

soon could we see these on the battlefield? These are some of the questions that are tackled in the chapter.

Overall, the purpose of Chap. 5 is to assess the magnitude of potential harm, as well as the imminence of potential misuse, of genome editing technologies. The first half of Chap. 5 includes an assessment of the magnitude of potential harm of using genome editing technologies to make existing pathogens more dangerous, weaponize the human microbiome, weaponize gene drives, and augment humans, including super soldiers and the general population. Following these assessments, the second half of Chap. 5 contains an assessment of the imminence of potential misuse of genome editing. It is noted that the magnitude of potential harm and imminence of potential misuse are both parameters specified within Tucker's *Decision Framework for Governance of Dual-Use Technologies* to assess the risk of misuse of technologies with dual-use potential. The other two parameters that are used to assess the risk of misuse, including accessibility and ease of misuse, are covered in Chap. 4.

5.1 Parameter 3: Magnitude of Potential Harm

5.1.1 Methodology

This subsection assesses the magnitude of potential harm that could result from using genome editing technologies to make pathogens more dangerous, weaponize the human microbiome, weaponize gene drives, and augment humans. First, an assessment of using genome editing technologies to make pathogens more dangerous is conducted by secondary data analysis of a study carried out in 2015, and published in 2016, by biodefense and biosecurity analysts at Gryphon Scientific, LLC on the *Risk and Benefit Analysis of Gain of Function Research* (hereafter, the Gryphon study). The purpose of the Gryphon study was to offer the National Science Advisory Board for Biosecurity (NSABB) recommendations for a general conceptual approach that could be used to evaluate gain of function (GoF) studies and to help the U.S. government formulate policy on such studies.[9] As part of the Gryphon study, analysts assessed the biosecurity risk from malicious acts that target GoF laboratories and result in a loss of containment event. They used a semi-quantitative approach and compared malicious acts on targeting laboratories with wild type strains to those with GoF strains.[10] Additionally, analysts assessed the biosecurity risk from GoF information in the literature.

Second, a goal of the assessment on the weaponization of the human microbiome is to understand the current state-of-understanding of the human microbiome and which genome editing technologies are being used in human microbiome research. By understanding how advancements in genome editing are being applied to

[9] *Risk and Benefit Analysis of Gain of Function Research, Final Report*, 1.

[10] *Risk and Benefit Analysis of Gain of Function Research, Final Report*, 202.

research activities on the human microbiome, biodefense and biosecurity experts are better positioned to assess the extent that these technologies and research activities could be conversely misused to cause harm. A full examination of every research project undertaken on the human microbiome and genome editing was not possible within the context of this study. Instead, data were gathered on human microbiome projects that are funded by the NIH, which is one of the prominent funders of this type of work. The NIH is a federal medical research agency part of the United States Department of Health and Human Services.[11] It is regarded as "the primary federal agency conducting and supporting basic, clinical, and translational medical research, and is investigating the causes, treatments, and cures for both common and rare diseases."[12]

The NIH Research Portfolio Online Reporting Tools (RePORT) is a website that "provides access to reports, data, and analyses of NIH research activities."[13] The NIH RePORTER, which stands for the RePORT Expenditures and Results, is a searchable system, available through RePORT, that stores data on NIH-funded research projects.[14] Searches were conducted on NIH RePORTER to quantitatively assess how many projects involve ZFN, TALEN, and CRISPR. Data were also gathered from PubMed to quantitatively assess how many publications involve the microbiome and ZFN, TALEN, and CRISPR. Furthermore, there are many anatomical areas, such as the gut, that could be targeted by nefarious actors. Data were obtained from a review article on NIH funded human microbiome research activities to determine which anatomical areas are being researched—the understanding of an anatomical area typically improves the more it is researched.[15]

Third, an aim of the book is to provide a better account over the type of fears that exist regarding the weaponization of gene drives. Within the context of the book, it was not possible to conduct an in-depth empirical study on the magnitude of potential harm from using genome editing technologies to weaponize gene drives. Instead, analysis was gathered from scientists who work with gene drives and biodefense experts who evaluated gene drives to provide a better account of the extent of concerns over weaponized gene drives. One of the reports that was analyzed includes the 2016 document by an expert committee on gene drives that

[11] "Who We Are," National Institutes of Health (NIH), accessed April 8, 2019, https://www.nih.gov/about-nih/who-we-are

[12] "NIH to Launch Genome Editing Research Program," National Institutes of Health (NIH), January 23, 2018, accessed December 15, 2018, https://www.nih.gov/news-events/news-releases/nih-launch-genome-editing-research-program

[13] NIH Office of Extramural Research, *Research Portfolio Online Reporting Tools (RePORT)* (US Department of Health and Human Services, 2010), accessed December 10, 2018, https://report.nih.gov/brochure/index.html

[14] NIH Office of Extramural Research.

[15] "A Review of 10 Years of Human Microbiome Research Activities at the US National Institutes of Health, Fiscal Years 2007-2016," *Microbiome* 7 (February 26, 2019): 31, doi:10.1186/s40168-019-0620-y.

was convened by the NASEM, titled *Gene Drives on the Horizon: Advancing Science, Navigating Uncertainty, and Aligning Research with Public.*[16]

Lastly, this assessment sought to provide a more realistic vision of human augmentation, and, specifically, the types of modifications that actors are likely to target through genome editing technologies in super soldiers and the general population. There are a vast number of traits that could theoretically be targeted for modification, but it would be difficult within the context of this study to determine the extent in which genome editing technologies could be applied in all these cases. Especially in Western nations, human augmentations beyond a typical or normal range of function are viewed as unethical, and the use of genome editing-based therapeutics are limited to treatments for infections, cancer, and other diseases. Despite these limitations, a critical first step in understanding the magnitude of potential harm of super soldiers is to identify the human traits that actors would strategically want to target for modification to gain battlefield superiority. The DOD Biotechnologies for Health and Human Performance Council (BHPC) developed a list of nine desirable capabilities for the warfighter.[17] The capabilities are analyzed here, within the context of genome editing, to provide a prioritized list of traits that actors are most likely to target using genome editing technologies.

5.1.2 Making Pathogens More Dangerous

There are several attributes of agents that nefarious actors may want to target using genome editing technologies to create a more dangerous biological weapon (see Chap. 3). In a large-scale and worst-case scenario, advancements in genome editing technologies could be used to modify a pathogen to cause a worldwide outbreak of disease in hundreds-of-thousands to millions of people. This would be *highly* dependent on the pathogen and attribute(s) that are selected for modification. For instance, 5–10% of the world population is annually infected by seasonal influenza, and an estimated 250,000 to 500,000 people die as a result.[18] Results from the Gryphon study suggest that enhanced transmissibility, ability to evade host immunity, and resistance to antivirals in seasonal influenza virus significantly increase the risk that an outbreak of disease could happen. Furthermore, the outbreak could

---

[16] National Academies of Sciences, Engineering, and Medicine, *Gene Drives on the Horizon: Advancing Science, Navigating Uncertainty, and Aligning Research with Public Values* (Washington, DC: The National Academies Press, 2016), 70, https://doi.org/10.17226/23405

[17] In 2018, the DOD BHPC conducted a year-long study to project the future of human/machine enhancement technologies over the next three decades. Human/machine technologies are machines that are physically integrated with humans, such as neural implants that would allow brain-computer interfacing (BCI) between the soldier and an external device, like an unmanned aerial vehicle (UAV)—in other words, soldiers using their thoughts to control UAVs. Peter Emanuel et al., "CCDC CBC-TR-1599, Cyborg Soldier 2050," 2.

[18] *Risk and Benefit Analysis of Gain of Function Research, Final Report*, 56.

escape local control and become widespread, or even global.[19] This risk increases even higher through the enhancement of pathogenicity of seasonal influenza virus.[20]

Unlike seasonal influenza, the risk for pandemic influenza virus is highly dependent on the comparator strain that is used in an assessment. For instance, enhanced pathogenicity of the 1918 H1N1 pandemic influenza virus, which is already highly pathogenic, does not significantly increase risk. Compared to 1918 H1N1, 1957 H2N2 pandemic influenza virus was less pathogenic. The enhancement of the pathogenicity of 1957 H2N2 to the level of pathogenicity of the 1918 H1N1, however, significantly increases risk. Finally, enhancement of the pathogenicity of avian influenza virus only modestly increases risk; whereas enhancement of transmissibility in mammals, with the potential to spread to humans, is very concerning and has the highest risk. Enhanced transmissibility of avian influenza virus that is spread by person-to-person contact could impact millions of people around the world.[21] An analysis of publications on genome editing technologies and influenza viruses reveals that most studies involved CRISPR (100 publications), one study involved TALEN, and no studies involved ZFN.[22] These results suggest that actors who want to target influenza virus are most likely to select CRISPR.

Next, it is important to consider the potential biosecurity risk from research on pathogens and genome editing technologies that generates GoF information. For instance, legitimate studies have conducted research on understanding the pathways in the cellular immune response to influenza virus and are working on detecting host factors for influenza virus. Scientists have suggested possible human genes that could be targeted with CRISPR to alter the sensitivity of cells to infection with influenza. While this knowledge could be applied to develop new influenza therapeutics, it could also be misused to use CRISPR to make humans more susceptible and permissive to influenza virus.[23] As the Gryphon study revealed, the ability to evade host immunity, even in seasonal influenza, could potentially lead to a global outbreak of disease. While this may be the case, the Gryphon study ultimately concluded that little information risk comes from GoF research that involves the influenza virus.[24] Table 5.1 provides an overall score for the assessment of the magnitude of potential harm that couldoccur as a result of making pathogens more dangerous via genome editing.

---

[19] *Risk and Benefit Analysis of Gain of Function Research, Final Report*, 5.

[20] *Risk and Benefit Analysis of Gain of Function Research, Final Report*, 210.

[21] *Risk and Benefit Analysis of Gain of Function Research, Final Report*, 5, 215.

[22] On March 10, 2022, a search was conducted on PubMed for publications that contained the terms "influenza" and "CRISPR," "TALEN," or "ZFN" in the title, abstract, or text. Publications between 2013 and 2021 were considered.

[23] Daria Prokhorova et al., "Application of the CRISPR/Cas9 System to Study Regulation Pathways of the Cellular Immune Response to Influenza Virus," *Viruses* 14, no. 2 (February 21, 2022): 437, doi:10.3390/v14020437.

[24] *Risk and Benefit Analysis of Gain of Function Research, Final Report*, 5.

Table 5.1 Overview score for the magnitude of potential harm from making pathogens more dangerous via genome editing

	ZFN	TALEN	CRISPR
Magnitude of potential harm from making pathogens more dangerous	Low	Low	Medium

5.1.3 Weaponizing the Human Microbiome

Humans are highly vulnerable to an attack on the microbiome. This is a relatively new application for biological weapons, with numerous anatomical areas that could be targeted, that we are not prepared to defend. Even so, the human microbiome is vast and relatively little is known about it, which will make targeting difficult. Furthermore, targeting more complex pathways, such as the gut-brain interface, will be hard. This subsection first explores the current state, as well as the trajectory, of human microbiome research to provide a foundation for understanding the types of weapons that could be made to target the human microbiome. Next, this subsection includes an analysis of the likelihood that actors could use genome editing technologies to target the human microbiome. Lastly, this subsection includes an assessment of the vulnerability of the population to an attack from a biological weapon that targets this anatomical area.

The human microbiome consists of the communities of microorganisms and their genomes that live on and within our bodies in regions such as the gut, oral cavity, nasopharyngeal space, and skin.[25] The microbiome is a broad term that is sometimes confused with the microbiota. The microbiota more narrowly refers to a unique collection of microorganisms (eukaryotes, archaea, bacteria, and viruses) that live together in a specific site in or on the body. The microbiome more broadly refers to the entire habitat, which includes the microorganisms and their genomes together with the surrounding environmental conditions. Scientists use the term microbiota when referring just to the microorganisms of the microbiome.[26]

For decades, research on the human microbiome has been scarce and scientists have known very little about this field. Given the lack of understanding and difficulty in precisely targeting this area, weaponization of the human microbiome in the past was unlikely to occur. As we turned the corner of the twenty-first century, however, research on the human microbiome began to exponentially increase. Two landmark projects that greatly improved the understanding of the human microbiome include the Human Microbiome Project (HMP), funded by NIH, and the Metagenomics of

[25] National Academies of Sciences, Engineering, and Medicine, *Biodefense in the Age of Synthetic Biology*, 71.

[26] Xu-Bo Qian et al., "A Guide to Human Microbiome Research: Study Design, Sample Collection, and Bioinformatics Analysis," *Chinese Medical Journal* 133, no. 15 (August 5, 2020): 1844–55, doi:10.1097/CM9.0000000000000871.

the Human Intestinal Tract (MetaHIT) project, funded by the European Commission. HMP and MetaHIT were initiated in 2007 and 2008, respectively.[27] These studies were enabled by advances in metagenomic sequencing, which allowed for the study of the entire microbial community at once.

The growth in knowledge about the human microbiome, coupled with advancements in genome editing technologies in the 2010s, began to worry biological weapons experts. There are different approaches that nefarious actors could take to target the human microbiome to cause harm. In one approach, harmful cargo could be delivered to humans through the microbiome. More specifically, commensal (beneficial) bacteria of the human microbiome could be modified to make a harmful biochemical to hurt humans.[28] In another approach, knowledge of the microbiome could be applied to modify pathogens that more effectively propagate in populations (e.g., modify a pathogen to become better established in the human host). Furthermore, since the microbiome influences human immunity, nefarious actors might try to purposefully cause an imbalance of an otherwise healthy microbiome through engineered dysbiosis, which would reduce human health and performance.[29]

When given the choice between ZFN, TALEN, and CRISPR, nefarious actors are most likely to select CRISPR to target the human microbiome given that CRISPR is the tool-of-choice for legitimate research on this subject. Most NIH funded projects on the human microbiome, which also include genome editing, involve CRISPR (339 projects), and very few projects involve TALEN (7 projects) or ZFN (3 projects).[30] Additionally, a review of publications on PubMed reveals that most publications on the microbiome, which also include genome editing, involve CRISPR (98%), and very few publications involve TALEN or ZFN (2%). Publications on the microbiome and CRISPR, which have been exponentially increasing since 2016, are on track to continue to grow.[31]

Certain anatomical areas of the human microbiome are more likely than others to be targeted with genome editing technologies; however, the human microbiome is

[27] "Human Microbiome Project," Human Microbiome Project (HMP) Data Analysis and Coordination Center (DACC), accessed February 24, 2022, https://hmpdacc.org/hmp/overview/; "Metagenomics of the Human Intestinal Tract," European Comission, CORDIS, July 16, 2019, accessed February 24, 2022, https://cordis.europa.eu/project/id/201052

[28] National Academies of Sciences, Engineering, and Medicine, *Biodefense in the Age of Synthetic Biology*, 71–72; Kirkpatrick et al., "Editing Biosecurity: Needs and Strategies for Governing Genome Editing," 52–53.

[29] National Academies of Sciences, Engineering, and Medicine, *Biodefense in the Age of Synthetic Biology*, 71–72; Kirkpatrick et al., "Editing Biosecurity: Needs and Strategies for Governing Genome Editing," 52–53.

[30] Data gathered March 6, 2022 from "NIH Research Portfolio Online Reporting Tools Expenditures and Results." Search terms include "Human Microbiome" plus "zinc finger nuclease," "TALEN," and "CRISPR," for 2014-2021.

[31] On March 1, 2022, a search was conducted on PubMed for publications that contained the terms "microbiome" and "CRISPR," "TALEN," or "ZFN" in the title, abstract, or text. Publications between 2010 and 2022 were considered. The search yielded 171 publications for CRISPR, 3 publications for TALEN, and 1 publication for ZFN.

complex and our understanding of it is still limited. A better understanding of a particular microbiome increases the likelihood that actors will be successful in targeting it. In general, most research has been on the gut microbiome. This is the case for NIH funded studies involving animal models—83% of which focused on the microbiomes associated with the gastrointestinal (GI) tract. Additionally, for NIH funded studies involving human cohorts, most focused on microbiomes associated with the GI tract (40%) and multiple body regions that often involved the GI tract (14%). Other high-focused microbiome areas of studies on human cohorts were the urogenital tract (16%), lung (11%), and oral cavity (11%).[32] Even if these anatomical areas are targeted, scientists do not have a good understanding of the interaction between microbiomes that are in different regions of the body, such as the skin and gut.[33] Modifications in one area of the microbiome may bring about unpredictable and unintended consequences elsewhere in the body.

Nevertheless, China has already authorized a therapy aimed at the gut microbiome for the treatment of Alzheimer's disease. The scientists who developed the treatment uncovered a possible link between the gut and brain, but they were unable to clearly elucidate the pathway through which it worked.[34] While this microbiome-based therapy demonstrates that it may be possible to control cognitive function through gut microbiota, there are mixed reviews as to whether the drug is effective in treating Alzheimer's. The use of the drug is controversial, and further clinical testing is expected to take place in the United States.

While CRISPR is technically capable of modifying the human microbiome, a key barrier to targeting the human microbiome continues to be a lack of knowledge about it. Scientists are still unable to agree on what makes up a healthy human microbiome or what constitutes an unhealthy human microbiome. Initially, research on the human microbiome worked on cataloging the microorganisms that live on and within us. The future of human microbiome research will seek to better understand the complex and mutable microbe-microbe relationships and human-microbe relationships.[35] The human microbiome is considered mutable because it differs with age; amount of interaction with people, animals, and the environment; consumption of foods, antibiotics, and other drugs; in response to stressors and disease; and possibly even an individual's genetic makeup.[36] In other words, there is no standard, one-size-fits-all microbiome.

---

[32] "A Review of 10 Years of Human Microbiome Research Activities at the US National Institutes of Health, Fiscal Years 2007-2016."

[33] Lita Proctor, "What's Next for the Human Microbiome?," *Nature* 569 (May 30, 2019): 624.

[34] Xinyi Wang et al., "Sodium Oligomannate Therapeutically Remodels Gut Microbiota and Suppresses Gut Bacterial Amino Acids-Shaped Neuroinflammation to Inhibit Alzheimer's Disease Progression," *Cell Research* 29, no. 10 (October 2019): 787–803, doi:10.1038/s41422-019-0216-x.

[35] Lita Proctor, "What's Next for the Human Microbiome?," 624.

[36] Nature Video, "The Microbes That Live with Us from Cradle to Grave," YouTube, June 19, 2019, accessed January 31, 2022, https://www.youtube.com/watch?v=c_ZRZkU-FEw

Table 5.2 Overview score for the magnitude of potential harm from weaponizing the human microbiome via genome editing

	ZFN	TALEN	CRISPR
Magnitude of potential harm from weaponizing the human microbiome	Low	Low	High

Nefarious actors may use the diversity of the human microbiome to their advantage. The community composition and density of a microbiome greatly varies within an anatomical area. For instance, there are denser populations of microbiota in the upper areas of the respiratory tract compared to the lower regions of the respiratory tract.[37] Additionally, native bacteria that live in the mucus layer of the upper colon are different from the bacteria that reside deeper in the intestinal folds. Actors will have more control over where they want to deliver a function within a body region by knowing which strain colonizes that area.[38] Still, the deployment of biological weapons that target the human microbiome will likely be challenged by issues with delivery. Table 5.2 provides an overall score for the assessment of the magnitude of potential harm that could occur as a result of weaponizing the human microbiome via genome editing.

5.1.4 *Weaponizing Gene Drives*

Shortly after scientists demonstrated CRISPR could be used as a genome editing tool, Kevin Esvelt and colleagues suggested the potential for CRISPR to create gene drives.[39] Although gene drives have been researched for over 50 years, this was the first time someone suggested using CRISPR to create a gene drive.[40] In response, David Gurwitz addressed possible dual-use concerns—thus igniting the debate over the misuse of CRISPR-based gene drives. Gurwitz warned, "just as gene drives can make mosquitoes unfit for hosting and spreading the malaria parasite, they could conceivable be designed with gene drives carrying cargo for delivering lethal bacteria toxins to humans."[41] The remainder of the subsection contains a review of the potential benefits of gene drives, followed by an analysis of the magnitude of

[37] Elakshi Dekaboruah et al., "Human Microbiome: An Academic Update on Human Body Site Specific Surveillance and Its Possible Role," *Archives of Microbiology* 202, no. 8 (October 1, 2020): 2147–67, doi:10.1007/s00203-020-01931-x.

[38] Jyoti Madhusoodanan, "Editing the Microbiome," *PNAS* 117, no. 7 (February 18, 2020), accessed March 3, 2022, https://www.pnas.org/doi/abs/10.1073/pnas.2000108117

[39] Esvelt et al., "Concerning RNA-Guided Gene Drives for the Alteration of Wild Populations."

[40] National Academies of Sciences, Engineering, and Medicine, *Gene Drives on the Horizon: Advancing Science, Navigating Uncertainty, and Aligning Research with Public Values*, 1, 14, 20, 26.

[41] Gurwitz, "Gene Drives Raise Dual-Use Concerns," 1010.

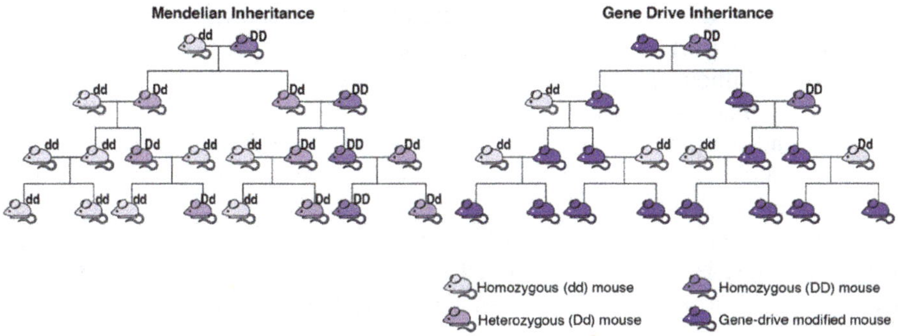

Fig. 5.1 Normal Inheritance vs. Gene Drive Inheritance. During normal, Mendelian, inheritance (left) there is a 50% chance the offspring will inherit the gene. Here, a homozygous recessive (dd) mouse mates with a homozygous dominant (DD) mouse, and they have heterozygous (Dd) offspring. Each parent passes on one copy of each chromosome to the offspring. The chances of inheriting the dominant gene (D) does not exceed 50%. Alternatively, during gene drive inheritance (right), the chances of spreading the targeted genetic element to offspring is over 50%. Here, the targeted genetic element is spread through a population of mice using a gene drive. When the gene-drive modified mouse mates with the homozygous dominant mouse, the gene drive is passed to all their offspring. The following genotypes of mice are depicted: dd, Dd, DD, or gene drive (Used with permission of The National Academies Press, from *Gene Drives on the Horizon: Advancing Science, Navigating Uncertainty, and Aligning Research with Public Values*, National Academies of Sciences, Engineering, and Medicine, 2016; permission conveyed through Copyright Clearance Center, Inc.)

potential harm and dual-use concerns of gene drives, and finally the limitations and challenges of gene drives.

Under normal conditions, heritable traits have a 50% chance that they will be passed to the next generation. Gene drives increase the speed that a gene could be spread through a community by driving these odds to above 50%—in other words, up to 100% of the offspring could inherit the gene. This is a dual edge sword because the gene drives that could be created to benefit us are the same that could be slightly modified to hurt us. Figure 5.1 depicts the difference between spreading an altered gene through a population of mosquitoes through normal, Mendelian, inheritance and gene drive-based inheritance.

There are many human, agricultural, and environmental benefits of gene drives. Gene drives could improve public health by combatting mosquito-borne diseases, such as malaria and dengue, and other infectious diseases, such as those transmitted by deer ticks and parasitic flatworms.[42] Gene drives could interrupt the transmission

[42]National Academies of Sciences, Engineering, and Medicine, 65–67.

chains of many infectious diseases and significantly decrease, or even eradicate, outbreaks.[43] Gene drives also have the potential to control weeds such as witchweed, which causes billions of dollars of crop losses each year in developing countries like Africa.[44] Furthermore, gene drives may be able to reverse "the development of pesticide and herbicide resistance."[45] Additionally, gene drives could be used to aid conservation efforts, help species that are threatened or endangered, and control or get rid of invasive species.[46]

While there are several potential benefits of gene drives, there are also many potential harms. Harm might be caused by the unintentional use of these drives. Unintentional harm could result from laboratory accidents or from unforeseen, negative effects of gene drives on target species, non-target species, or ecosystem. To illustrate the latter cases, suppose mosquitoes (i.e., the target species) are modified to no longer host dengue virus. Modified mosquitoes may unknowingly be susceptible to another pathogen that harms humans. Likewise, dengue virus could evolve into an unexpected variation that presents a different health hazard to humans. If, instead of modifying the mosquitoes, the mosquitoes were eliminated altogether then there may be unintended effects on non-target organisms—the gene drive causing suppression could be transferred to non-target organisms, or non-target organisms that are also disease vectors could fill the gap in the ecosystem left by the elimination of mosquitoes.[47]

There are many ways in which gene drives could be manipulated by nefarious actors to intentionally cause harm. A terrorist may want to use gene drives to spread disease-causing pathogens. In another scenario, a nefarious actor may want to magnify the threat from a mosquito that already carries an infectious disease (e.g., malaria) and use a gene drive to drive a gene that codes for a deadly toxin through the mosquito population. Furthermore, the nefarious actor could target genes to expand the habitat of the mosquito into non-tropical areas, which would increase the range that the organism can spread disease.[48]

While gene drives could be misused, there are inherent restrictions that make application of gene drives difficult. Gene drives are limited to sexually reproducing organisms (e.g., insects and rodents), so populations of organisms such as viruses and bacteria cannot be engineered with gene drives. Moreover, it can take "dozens of

---

[43] Sonia Ben Ouagrham-Gormley and Kathleen M. Vogel, "Gene Drives: The Good, the Bad, and the Hype," Bulletin of the Atomic Scientists, October 14, 2016, accessed April 20, 2018, https://thebulletin.org/gene-drives-good-bad-and-hype10027

[44] National Academies of Sciences, Engineering, and Medicine, *Gene Drives on the Horizon: Advancing Science, Navigating Uncertainty, and Aligning Research with Public Values*, 67.

[45] Kenneth A. Oye et al., "Regulating Gene Drives," *Science* 345, no. 6197 (July 17, 2014): 626, doi:10.1126/science.1254287.

[46] National Academies of Sciences, Engineering, and Medicine, *Gene Drives on the Horizon: Advancing Science, Navigating Uncertainty, and Aligning Research with Public Values*, 70–72; Ouagrham-Gormley and Vogel, "Gene Drives"; Oye et al., "Regulating Gene Drives," 626.

[47] National Academies of Sciences, Engineering, and Medicine, *Gene Drives on the Horizon: Advancing Science, Navigating Uncertainty, and Aligning Research with Public Values*, 68.

[48] Ouagrham-Gormley and Vogel, "Gene Drives."

Table 5.3 Overview score for the magnitude of potential harm from weaponizing gene drives via genome editing

	ZFN	TALEN	CRISPR
Magnitude of potential harm from weaponizing gene drives	Low	Low	Medium-to-high

generations" after a gene drive is introduced into a population for it to become widespread; this may take up to a year or even hundreds of years depending on the length of the generation time of the organism.[49] In a 2016 report by an expert committee on gene drives that convened by the NASEM, said gene drives "would not be effective on humans" due to the long generation time of humans.[50]

The 2018 NASEM report on synthetic biology added that it would take "thousands of years" for human gene drives to spread in the human population.[51] Additionally, the 2018 *Editing Biosecurity* report suggested that gene drives would not have a meaningful impact on humans due to long generation times.[52] In reference to crops and livestock, the 2016 NASEM report on gene drives said that gene drives may have a "limited effect."[53] Ultimately, the 2016 NASEM report concluded that "gene drives may be challenging or not possible to develop in many organisms."[54] Table 5.3 provides an overall score for the assessment of the magnitude of potential harm that couldoccur as a result of weaponizing gene drives via genome editing.

5.1.5 *Augmenting Humans: Super Soldiers and the General Population*

Actors may want to use genome editing technologies to augment traits in soldiers to create a stronger military force or in everyday people to create a smarter population with an economic edge. The modification of human DNA using genome editing technologies, however, is not an easy undertaking and analysts should be careful not to underestimate the challenges that actors will face. This subsection first reviews the general approach to engineering traits in humans, through either germline gene

[49]Oye et al., "Regulating Gene Drives," 626.

[50]National Academies of Sciences, Engineering, and Medicine, *Gene Drives on the Horizon: Advancing Science, Navigating Uncertainty, and Aligning Research with Public Values*, 70.

[51]National Academies of Sciences, Engineering, and Medicine, *Biodefense in the Age of Synthetic Biology*, 101.

[52]Kirkpatrick et al., "Editing Biosecurity: Needs and Strategies for Governing Genome Editing," 55.

[53]National Academies of Sciences, Engineering, and Medicine, *Gene Drives on the Horizon: Advancing Science, Navigating Uncertainty, and Aligning Research with Public Values*, 70.

[54]National Academies of Sciences, Engineering, and Medicine, 18.

therapy or somatic gene therapy, to provide a basis for this assessment. Next, this subsection includes an analysis of which traits actors are most likely to improve, beyond what is considered normal, in soldiers and the general population.

There are two approaches to gene therapy that actors could take to modify humans. First, an actor could target the human germline, termed germline gene therapy. Second, an actor could target somatic cells, termed somatic gene therapy. Gene therapy is described as "an experimental treatment that involves introducing genetic material (DNA or RNA) into a person's cells for therapeutic purposes."[55] Germline cells are described as "cells with the capacity to be involved in forming a new individual and to have their genetic material passed on to a new generation."[56] These cells not only give rise to more germ cells but also differentiate into non-reproductive, body cells known as somatic cells.[57] In short, modifications made to the germline are heritable changes, whereas modifications to the somatic cells are not passed on to future generations.

At this point, actors who seek to augment humans are more likely to use CRISPR over ZFN or TALEN. As revealed later in this chapter, there are more clinical studies that involve CRISPR-based therapies (26 studies) than there are clinical studies that involve ZFN-based therapies (15 studies) or TALEN-based therapies (6 studies). Nevertheless, there may be a few specific applications, such as targeting heterochromatin, in which a predecessor genome editing technology outcompetes CRISPR. The chromatin in eukaryotic cells is organized into more compact regions (heterochromatin) and more open regions (euchromatin). Research suggests that TALEN is five times more efficient in editing heterochromatin than CRISPR-Cas9. While heterochromatin DNA is usually not as active and has relatively low gene density, loss of heterochromatin has been associated with aging and other human diseases.[58]

Although CRISPR is already used in clinical trials to treat disease, the modification of cells may not be as straightforward as scientists initially thought. Scientists have experienced issues with dysfunctional p53, a tumor suppressor gene, in CRISPR modified cells. The resultant cells were referred to as "ticking time bombs" that could develop into cancer, and scientists were advised to thoroughly check modified cells for problems with p53 before reintroducing them into

---

[55] Koray Özduman, Guido Wollmann, and Joseph M. Piepmeier, "CHAPTER 57 - Gene Therapy for Meningiomas," in *Meningiomas*, ed. M. Necmettin Pamir, Peter M. Black, and Rudolf Fahlbusch (Philadelphia: W.B. Saunders, 2010), 681, doi:10.1016/B978-1-4160-5654-6.00057-X.

[56] National Academies of Sciences, Engineering, and Medicine, *Human Genome Editing: Science, Ethics, and Governance*, 70.

[57] Germline editing can be achieved by using precursors of sperm and egg (gametes), gametes, zygotes, and early embryos. National Academies of Sciences, Engineering, and Medicine, 70, 111.

[58] Cas9 was hindered by local searches on non-specific sites and became trapped in heterochromatin structures that were more compact. However, Cas9 was more efficient in euchromatin regions. Surbhi Jain et al., "TALEN Outperforms Cas9 in Editing Heterochromatin Target Sites," *Nature Communications* 12, no. 1 (January 27, 2021): 606, doi:10.1038/s41467-020-20,672-5; Andrés Penagos-Puig and Mayra Furlan-Magaril, "Heterochromatin as an Important Driver of Genome Organization," *Frontiers in Cell and Developmental Biology* 8 (September 18, 2020), accessed March 14, 2022, https://www.frontiersin.org/article/10.3389/fcell.2020.579137

patients.[59] Additionally, scientists may have severely underestimated the amount of collateral damage that CRISPR has on DNA. Experts warn that damage can occur far from the target site and can be difficult to detect through standard DNA tests.[60] Actors who do not proceed with extreme caution and have the knowledge, skills, and resources to ensure CRISPR-based therapeutics are safe prior to giving them to individuals could seriously hurt or kill someone.

5.1.5.1 Super Soldiers

Humans are limited by their own natural abilities on the battlefield. Military forces could overcome the natural limitations of the soldier by modifying the human genome via genome editing technologies. Nation-states may want to genetically enhance their militaries to improve the cognitive and physical abilities of soldiers. The goal for military forces would be to make their troops more capable, more lethal, and better able to survive.[61] In short, military forces that are genetically augmented would have more overall battlefield superiority compared to their nonengineered counterparts.

In 2019, the DOD Biotechnologies for Health and Human Performance Council (BHPC) published a list of nine capabilities that could improve the warfighter on the battlefield. The prioritized list, from most to least impactful, is as follows: "1. situational awareness, 2. strength and speed, 3. imaging and sight, 4. communication, 5. physiology (endurance/sleep/health), 6. virtual (avatar) control, 7. attention and memory, 8. learning, and 9. olfaction (sense of smell)."[62] As a trait becomes more impactful (i.e., is higher on the list), the chances increase that a nation-state would want to target it for augmentation using genome editing technologies. In other words, it is most likely for a nation-state to use genome editing to increase a soldier's situational awareness, and it is least likely for a nation-state to use genome editing to improve a soldier's sense of smell.

It is important to understand the threat within the context of other options, besides genome editing, that are available to achieve human enhancements. Other options, such as pharmaceuticals and wearable technologies, provide viable alternatives especially if nation-states have already invested time and money in researching and developing them for weapons purposes. For instance, there are already visual enhancements to correct vision back to a normal, such as glasses, contacts, and laser

[59] Sharon Begley, "A Serious New Hurdle for CRISPR: Edited Cells Might Cause Cancer, Two Studies Find," STAT, June 11, 2018, accessed April 10, 2022, https://www.statnews.com/2018/06/11/crispr-hurdle-edited-cells-might-cause-cancer/

[60] Sharon Begley, "Potential DNA Damage from CRISPR Has Been 'seriously Underestimated,' Study Finds," STAT, July 16, 2018, accessed April 10, 2022, https://www.statnews.com/2018/07/16/crispr-potential-dna-damage-underestimated/

[61] Peter Emanuel et al., "CCDC CBC-TR-1599, Cyborg Soldier 2050," 8.

[62] Peter Emanuel et al., 2.

eye surgery. There are also visual enhancements that move beyond this baseline, such as wearable night vision devices.

Nation-states are less likely to invest in enhancing a trait, such as night vision, if they have already spent considerable time and effort in developing an alternative technology that meets its needs. Huge investments have been made in night vision technology since World War II.[63] In 2021, the U.S. Army unveiled its new Enhanced Night Vision Goggle–Binocular system that improves a soldier's situational awareness and depth perception on the battlefield. The device works by using white phosphor, instead of traditional green phosphor, to improve contrast and thermal imaging to improve target recognition with an augmented reality overlay. The Program of Record is valued up to $422 million.[64]

Nevertheless, technologies have their limitations, and so it is critical to ask whether genome editing could be used to better enhance a trait over existing technologies. Wearable technologies can fall off and break. Pharmaceuticals wear off and can require repeated applications—they can even become addictive and cause long-term harm. Augmentation through genome editing may provide a better long-term solution. Furthermore, genome editing may be used in combination with other existing and emerging technologies, such as nanotechnology and artificial intelligence, to achieve a higher level of augmentation.

5.1.5.2 General Population

In addition to augmenting soldiers, actors may want to use genome editing to augment traits in the general population. This would generally have a greater impact on nation-states with larger populations of people who could be augmented rather than non-state actors functioning as individuals or small groups. Actors are most likely to target traits that enhance the intelligence and immunity, while reducing the rate of aging, of the general population. Although there are several other traits that actors might want to improve—such as sight, hearing, chemosenses (i.e., chemical sensing), touch, heat sensing, memory span, memory content, heat tolerance, locomotion, ocean depth, altitude, and voice—there are already chemical and physical means that are available to enhance these traits.

The augmentation of traits via genome editing technologies in the general population could allow countries to gain a long-term advantage, especially economic, which could lead to a future imbalance of power. Research suggests that there is a positive link between intelligence and the economy. Previous studies have concluded that countries with higher intelligence quotient (IQ) levels tend to have

---

[63]Emily Matchar, "Seeing in the Dark: The History of Night Vision," Smithsonian Magazine, May 19, 2017, accessed January 15, 2022, https://www.smithsonianmag.com/innovation/seeing-dark-history-night-vision-180963357/

[64]Dalvin Brown, "The Army's Latest Night-Vision Tech Looks like Something out of a Video Game," *Washington Post*, May 24, 2021, accessed January 15, 2022, https://www.washingtonpost.com/technology/2021/05/24/army-night-vision-goggles/

Table 5.4 Overview score for the magnitude of potential harm from augmenting humans via genome editing

	ZFN	TALEN	CRISPR
Magnitude of potential harm from augmenting humans	Low	Medium	High

faster rates of economic growth.[65] A study of 74 countries revealed that a one-point increase in IQ was correlated with a 4% increase in the growth of a country's economic welfare (i.e., economic well-being). A gain of 2 or 3% IQ will considerably affect a country's gross domestic product (GDP)."[66] Table 5.4 provides an overall score for the assessment of the magnitude of potential harm that could occur asa result of augmenting humans, both super soldiers and the general population, via genome editing.

5.2 Parameter 4: Imminence of Potential Misuse

5.2.1 Methodology

Parameter 4, imminence of potential misuse, is assessed by measuring the ability for genome editing technologies to be exploited based on the maturity of the technologies. Maturity is determined by obtaining and analyzing data on scientific publications, clinical studies, and patents involving ZFN, TALEN, and CRISPR. Additionally, data are gathered and analyzed on companies that formed, and even went public, based on a genome editing technology. Lastly, the global market size projections are calculated based on industry market research reports. Parameter 4 is mostly studied using quantitative methods.

Data for scientific publications was obtained from the Web of Science Core Collection. The Web of Science is a platform that allows searching across several databases—totaling over 30,000 journals.[67] Searches on the Web of Science were

[65] Erich Weede and Sebastian Kämpf, "The Impact of Intelligence and Institutional Improvements on Economic Growth," *Kyklos* 55, no. 3 (August 2002): 361–80, doi:10.1111/1467-6435.00191; Garett Jones and W. Joel Schneider, "Intelligence, Human Capital, and Economic Growth: A Bayesian Averaging of Classical Estimates (BACE) Approach," *Journal of Economic Growth* 11, no. 1 (March 1, 2006): 71–93, doi:10.1007/s10887-006-7407-2; Heiner Rindermann, "Relevance of Education and Intelligence at the National Level for the Economic Welfare of People," *Intelligence* 36, no. 2 (March 1, 2008): 127–42, doi:10.1016/j.intell.2007.02.002.

[66] Kirsten Gronlund, "Genome Editing and the Future of Biowarfare: A Conversation with Dr. Piers Millett," Future of Life Institute, October 12, 2018, accessed October 4, 2019, https://futureoflife.org/2018/10/12/genome-editing-and-the-future-of-biowarfare-a-conversation-with-dr-piers-millett/

[67] Tracy Matthews, "Web of Science Platform: Introduction," Clarivate Analytics, accessed December 7, 2018, //clarivate.libguides.com/webofscienceplatform/introduction; "Web of Science," Clarivate Analytics, accessed December 7, 2018, https://clarivate.com/products/web-of-science/

conducted in the title, abstract, author keywords, and keywords plus fields through the publication year 2018 for the following terms: ("ZF nuclease$" OR "zinc finger nuclease$"), ("TALE nuclease$" OR "transcription activator like effector nuclease $" OR "TALEN"), and ("CRISPR$" OR "clustered regularly interspaced short palindromic repeat$).[68] Search results were returned for a variety of publication types such as articles, reviews, meeting abstracts, book chapters, and proceedings papers. Results were assessed to determine the number of publications, global distribution of publications, and funding acknowledgments of publications. As described below, additional data was obtained from NIH RePORTER to determine the amount that the NIH—the organization with the most financial acknowledgements for ZFN, TALEN, and CRISPR publications on the Web of Science—invests in genome editing.

Searches were conducted in NIH RePORTER for project titles and abstracts that contained the following: ("ZF nuclease" OR "zinc finger nuclease" OR "ZFN"), ("TALE nuclease" OR "transcription activator like effector nuclease" OR "TALEN"), and ("CRISPR" OR "clustered regularly interspaced short palindromic repeat").[69] The data were analyzed to determine the overall amount that the NIH invests into ZFN, TALEN, and CRISPR research. Additionally, the data were analyzed to determine the amount of funds awarded by the NIH under two programs—the Small Business Innovation Research (SBIR) program and the Small Business Technology Transfer (STTR) program. The SBIR and STTR are congressionally established programs that award much needed startup and development stage funds to United States small businesses that pursue research or research and development (R/R&D) with commercialization potential.[70] It should be noted that it is difficult to gather data on the amount that other governments (i.e., not the United States) are investing in CRISPR. Data is gathered here on the funding organizations for all the genome editing publications on the Web of Science, and the top 5 funding organizations are identified.

Next, clinical studies that involve ZFN, TALEN, and CRISPR were assessed to determine the number of clinical studies, time to move from the laboratory to clinical studies, maturity of clinical trial (i.e., the clinical trial phase), and types of conditions studied in clinical studies. Data on clinical studies was obtained through ClinicalTrials.gov. ClinicalTrials.gov is operated by the United States National

[68] Searches conducted in Web of Science on May 8, 2019. "$" represents a wildcard. Results for TALEN prior to 2010 were irrelevant and excluded; all but one result from 2010 were excluded.

[69] Searches conducted on May 7, 2019. Use of "-" in search terms did not affect search results. "NIH Research Portfolio Online Reporting Tools Expenditures and Results," NIH RePORTER, accessed May 7, 2019, https://projectreporter.nih.gov/reporter.cfm

[70] "NIH Small Business Innovation Research (SBIR) and Small Business Technology Transfer (STTR) Programs," accessed December 11, 2018, https://sbir.nih.gov/; "Types of Grant Programs | Grants.Nih.Gov," May 16, 2018, accessed December 11, 2018, https://grants.nih.gov/grants/funding/funding_program.htm; "About SBIR | SBIR.Gov," accessed December 11, 2018, https://www.sbir.gov/about/about-sbir; "About STTR | SBIR.Gov," accessed December 11, 2018, https://www.sbir.gov/about/about-sttr

Library of Medicine at the NIH. The web-based resource was selected as a data source because it is a comprehensive source that is updated with information on clinical studies from the United States and over 200 other countries around the world.[71] Searches were conducted in ClinicalTrials.gov for clinical studies, which includes observational studies and interventional studies (i.e., clinical trials), involving ZFN ("ZFN" and "zinc finger nuclease"), "TALEN," and "CRISPR." Additional studies involving these technologies were identified in review articles, and data on the studies were retrieved from ClinicalTrials.gov.[72] Data were obtained on 15 ZFN, 6 TALEN, and 26 CRISPR clinical studies.

Data on patents were obtained from the Lens. The Lens is a comprehensive public search system that hosts over 115 million patent documents that originate from jurisdictions around the world.[73] Every 3–4 weeks, the Lens is updated with data from major patent data sources including the United States Patent and Trademark Office, European Patent Office (EPO), IP Australia, and United Nations World Intellectual Property Organization (WIPO).[74] Searches were conducted for patent application documents and granted patent documents in the Lens for the following terms: ("zinc finger nuclease" OR "ZF nuclease"), ("TALE nuclease" OR "transcription activator like effector nuclease"), and ("CRISPR" OR "clustered regularly interspaced short palindromic repeat"). In addition to conducting searches for these terms in the title and abstract, searches were also conducted in the claims. The claims field is described as "the most important part of a patent" in which the subject matter of the patent is clearly indicated.[75] Data were analyzed to determine the number and location of patent documents.

Data on companies were obtained from Crunchbase and Owler. Crunchbase and Owler are online platforms that provide users with information on public and private companies; such as the total funding amount, number of investors, number of lead investors, and number of funding rounds.[76] Searches were conducted in both platforms for companies that have started based on ZFN, TALEN, and CRISPR technologies. The data were analyzed to determine the number of companies started each year and the amount of time that it took for the companies to go from private to

[71] "ClinicalTrials.Gov Background," ClinicalTrials.gov, accessed April 12, 2019, https://clinicaltrials.gov/ct2/about-site/background

[72] Gayong Shim et al., "Therapeutic Gene Editing: Delivery and Regulatory Perspectives," *Acta Pharmacologica Sinica* 38, no. 6 (June 2017): 738–53, doi:10.1038/aps.2017.2; Cynthia E. Dunbar et al., "Gene Therapy Comes of Age," *Science* 359, no. 6372 (January 12, 2018): 4, doi:10.1126/science.aan4672.

[73] "The Lens - Free & Open Patent and Scholarly Search," accessed December 5, 2018, https://www.lens.org/lens

[74] "About The Lens," accessed May 20, 2019, https://about.lens.org/

[75] "Support Center » What Are Claims?," Lens.org, accessed May 21, 2019, https://support.lens.org/help-resources/claims/what-are-claims/

[76] "Who We Are," Crunchbase, accessed February 10, 2019, https://about.crunchbase.com/about-us/; "About Owler," Owler, accessed February 10, 2019, https://www.owler.com/about/what-is-owler

public. Lastly, data on the size of the global genome editing market were obtained through reports from industry firms that specialize in market research.

5.2.2 *Ability to Be Exploited Based on Maturity*

5.2.2.1 Scientific Publications

In general, the number of CRISPR publications far exceeds the number of ZFN and TALEN publications. Figure 5.2 depicts the number of publications available in the Web of Science for ZFN, TALEN, and CRISPR between 2002 and 2018. The rise of ZFN records after 2008 coincides with two events: the decrease in price of ZFN tools available to purchase from Sigma-Aldrich and the emergence of publicly available ZFN protocols. The number of TALEN publications gradually increased after the tool emerged in 2010. Unlike ZFN and TALEN, as soon as CRISPR became available as a genome editing tool in 2012, the number of CRISPR publications exponentially increased. The sharp rise in CRISPR publications in 2014 was met with the beginning of the decline of ZFN and TALEN publications.

ZFN, TALEN, and CRISPR publications are written by researchers at institutions around the world. ZFN publications (Fig. 5.3) are slightly more widespread than TALEN publications (Fig. 5.4). Between 2002 and 2018, ZFN publications were affiliated with 70 countries, and TALEN publications were only affiliated with 60 countries. CRISPR publications (Fig. 5.5) are still the most widespread with affiliations in 116 countries. Despite the vast number of countries where CRISPR publications are affiliated, half of CRISPR publications are from affiliations with only two countries: the United States and China.

By far, the major funding organization for studies that were published in the Web of Science for ZFN, TALEN, and CRISPR, combined, was the United States National Institutes of Health (NIH) with 5347 publications. Following the NIH, the National Natural Science Foundation of China was highly cited as a funding organization for 1934 publications. The National Natural Science Foundation of China, which is modeled after the National Science Foundation of the United States, was established in China as a major funder of basic research, and, more recently, its focus has expanded to application-based research.[77] Next, Deutsche Forschungsgemeinschaft, known as the German Research Foundation, which is a self-governing, German research funding organization that funds research for higher education and other public research institutions, was cited as the funding organization in 627 publications. The budget of the DFG mostly originates from the federal

[77] Micah Springut, Stephen Schlaikjer, and David Chen, *China's Program for Science and Technology Modernization: Implications for American Competitiveness*, n.d., 30.

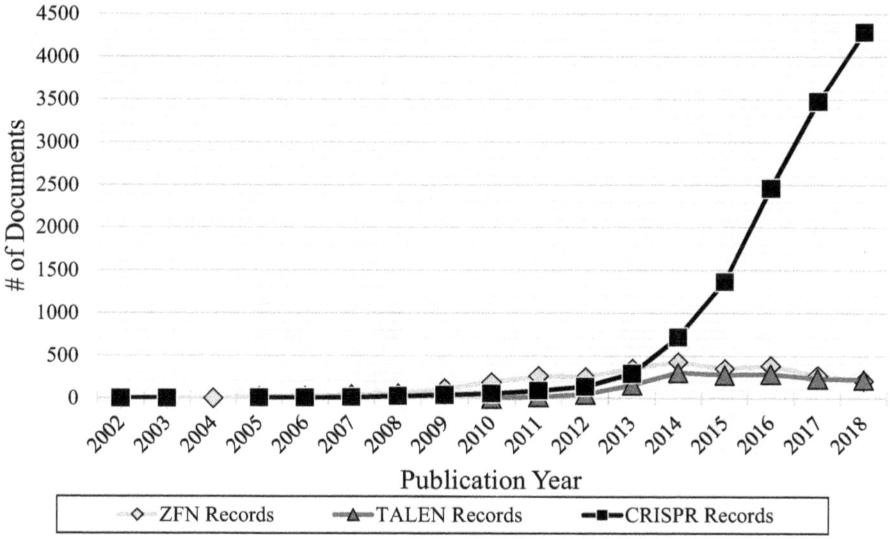

Fig. 5.2 Genome editing technology publications available globally (2002–2018). Search results were returned for a variety of publication types such as articles, reviews, meeting abstracts, book chapters, and proceedings papers (Data for the figure were obtained from the Web of Science on May 8, 2019)

government (approximately two-thirds) and German states (approximately one-third).[78]

The National Science Foundation, which is a United States independent federal agency that is the funding source for about one-quarter of federally funded basic research carried out at colleges and universities in the United States, was cited as the funding organization for 609 publications.[79] Lastly, the Japan Society for the Promotion of Science, which is an independent administrative institution that administers scientific and academic programs in Japan, was cited as the funding organization for 597 publications. The Japan Society for the Promotion of Science is almost 100% subsidized by the government of Japan.[80] In all instances, these funding organizations were cited in more CRISPR publications than in TALEN and ZFN publications, combined.

Additional analysis of data from the NIH reveals the federal agency spends much more on CRISPR research than ZFN or TALEN research. By the end of fiscal year

[78] "Deutsche Forschungsgemeinschaft (DFG) - Research in Germany," accessed May 14, 2019, https://www.research-in-germany.org/en/research-funding/funding-organisations/deutsche-forschungsgemeinschaft-(dfg).html.

[79] "About NSF - Overview | NSF - National Science Foundation," accessed December 14, 2018, https://www.nsf.gov/about/

[80] "Purpose & Functions,History | About US | Japan Society for the Promotion of Science," accessed May 15, 2019, https://www.jsps.go.jp/english/aboutus/index2.html

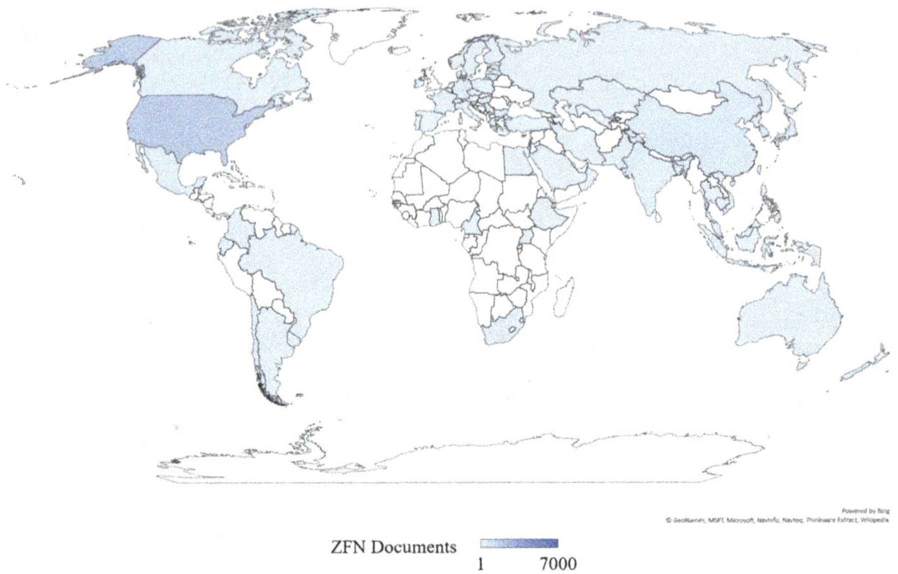

Fig. 5.3 Global spread of ZFN publications (2002–2018). The top five countries with the most publication affiliations are the United States (1519 publications), China (346 publications), Germany (327 publications), Japan (204 publications), and England (169 publications) (Data for the figure were obtained from the Web of Science on May 8, 2019. The map was generated in Microsoft Excel, powered by Bing, © GeoNames, MSFT, Microsoft, NavInfo, Navteq, Thinkware Extract, Wikipedia)

(FY) 2018, the NIH spent nearly $2 billion overall on CRISPR-related projects, which is over six-times what the NIH spent on ZFN- and TALEN-related projects combined. ZFN funding picked up some in FY 2006 and gained more traction after FY 2008; however, it fell again after peaking in FY 2015. Similarly, TALEN funding picked up between FY 2011 and FY 2015 and then fell. Unlike ZFN and TALEN funding, CRISPR funding began to take off in FY 2013 and exponentially increased thereafter. While funding for CRISPR continues to increase, funding for ZFN and TALEN projects remains low (Fig. 5.6).

Not only does NIH provide more overall funding for CRIPSR projects, but it also provides more funding for CRISPR R/R&D with strong commercialization potential—which are funded through the SBIR and STTR programs (Fig. 5.7). Between FY 2002 and FY 2018, NIH's SBIR and STTR programs invested nearly $23 million into CRISPR R/R&D, which is almost three-times the amount invested into ZFN and TALEN R/R&D combined. Prior to FY 2014, NIH sparsely invested in genome editing R/R&D with strong commercialization potential—only three projects involving ZFN received SBIR and STTR funding. In FY 2014, NIH increased its SBIR and STTR investments in ZFN, TALEN, and CRISPR R/R&D. Even so, between FY 2014 and FY 2018, SBIR and STTR investments in ZFN and TALEN R/R&D did not annually exceed $2.5 million. On the other hand, SBIR and STTR investment in CRISPR R/R&D began to rapidly increase after FY 2016. For

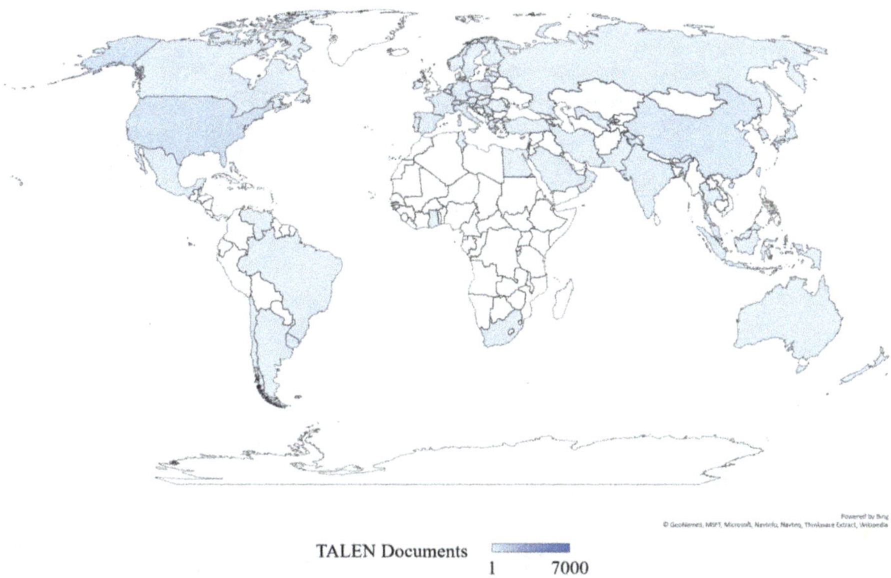

Fig. 5.4 Global spread of TALEN publications (2002–2018). The top five countries with the most document affiliations are the United States (620 publications), China (337 publications), Japan (211 publications), Germany (171 publications), and France (77 publications) (Data for the figure were obtained from the Web of Science on May 8, 2019. The map was generated in Microsoft Excel, powered by Bing, © GeoNames, MSFT, Microsoft, NavInfo, Navteq, Thinkware Extract, Wikipedia)

FY 2018, NIH's SBIR and STTR programs investment in CRISPR R/R&D reached over $10.6 million for 25 projects.

5.2.2.2 Clinical Studies

Overall, there are more clinical studies involving CRISPR (26 studies) than there are those involving ZFN (15 studies) or TALEN (6 studies). The first clinical trial for ZFN started in 2009 (NCT00842634)—13 years after ZFN was first designed (Fig. 5.8). A few more clinical trials involving ZFN launched in the years after the first clinical trial, and, since 2015, two clinical trials involving ZFN have started each year. In contrast to ZFN, the first clinical trial involving TALEN started in 2016, which is only 6 years after TALEN was first designed (NCT02808442). Between 2016 and 2018, one to three clinical trials involving TALEN started each year.

Unlike ZFN and TALEN, it only took 3 years to move CRISPR from the laboratory to clinical studies. While this is the case, the two studies involving CRISPR that started in 2015, both of which were in the United States, were observational studies and did not actually involve any intervention (NCT02964481 and NCT03332030). The first clinical trial (i.e., first interventional study) involving CRISPR, which treated metastatic non-small cell lung cancer, started in 2016 in

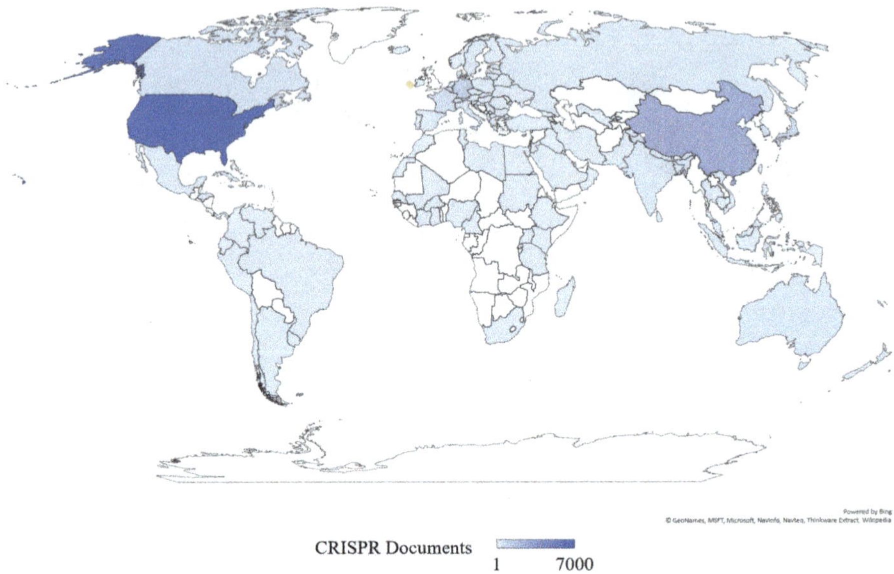

Fig. 5.5 Global spread of CRISPR publications (2002–2018). The top five countries with the most publication affiliations are the United States (6163 publications), China (2429 publications), Germany (1077 publications), Japan (897 publications), and England (832 publications) (Data for the figure were obtained from the Web of Science on May 8, 2019. The map was generated in Microsoft Excel, powered by Bing, © GeoNames, MSFT, Microsoft, NavInfo, Navteq, Thinkware Extract, Wikipedia)

China (NCT02793856). The three other clinical trials that were scheduled to start in 2016 in China were withdrawn because there was no funding (NCT02863913, NCT02867345, and NCT02867332). Despite the setback, all clinical trials involving CRISPR that started the next year, in 2017, took place in China.

Compared to China, clinical trials involving CRISPR were off to a slower start in the United States. The first clinical trial that was set to begin in the United States in May 2018 was stopped before enrollment (NCT03538613). The National Cancer Institute, the sponsor of NCT03538613, did not move forward with the execution of the study. In an interview, it was reported that the study team for NCT03538613 decided to pursue a "more promising treatment."[81] The first clinical trial involving CRISPR that took place in the United States started on September 5, 2018 (NCT03399448). The clinical trial, NCT03399448, began more than 2 years after

[81] Andrew D. Smith, "Controversy Adds Caution to CRISPR Editing in United States," OncLive, January 17, 2019, accessed April 12, 2019, https://www.onclive.com/publications/Oncology-live/2019/vol-20-no-2/controversy-adds-caution-to-crispr-editing-in-united-states?p=3

[82] Sara Reardon, "First CRISPR Clinical Trial Gets Green Light from US Panel," *Nature News*, accessed April 15, 2019, doi:10.1038/nature.2016.20137.

[83] "Recombinant DNA Advisory Committee," Office of Science Policy, accessed April 15, 2019, https://osp.od.nih.gov/biotechnology/recombinant-dna-advisory-committee/

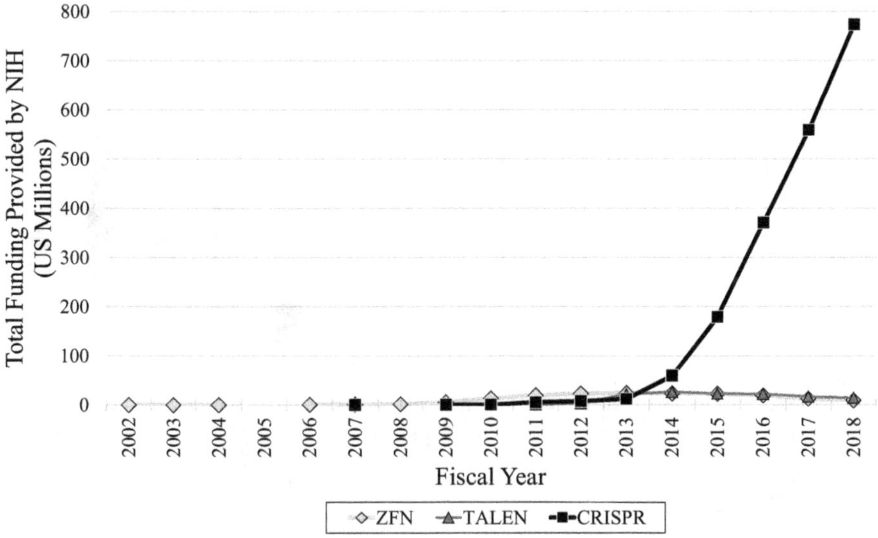

Fig. 5.6 NIH funding for genome editing projects (FY 2002–2018) (Data gathered May 7, 2019 from "NIH Research Portfolio Online Reporting Tools Expenditures and Results")

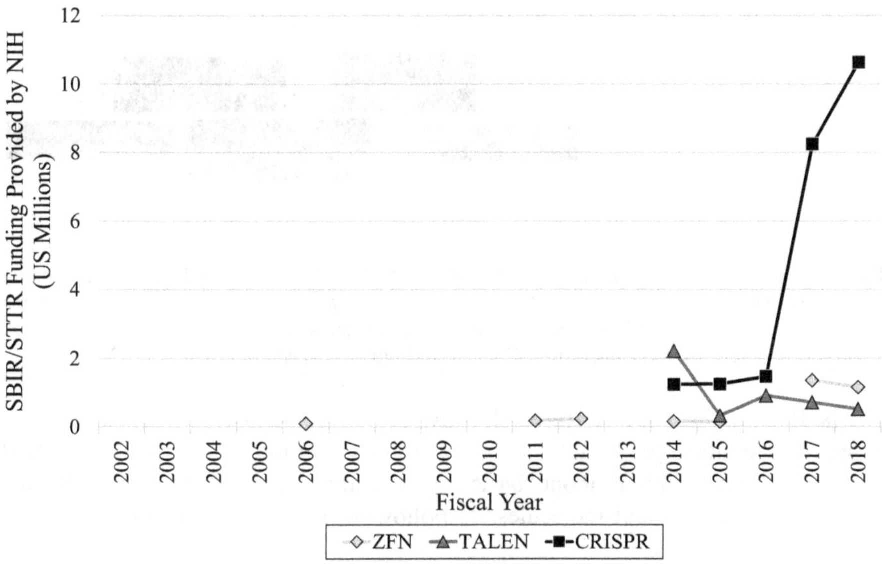

Fig. 5.7 NIH's SBIR and STTR programs investment into genome editing projects with potential for commercialization (2002–2018) (Data gathered May 7, 2019 from "NIH Research Portfolio Online Reporting Tools Expenditures and Results")

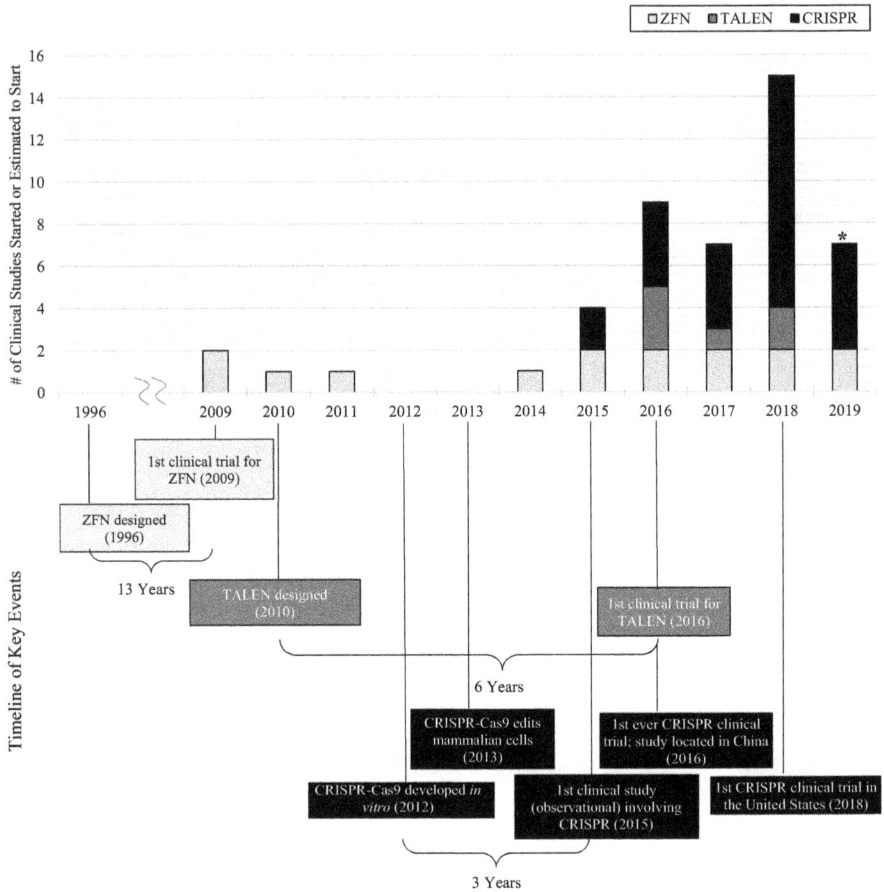

Fig. 5.8 Clinical studies involving ZFN, TALEN, and CRISPR. The figure depicts the number of clinical studies involving ZFN, TALEN, and CRISPR that started, or were scheduled to start, per year as well as a timeline of key events. Asterisk, Clinical studies are depicted as of April 10, 2019 (Data for the figure were obtained from ClinicalTrials.gov on April 10, 2019)

NIH's Recombinant DNA Research Advisory Committee (RAC) reviewed and approved the study proposal.[82] As an advisory committee, RAC offers the NIH director recommendations about "basic and clinical research involving recombinant or synthetic nucleic acid molecules."[83] Following RAC approval, the study also

[82]Sara Reardon, "First CRISPR Clinical Trial Gets Green Light from US Panel," *Nature News*, accessed April 15, 2019, doi:10.1038/nature.2016.20137.

[83]"Recombinant DNA Advisory Committee," Office of Science Policy, accessed April 15, 2019, https://osp.od.nih.gov/biotechnology/recombinant-dna-advisory-committee/

needed approval from the United States Food and Drug Administration (FDA) as well as review committees at the institution that sponsored the study—in this case, the University of Pennsylvania.[84]

In October 2018, shortly after the first clinical trial involving CRISPR began in the United States, Vertex Pharmaceuticals and CRISPR Therapeutics informed the public that the FDA removed the clinical hold and approved the Investigational New Drug application for the drug CTX001 in the treatment of sickle cell disease. CTX001 is a CRISPR-based therapy, co-developed by Vertex Pharmaceuticals and CRISPR Therapeutics, for adults with diseases that involve a deficiency in hemoglobin production, including β-thalassemia and sickle cell disease. While a clinical trial, NCT03745287, that evaluated CTX001 in participants with sickle cell disease began in the United States a month after the FDA lifted the hold, the drug developers had pursued Clinical Trial Applications in multiple countries for CTX001—the strategy paid off and allowed drug developers to have already initiated another clinical trial, NCT03655678, abroad before the FDA even removed the clinical hold. In September 2018, the clinical trial, NCT03655678, began in Canada, Germany, and the United Kingdom to evaluate CTX001 in participants with β-thalassemia.[85]

Phases are used to indicate the stage of clinical trials that involve testing drug products. Not applicable (N/A) indicates trials without phases. Early phase 1, previously known as phase 0, refers to exploratory trials carried out prior to phase 1; these trials may speed up drug development by studying the effect of small doses of the drug on people. Phase 1 refers to trials studying the safety of the drug. Phase 2 indicates trials studying the effectiveness of the drug in persons with a disease or condition. Phase 1/2 indicates trials that have combined phases 1 and 2—perhaps to answer research questions faster or with fewer participants. Phase 3 refers to trials further studying safety, effectiveness, and risk-benefit by examining multiple populations, varied doses, and drug combinations. Finally, phase 4 indicates trials involving an FDA-approved drug in which more information is obtained on how well the drug products its desired result, safety, and longer-term risk-benefit.[86]

---

[84] Review committees from the University of Pennsylvania include “the Institutional Biosafety Committee (IBC), Conflict of Interest Standing Committee, and Institutional Review Board (IRB).” Françoise Baylis and Marcus McLeod, “First-in-Human Phase 1 CRISPR Gene Editing Cancer Trials: Are We Ready?,” *Current Gene Therapy* 17, no. 4 (August 2017): 309, doi:10.2174/1566523217666171121165935.

[85] “CRISPR Therapeutics and Vertex Announce FDA Has Lifted the Clinical Hold on the Investigational New Drug Application for CTX001 for the Treatment of Sickle Cell Disease,” Vertex Pharmaceuticals, accessed October 22, 2018, https://investors.vrtx.com/news-releases/news-release-details/crispr-therapeutics-and-vertex-announce-fda-has-lifted-clinical; “Vertex and CRISPR Therapeutics to Co-Develop and Co-Commercialize CTX001 as CRISPR/Cas9 Gene Edited Treatment for Sickle Cell Disease and β-Thalassemia,” Vertex Pharmaceuticals, December 12, 2017, accessed October 22, 2018, https://investors.vrtx.com/news-releases/news-release-details/vertex-and-crispr-therapeutics-co-develop-and-co-commercialize

[86] “ClinicalTrials.Gov Protocol Registration Data Element Definitions for Interventional and Observational Studies,” accessed April 17, 2019, https://prsinfo.clinicaltrials.gov/definitions.html;

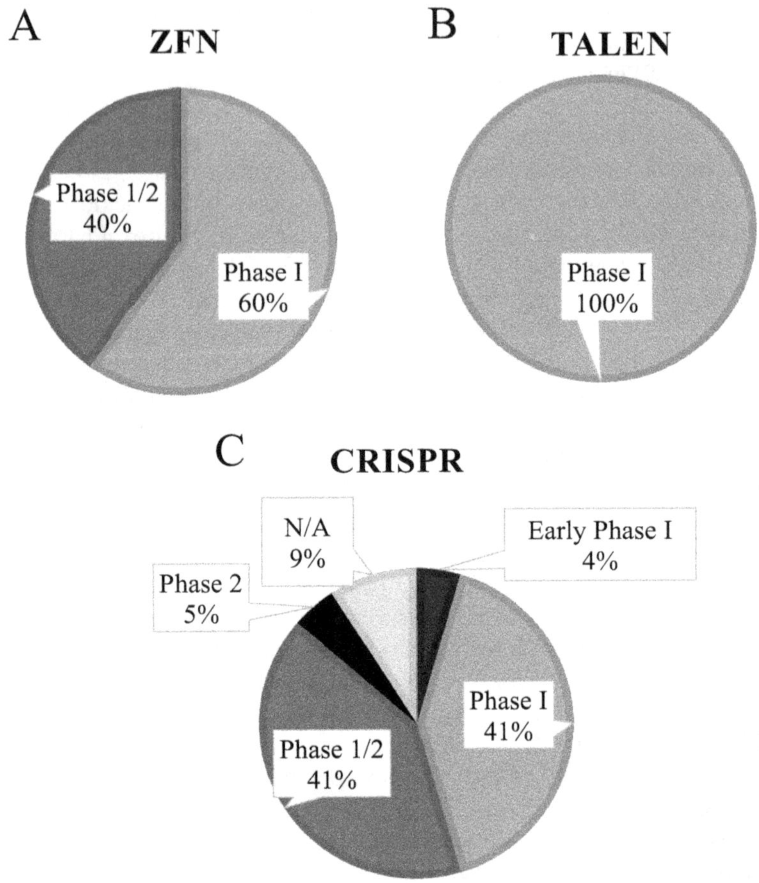

Fig. 5.9 Phases of clinical trials involving (**a**) ZFN, (**b**) TALEN, and (**c**) CRISPR (Data for the figure were obtained from ClinicalTrials.gov on April 10, 2019)

Even though clinical trials involving CRISPR only began in 2016, about half (46%) of the trials were already in phase 1/2 and phase 2 by mid-2019 (Fig. 5.9). Clinical trials that are phase 1/2 or phase 2 study safety, side effects and optimal dosage of the drug (phase 1) as well as how well the drug works in patients who have the disease or condition (phase 2). Note that phase 1/2 studies combine phases 1 and 2 into one trial. Like CRISPR, a little less than half (40%) of clinical trials involving ZFN are phase 1/2. The remainder of clinical trials involving ZFN and CRISPR are

"Glossary of Common Site Terms - ClinicalTrials.Gov," accessed April 17, 2019, https://clinicaltrials.gov/ct2/about-studies/glossary; "NCI Dictionary of Cancer Terms: Phase I/II Clinical Trial," nciAppModulePage, National Cancer Institute, February 2, 2011, accessed April 15, 2019, https://www.cancer.gov/publications/dictionaries/cancer-terms; "What Are Clinical Trials? | ECMC," accessed April 15, 2019, https://www.ecmcnetwork.org.uk/what-are-clinical-trials

phase I or earlier—except for 9% of CRISPR trials that are without phases. While several clinical trials involving ZFN and CRISPR are in phases 1/2 and 2, all clinical trials involving TALEN are in phase 1 and had not progressed to the next phase as of mid-2019.

In general, clinical studies involving CRISPR study a wider variety of conditions than clinical studies involving ZFN and TALEN (Fig. 5.10). Clinical studies involving ZFN tend to concentrate on inherited metabolic diseases (Mucopolysaccharidosis type I and Mucopolysaccharidosis type II, abbreviated MPS I and MPS II, respectively), inherited blood diseases (Beta-Thalassemia, Hemophilia B, and sickle cell disease), and infection from human immunodeficiency virus (HIV) as well as Human Papillomavirus-Related Malignant Neoplasm. Compared to TALEN, clinical studies involving TALEN are narrowly focused on Human Papillomavirus-Related Malignant Neoplasm as well as leukemia, lymphoma, and other lymphoid malignancies. So far, clinical studies involving CRISPR consider all the conditions (except MPS I, MPS II, and Hemophilia B) as those involving ZFN and TALEN. Additionally, clinical studies involving CRISPR are studying more types of cancers as well as more inherited and congenital diseases.

5.2.2.3 Patents

Since 2003, over 9000 patent documents for ZFN, TALEN, and CRISPR have been published (Fig. 5.11). Over half of these patent documents are for CRISPR (5310 documents) and the rest are for ZFN (2038 documents) and TALEN (1923 documents). Unlike the number of ZFN and TALEN patent documents, which has only slightly increased each year, the number of patent documents published for CRISPR has increased exponentially since 2012. Many of these documents are patent application documents (ZFN: 83%, TALEN: 88%, and CRISPR: 91%), and the rest of the documents are granted patent documents (ZFN: 17%, TALEN: 12%, and CRISPR: 9%).

For ZFN, TALEN, and CRISPR, the top five jurisdictions of patent documents are Australia, China, European Patent Office (EPO), Patent Cooperation Treaty (PCT) administered by the WIPO, and United States (Fig. 5.12). The jurisdictions with the most patent documents for all three technologies are the PCT WIPO and United States. Under the PCT, "it possible to seek patent protection for an invention simultaneously in a large number of countries by filing a single 'international' patent application instead of filing several separate national or regional patent applications. The granting of patents remains under the control of the national or regional patent Offices in what is called the 'national phase.'"[87] While the United States has a greater percentage of patent documents than the PCT WIPO for ZFN, the reverse is true for TALEN and CRISPR. As newer technologies emerge, this trend is

[87] "PCT FAQs," World Intellectual Property Organization, accessed July 7, 2019, https://www.wipo.int/pct/en/faqs/faqs.html

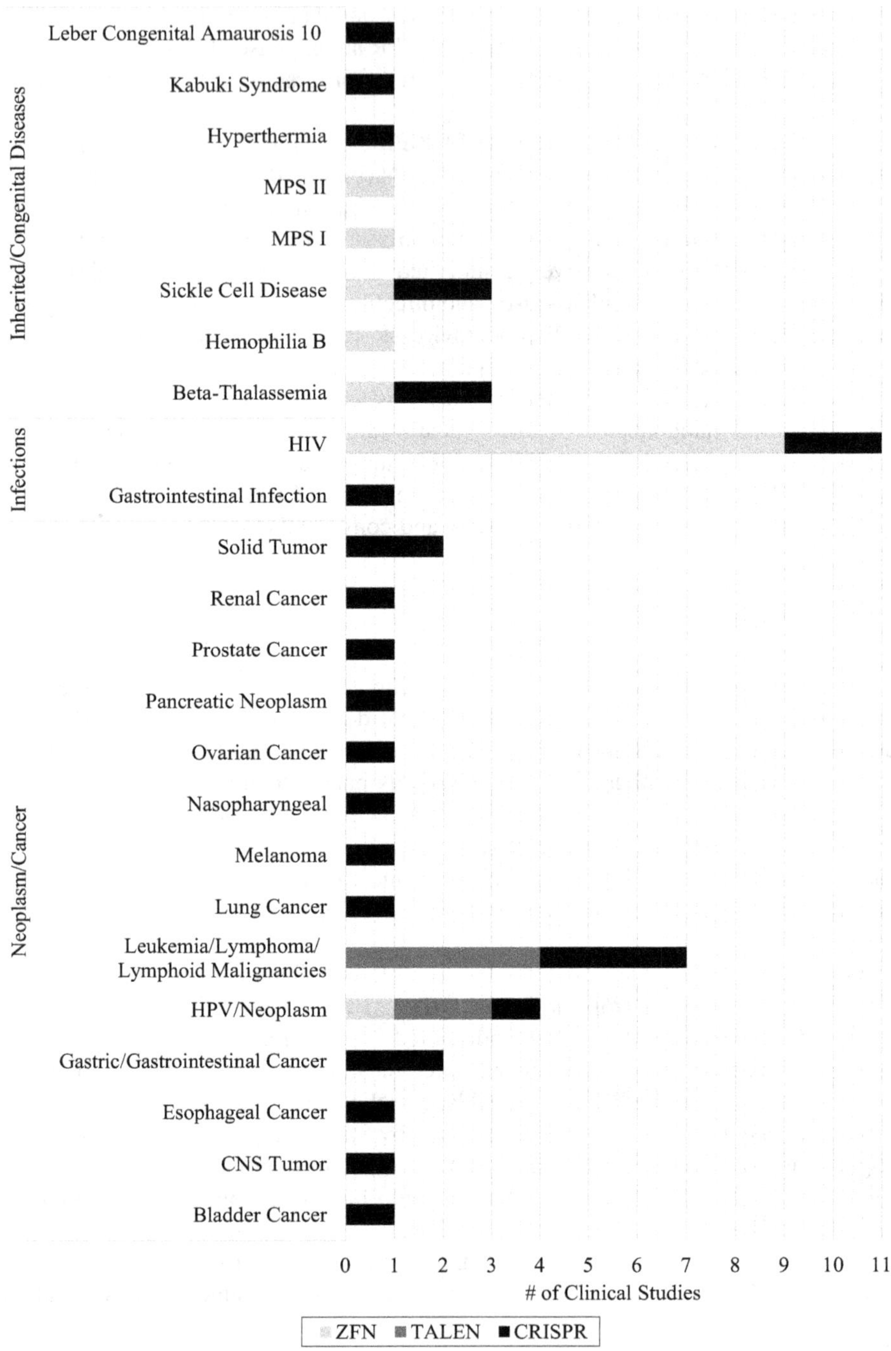

Fig. 5.10 Types of Conditions Studied in Clinical Studies Involving ZFN, TALEN, and CRISPR. Some clinical studies involve multiple conditions. *MPS* mucopolysaccharidosis, *HIV* human immunodeficiency virus, *HPV* human papillomavirus, *CNS* central nervous system (Data for the figure were obtained from ClinicalTrials.gov on April 10, 2019)

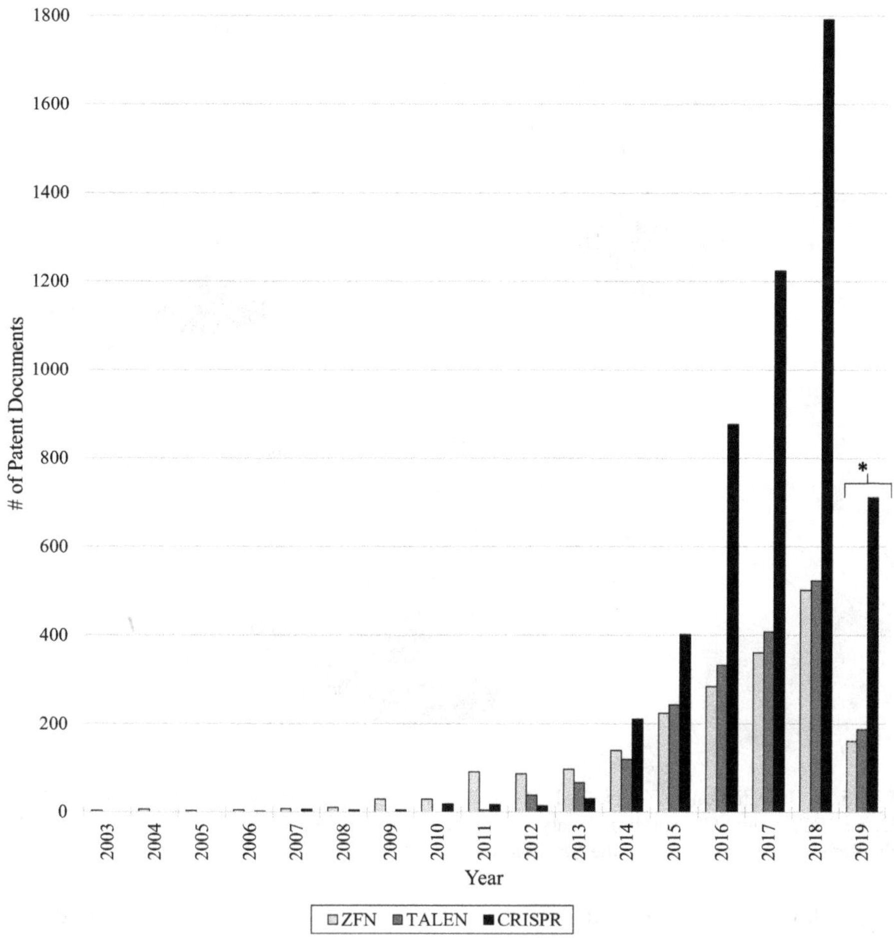

Fig. 5.11 Number of global patent documents published per year for ZFN, TALEN, and CRISPR. Asterisk, Data for 2019 is provided from January 1, 2019 through May 21, 2019 (Data for the figure were obtained from the Lens on May 21, 2019)

increasingly more pronounced. This suggests that patent applicants seeking invention rights for TALEN and, especially, CRISPR are pursuing a more international approach than patent applicants seeking invention rights for ZFN.

The jurisdictions of the remaining patent documents are Australia, China, EPO, and others including Brazil, Canada, Chile, Columbia, Cyprus, Eurasian Patent Office, Hong Kong, Israel, Japan, Mexico, New Zealand, Poland, Portugal, Republic of Korea, Russia, Singapore, Spain, Sweden, Taiwan, United Kingdom, and Uruguay. In Australia, although the percentage of CRISPR patent documents is lower than ZFN and TALEN patent documents, the total number of patent documents for CRISPR (439 documents) is still higher than those for ZFN (284 documents) and TALEN (244 documents). In Eastern Asia, there is a substantial increase in CRISPR

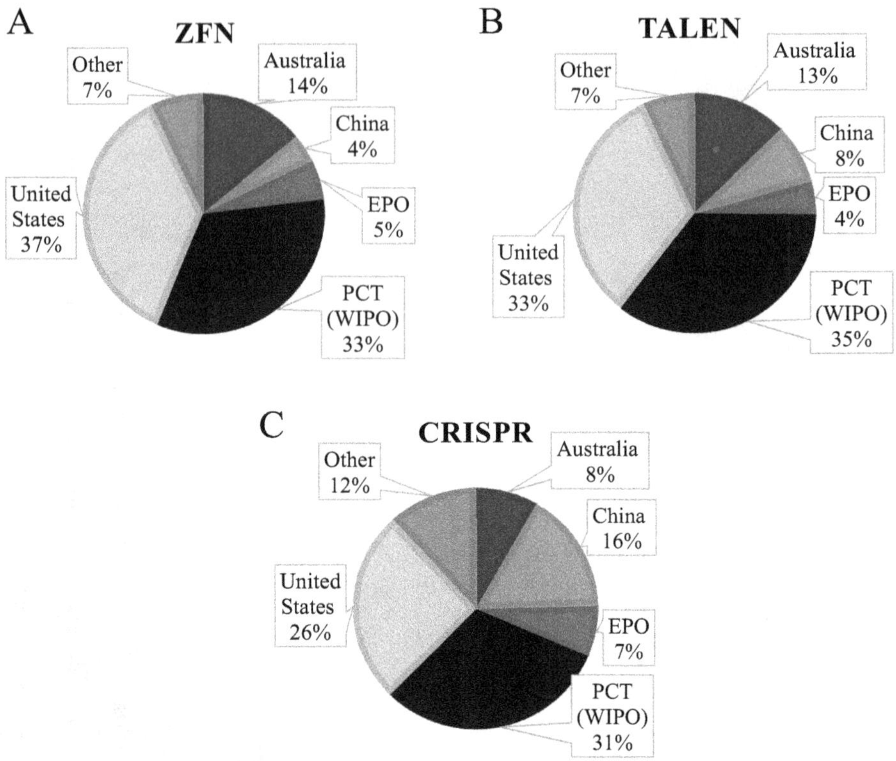

Fig. 5.12 Jurisdictions of ZFN, TALEN, and CRISPR patent documents. *EPO* European Patent Office, *PCT* Patent Cooperation Treaty, *WIPO* World Intellectual Property Organization (Data for the figure were obtained from the Lens on May 21, 2019)

patent documents. Compared to ZFN and TALEN, combined, there are 3.7-fold, 2.2-fold, and 2.2-fold more CRISPR patent documents in China, Japan, and Republic of Korea, respectively. Finally, in the EPO, there is a slight increase in the percentage of CRISPR patent documents compared to the percentage of ZFN and TALEN patent documents.[88]

5.2.2.4 Companies

Many more CRISPR-focused companies have started up than ZFN- or TALEN-focused companies (Fig. 5.13). Sangamo Biosciences (now Sangamo Therapeutics), and Phytodyne were founded in 1995 and 2000, respectively, based on zinc-finger

[88] As of March 1, 2019, the EPO provided coverage to 38 member states, two extension states, and four validation states across Europe. European Patent Office, "The EPO at a Glance," accessed May 29, 2019, https://www.epo.org/about-us/at-a-glance.html

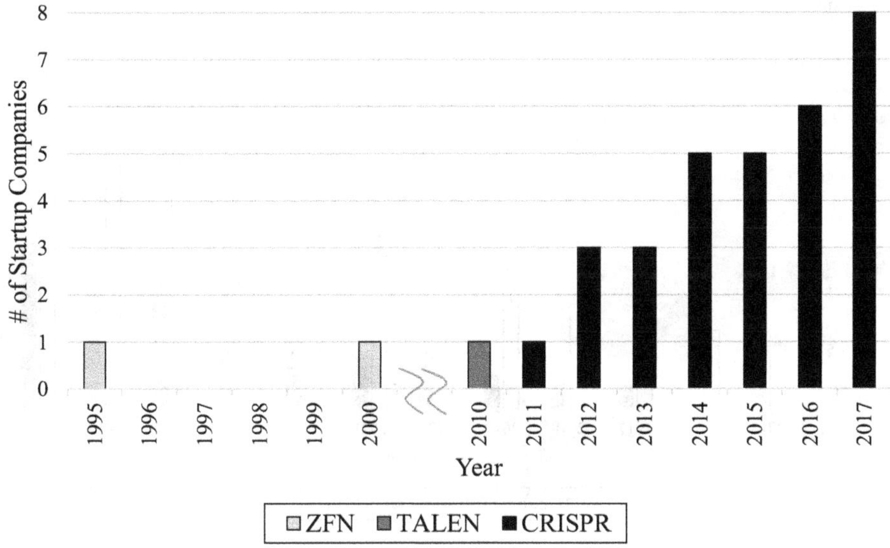

Fig. 5.13 Number of genome editing focused startup companies (1995–2017) (Data on startup companies were obtained from Crunchbase and Owler on November 30, 2018)

technology. While Sangamo is still in operation, Phytodyne closed in 2004 due to a lack of funding and inability to reach a deal with Sangamo to access Sangamo's proprietary proteins.[89] Calyxt (a separate unit in Cellectis), was formed in 2010 based on TALEN technology. Unlike zinc-fingers and TALEN, CRISPR-focused startup companies began in 2011 and many continue to form each year. By 2017, over $1 billion had been invested into CRISPR-focused startups, and, on average, nearly one new CRISPR-focused company was springing up every 1–2 months. This trend suggests that CRISPR-focused companies will continue to form at an increasing rate, while ZFN- and TALEN-focused startup companies are not likely to form in the future.

Not only are startups forming that are focused on CRISPR, but companies that have formed based on CRISPR are also going public about twice as fast as ZFN- and TALEN-focused startups. Sangamo (zinc-fingers) and Calyxt (TALEN) took 5 and 7 years, respectively, to go public. On the other hand, three CRISPR human therapeutic companies—Editas Medicine, CRISPR Therapeutics, and Intellia Therapeutics—took 2–3 years to go public. Editas Medicine was formed in September 2013 and, only 2 years and 5 months later, was the first company to go public in February 2016. Similarly, CRISPR Therapeutics formed in late October 2013 and went public in October 2016. Lastly, Intellia Therapeutics formed in May 2014 and went public 2 years later in May 2016 (see Appendix E).

[89] Pearson, "Protein Engineering," 163.

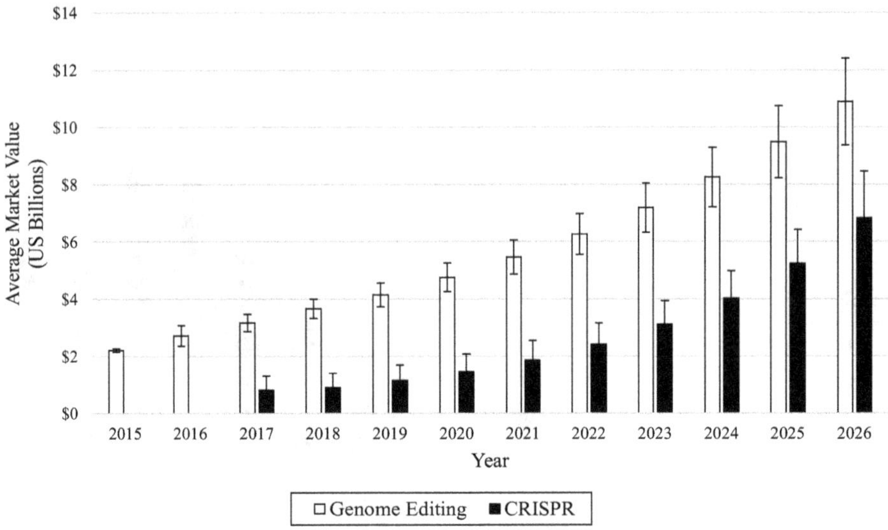

Fig. 5.14 Global genome editing and CRISPR market valuations. The figure depicts the average market value of genome editing. The market value of CRISPR, which is a portion of the genome editing market, is also illustrated (Data on the size of the global genome editing market were obtained through the following market research reports: Global Market Insight, Grand View Research, Markets and Markets, Maximize Market Research, Wise Guy Reports, Research and Markets, and Inkwood Research: "Global CRISPR And CRISPR-Associated (Cas) Genes Market 2018 by Manufacturers, Regions, Type and Application, Forecast to 2023, Description," Wise Guy Reports, September 6, 2018, accessed October 10, 2018, https://www.wiseguyreports.com/reports/3388744-global-crispr-and-crispr-associated-cas-genes-market; "CRISPR Market - Forecasts from 2018 to 2023, Description," Research and Markets, August 2018, accessed October 9, 2018, https://www.researchandmarkets.com/reports/4618285/crispr-market-forecasts-from-2018-to-2023; "GLOBAL CRISPR MARKET FORECAST 2018-2026," Inkwood Research, accessed September 11, 2018, https://www.inkwoodresearch.com/reports/global-crispr-market-forecast/; "Gene Editing Market Size – Industry Share Report 2024, Summary," Global Market Insights, October 2016, accessed October 7, 2018, https://www.gminsights.com/industry-analysis/gene-editing-market; "Genome Editing Market Size To Reach $8.1 Billion By 2025," Grand View Research, accessed October 9, 2018, https://www.grandviewresearch.com/press-release/global-genome-editing-market; "Genome Editing/Genome Engineering Market by Application, Technology & End User - 2022 | MarketsandMarkets," MarketsandMarkets, November 2017, https://www.marketsandmarkets.com/Market-Reports/genomeediting-engineering-market-231037000.html, https://www.marketsandmarkets.com/Market-Reports/genome-editing-engineering-market-231037000.html; "Genome Editing Market: Global Industry Analysis and Forecast 2026, Report Details," *Maximize Market Research* (blog), accessed October 10, 2018, https://www.maximizemarketresearch.com/market-report/genome-editing-market/2747/)

5.2.2.5 Global Market Size Projections

Public and private entities around the world are investing heavily into initiatives that use genome editing technologies to advance human therapeutics, agricultural biotech, biotech research, and other industrial biotech. Over the next decade, Russia

Table 5.5 Overview score for the ability of genome editing to be exploited based on maturity

	ZFN	TALEN	CRISPR
Ability to be exploited based on maturity	High to moderate	Moderate	High

plans to invest $1.7 billion into genome editing.[90] As Fig. 5.14 shows, the genome editing market is predicted to grow from $2.17 billion in 2015 to $12.5 billion in 2027 at an average Compound Annual Growth Rate (CAGR) of 14.43%. Additionally, the global CRISPR market, which is a subset of the genome editing market, is predicted to grow from $467 million in 2017 to $10.3 billion in 2027 at an average CAGR of 36.53%. The global CRISPR market (CAGR of 36.53%) is expected to grow over twice as fast as the global genome editing market (CAGR 14.43%). In 2017, CRISPR made up less than one-fifth (14.46%) of the total global genome editing market. By 2027, CRISPR is expected to make up over four-fifths (82.23%) of the genome editing market. The data suggest that CRISPR will continue to make up more and more of the total global genome editing market. Overall, the analysis of data on scientific publications, clinical studies, patents, companies, and global market size projections suggests that the ability to exploit genome editing technologies based on maturity is high to moderate for ZFN, moderate for TALEN, and high for CRISPR (Table 5.5).

5.3 Conclusion

The assessment on the magnitude of potential harm of genome editing technologies offered key insight into making pathogens more dangerous, weaponizing the human microbiome, weaponizing gene drives, and augmenting humans. First, analysis of the Gryphon study revealed that the modification of some traits using genome editing technologies will have more of an impact than other traits. Second, the analysis of the human microbiome revealed that it is vast and relatively little is known about it, which will make targeting difficult. Third, there are many concerning scenarios of misuse of gene drives; there are inherent restrictions that make application of gene drives difficult. Lastly, there are several traits that actors may want to target to create super soldiers or to augment the general population, and some of the traits (e.g., enhancing situational awareness) are more likely than others to be targeted.

The imminence of potential misuse is highest for CRISPR, followed by ZFN, and, finally, TALEN. There are nearly three-times more publications for CRISPR compared to ZFN and TALEN, combined. The NIH, which invests in R/R&D with strong commercialization potential through the SBIR and STTR programs, spent almost three-times more for CRISPR R/R&D compared to ZFN and TALEN

[90] Rodolphe Barrangou, "Taking CRISPR to New Heights," *The CRISPR Journal* 2, no. 3 (June 21, 2019), doi:10.1089/crispr.2019.29064.rba.

R/R&D, combined. In terms of clinical studies, there are over two-times as many clinical studies for ZFN compared to TALEN, and there are almost two-times as many clinical studies for CRISPR compared to ZFN. By mid-2019, all clinical studies involving TALEN were only in phase 1, and almost half of clinical studies involving ZFN and CRISPR were further along in phases 1/2 or 2. The number of CRISPR patent documents published each year is exponentially increasing, whereas the number of ZFN and TALEN patent documents are only linearly increasing. Lastly, while few companies were formed based on ZFN and TALEN technologies, new CRISPR-focused companies are forming every 1–2 months. The findings from the risk of misuse assessment, which was carried out in Chaps. 4 and 5, are presented in Chap. 6.

References

1. "A Review of 10 Years of Human Microbiome Research Activities at the US National Institutes of Health, Fiscal Years 2007-2016." *Microbiome* 7 (February 26, 2019): 31. doi:https://doi.org/10.1186/s40168-019-0620-y.
2. "About NSF - Overview | NSF - National Science Foundation." Accessed December 14, 2018. https://www.nsf.gov/about/.
3. Owler. "About Owler." Accessed February 10, 2019. https://www.owler.com/about/what-is-owler.
4. "About SBIR | SBIR.Gov." Accessed December 11, 2018. https://www.sbir.gov/about/about-sbir.
5. "About STTR | SBIR.Gov." Accessed December 11, 2018. https://www.sbir.gov/about/about-sttr.
6. "About The Lens." Accessed May 20, 2019. https://about.lens.org/.
7. Amoasii, Leonela, John C. W. Hildyard, Hui Li, Efrain Sanchez-Ortiz, Alex Mireault, Daniel Caballero, Rachel Harron, et al. "Gene Editing Restores Dystrophin Expression in a Canine Model of Duchenne Muscular Dystrophy." *Science* 362, no. 6410 (October 5, 2018): 86–91. doi:https://doi.org/10.1126/science.aau1549.
8. Antonio Regalado. "Engineering the Perfect Baby." MIT Technology Review, March 5, 2015. https://www.technologyreview.com/2015/03/05/249167/engineering-the-perfect-baby/.
9. Baylis, Françoise, and Marcus McLeod. "First-in-Human Phase 1 CRISPR Gene Editing Cancer Trials:Are We Ready?" *Current Gene Therapy* 17, no. 4 (August 2017): 309–19. doi:https://doi.org/10.2174/1566523217666171121165935.
10. ClinicalTrials.gov. "ClinicalTrials.Gov Background." Accessed April 12, 2019. https://clinicaltrials.gov/ct2/about-site/background.
11. "ClinicalTrials.Gov Protocol Registration Data Element Definitions for Interventional and Observational Studies." Accessed April 17, 2019. https://prsinfo.clinicaltrials.gov/definitions.html.
12. Research and Markets. "CRISPR Market - Forecasts from 2018 to 2023, Description," August 2018. https://www.researchandmarkets.com/reports/4618285/crispr-market-forecasts-from-2018-to-2023.
13. Vertex Pharmaceuticals. "CRISPR Therapeutics and Vertex Announce FDA Has Lifted the Clinical Hold on the Investigational New Drug Application for CTX001 for the Treatment of Sickle Cell Disease." Accessed October 22, 2018. https://investors.vrtx.com/news-releases/news-release-details/crispr-therapeutics-and-vertex-announce-fda-has-lifted-clinical.

14. Dalvin Brown. "The Army's Latest Night-Vision Tech Looks like Something out of a Video Game." *Washington Post*, May 24, 2021. https://www.washingtonpost.com/technology/2021/05/24/army-night-vision-goggles/.
15. Dekaboruah, Elakshi, Mangesh Vasant Suryavanshi, Dixita Chettri, and Anil Kumar Verma. "Human Microbiome: An Academic Update on Human Body Site Specific Surveillance and Its Possible Role." *Archives of Microbiology* 202, no. 8 (October 1, 2020): 2147–67. doi:https://doi.org/10.1007/s00203-020-01931-x.
16. "Deutsche Forschungsgemeinschaft (DFG) - Research in Germany." Accessed May 14, 2019. https://www.research-in-germany.org/en/research-funding/funding-organisations/deutsche-forschungsgemeinschaft-(dfg).html.
17. Dunbar, Cynthia E., Katherine A. High, J. Keith Joung, Donald B. Kohn, Keiya Ozawa, and Michel Sadelain. "Gene Therapy Comes of Age." *Science* 359, no. 6372 (January 12, 2018): eaan4672. doi:https://doi.org/10.1126/science.aan4672.
18. "Emerging Science and Technology Trends: 2017-2047." Office of the Deputy Assistant Secretary of the Army (Research & Technology), November 21, 2017.
19. Esvelt, Kevin M., Andrea L. Smidler, Flaminia Catteruccia, and George M. Church. "Concerning RNA-Guided Gene Drives for the Alteration of Wild Populations." *Elife* 3, no. e03104 (2014). doi:https://doi.org/10.7554/eLife.03401.
20. European Patent Office. "The EPO at a Glance." Accessed May 29, 2019. https://www.epo.org/about-us/at-a-glance.html.
21. Global Market Insights. "Gene Editing Market Size – Industry Share Report 2024, Summary," October 2016. https://www.gminsights.com/industry-analysis/gene-editing-market.
22. Maximize Market Research. "Genome Editing Market : Global Industry Analysis and Forecast 2026, Report Details." Accessed October 10, 2018. https://www.maximizemarketresearch.com/market-report/genome-editing-market/2747/.
23. Grand View Research. "Genome Editing Market Size To Reach $8.1 Billion By 2025." Accessed October 9, 2018. https://www.grandviewresearch.com/press-release/global-genome-editing-market.
24. MarketsandMarkets. "Genome Editing/Genome Engineering Market by Application, Technology & End User - 2022 | MarketsandMarkets," November 2017. https://www.marketsandmarkets.com/Market-Reports/genomeediting-engineering-market-231037000.html, https://www.marketsandmarkets.com/Market-Reports/genome-editing-engineering-market-231037000.html.
25. Wise Guy Reports. "Global CRISPR And CRISPR-Associated (Cas) Genes Market 2018 by Manufacturers, Regions, Type and Application, Forecast to 2023, Description," September 6, 2018. https://www.wiseguyreports.com/reports/3388744-global-crispr-and-crispr-associated-cas-genes-market.
26. Inkwood Research. "GLOBAL CRISPR MARKET FORECAST 2018-2026." Accessed September 11, 2018. https://www.inkwoodresearch.com/reports/global-crispr-market-forecast/.
27. "Glossary of Common Site Terms - ClinicalTrials.Gov." Accessed April 17, 2019. https://clinicaltrials.gov/ct2/about-studies/glossary.
28. Gronlund, Kirsten. "Genome Editing and the Future of Biowarfare: A Conversation with Dr. Piers Millett." Future of Life Institute, October 12, 2018. https://futureoflife.org/2018/10/12/genome-editing-and-the-future-of-biowarfare-a-conversation-with-dr-piers-millett/.
29. Gurwitz, David. "Gene Drives Raise Dual-Use Concerns." *Science* 345, no. 6200 (August 29, 2014): 1010–1010. doi:https://doi.org/10.1126/science.345.6200.1010-b.
30. Human Microbiome Project (HMP) Data Analysis and Coordination Center (DACC). "Human Microbiome Project." Accessed February 24, 2022. https://hmpdacc.org/hmp/overview/.
31. Jain, Surbhi, Saurabh Shukla, Che Yang, Meng Zhang, Zia Fatma, Manasi Lingamaneni, Shireen Abesteh, et al. "TALEN Outperforms Cas9 in Editing Heterochromatin Target Sites." *Nature Communications* 12, no. 1 (January 27, 2021): 606. doi:https://doi.org/10.1038/s41467-020-20672-5.

32. John Ratcliffe. "China Is National Security Threat No. 1." The Wall Street Journal, December 3, 2020. https://www.wsj.com/articles/china-is-national-security-threat-no-1-11607019599.
33. Jones, Garett, and W. Joel Schneider. "Intelligence, Human Capital, and Economic Growth: A Bayesian Averaging of Classical Estimates (BACE) Approach." *Journal of Economic Growth* 11, no. 1 (March 1, 2006): 71–93. doi:https://doi.org/10.1007/s10887-006-7407-2.
34. Jyoti Madhusoodanan. "Editing the Microbiome." *PNAS* 117, no. 7 (February 18, 2020). https://www.pnas.org/doi/abs/10.1073/pnas.2000108117.
35. Kirkpatrick, Jesse, Gregory D. Koblentz, Megan J. Palmer, Edward Perello, David A. Relman, and Sarah W. Denton. "Editing Biosecurity: Needs and Strategies for Governing Genome Editing," December 2018.
36. Lita Proctor. "What's Next for the Human Microbiome?" *Nature* 569 (May 30, 2019): 623–25.
37. Matchar, Emily. "Seeing in the Dark: The History of Night Vision." Smithsonian Magazine, May 19, 2017. https://www.smithsonianmag.com/innovation/seeing-dark-history-night-vision-180963357/.
38. Matthews, Tracy. "Web of Science Platform: Introduction." Clarivate Analytics. Accessed December 7, 2018. //clarivate.libguides.com/webofscienceplatform/introduction.
39. European Comission, CORDIS. "Metagenomics of the Human Intestinal Tract," July 16, 2019. https://cordis.europa.eu/project/id/201052.
40. National Academies of Sciences, Engineering, and Medicine. *Biodefense in the Age of Synthetic Biology*. Washington, DC: The National Academies Press, 2018. https://doi.org/10.17226/24890.
41. National Academies of Sciences, . *Gene Drives on the Horizon: Advancing Science, Navigating Uncertainty, and Aligning Research with Public Values*. Washington, DC: The National Academies Press, 2016. https://doi.org/10.17226/23405.
42. National Academies of Sciences, . *Human Genome Editing: Science, Ethics, and Governance*. Washington, DC: The National Academies Press, 2017. https://doi.org/10.17226/24623.
43. Nature Video. "The Microbes That Live with Us from Cradle to Grave." YouTube, June 19, 2019. https://www.youtube.com/watch?v=c_ZRZkU-FEw.
44. National Cancer Institute. "NCI Dictionary of Cancer Terms: Phase I/II Clinical Trial." NciAppModulePage, February 2, 2011. https://www.cancer.gov/publications/dictionaries/cancer-terms.
45. NIH Office of Extramural Research. *Research Portfolio Online Reporting Tools (RePORT)*. US Department of Health and Human Services, 2010. https://report.nih.gov/brochure/index.html.
46. NIH RePORTER. "NIH Research Portfolio Online Reporting Tools Expenditures and Results." Accessed May 7, 2019. https://projectreporter.nih.gov/reporter.cfm.
47. "NIH Small Business Innovation Research (SBIR) and Small Business Technology Transfer (STTR) Programs." Accessed December 11, 2018. https://sbir.nih.gov/.
48. National Institutes of Health (NIH). "NIH to Launch Genome Editing Research Program," January 23, 2018. https://www.nih.gov/news-events/news-releases/nih-launch-genome-editing-research-program.
49. Ouagrham-Gormley, Sonia Ben, and Kathleen M. Vogel. "Gene Drives: The Good, the Bad, and the Hype." Bulletin of the Atomic Scientists, October 14, 2016. https://thebulletin.org/gene-drives-good-bad-and-hype10027.
50. Oye, Kenneth A., Kevin Esvelt, Evan Appleton, Flaminia Catteruccia, George Church, Todd Kuiken, Shlomiya Bar-Yam Lightfoot, Julie McNamara, Andrea Smidler, and James P. Collins. "Regulating Gene Drives." *Science* 345, no. 6197 (July 17, 2014): 626–28. doi:https://doi.org/10.1126/science.1254287.
51. Özduman, Koray, Guido Wollmann, and Joseph M. Piepmeier. "CHAPTER 57 - Gene Therapy for Meningiomas." In *Meningiomas*, edited by M. Necmettin Pamir, Peter M. Black, and Rudolf Fahlbusch, 681–90. Philadelphia: W.B. Saunders, 2010. doi:https://doi.org/10.1016/B978-1-4160-5654-6.00057-X.
52. World Intellectual Property Organization. "PCT FAQs." Accessed July 7, 2019. https://www.wipo.int/pct/en/faqs/faqs.html.

53. Pearson, Helen. “Protein Engineering: The Fate of Fingers.” *Nature* 455, no. 7210 (September 11, 2008): 160–64. doi:https://doi.org/10.1038/455160a.
54. Penagos-Puig, Andrés, and Mayra Furlan-Magaril. “Heterochromatin as an Important Driver of Genome Organization.” *Frontiers in Cell and Developmental Biology* 8 (September 18, 2020). https://www.frontiersin.org/article/10.3389/fcell.2020.579137.
55. Peter Emanuel, Scott Walper, Diane DiEuliis, Natalie Klein, James B. Petro, and James Giordano. “Cyborg Soldier 2050: Human/Machine Fusion and the Implications for the Future of the DOD.” *CCDC CBC-TR-1599,* August 2019.
56. Prokhorova, Daria, Natalya Zhukova Eschenko, Anna Lemza, Mariia Sergeeva, Rinat Amirkhanov, and Grigory Stepanov. “Application of the CRISPR/Cas9 System to Study Regulation Pathways of the Cellular Immune Response to Influenza Virus.” *Viruses* 14, no. 2 (February 21, 2022): 437. doi:https://doi.org/10.3390/v14020437.
57. “Purpose & Functions,History | About US | Japan Society for the Promotion of Science.” Accessed May 15, 2019. https://www.jsps.go.jp/english/aboutus/index2.html.
58. Putin, Vladimir. “Youth 2030: The Image of the Future Panel Session.” Presented at the 19th World Festival of Youth and Students, Russia, October 2017. https://www.youtube.com/watch?v=9v3TNGmbArs.
59. Qian, Xu-Bo, Tong Chen, Yi-Ping Xu, Lei Chen, Fu-Xiang Sun, Mei-Ping Lu, and Yong-Xin Liu. “A Guide to Human Microbiome Research: Study Design, Sample Collection, and Bioinformatics Analysis.” *Chinese Medical Journal* 133, no. 15 (August 5, 2020): 1844–55. doi:https://doi.org/10.1097/CM9.0000000000000871.
60. Reardon, Sara. “First CRISPR Clinical Trial Gets Green Light from US Panel.” *Nature News*. Accessed April 15, 2019. doi:https://doi.org/10.1038/nature.2016.20137.
61. Office of Science Policy. “Recombinant DNA Advisory Committee.” Accessed April 15, 2019. https://osp.od.nih.gov/biotechnology/recombinant-dna-advisory-committee/.
62. Rindermann, Heiner. “Relevance of Education and Intelligence at the National Level for the Economic Welfare of People.” *Intelligence* 36, no. 2 (March 1, 2008): 127–42. doi:https://doi.org/10.1016/j.intell.2007.02.002.
63. *Risk and Benefit Analysis of Gain of Function Research, Final Report*. Takoma Park, MD: Gryphon Scientific, LLC, 2016.
64. Rodolphe Barrangou. “Taking CRISPR to New Heights.” *The CRISPR Journal* 2, no. 3 (June 21, 2019). doi:https://doi.org/10.1089/crispr.2019.29064.rba.
65. Sharon Begley. “A Serious New Hurdle for CRISPR: Edited Cells Might Cause Cancer, Two Studies Find.” STAT, June 11, 2018. https://www.statnews.com/2018/06/11/crispr-hurdle-edited-cells-might-cause-cancer/.
66. ———. “Potential DNA Damage from CRISPR Has Been ‘seriously Underestimated,’ Study Finds.” STAT, July 16, 2018. https://www.statnews.com/2018/07/16/crispr-potential-dna-damage-underestimated/.
67. Shim, Gayong, Dongyoon Kim, Gyu Thae Park, Hyerim Jin, Soo-Kyung Suh, and Yu-Kyoung Oh. “Therapeutic Gene Editing: Delivery and Regulatory Perspectives.” *Acta Pharmacologica Sinica* 38, no. 6 (June 2017): 738–53. doi:https://doi.org/10.1038/aps.2017.2.
68. Smith, Andrew D. “Controversy Adds Caution to CRISPR Editing in United States.” OncLive, January 17, 2019. https://www.onclive.com/publications/Oncology-live/2019/vol-20-no-2/controversy-adds-caution-to-crispr-editing-in-united-states?p=3.
69. Ministry of the Armed Forces. “Speech by Florence Parly, Minister of the Armed Forces, Introducing the Round Table ‘Ethics and Augmented Soldier’ at the Digital Defense Innovation Forum,” December 4, 2020. https://www.defense.gouv.fr/salle-de-presse/discours/discours-de-florence-parly-ministre-des-armees-introduisant-la-table-ronde-ethique-et-soldat-augmente-au-digital-forum-innovation-defense.
70. Springut, Micah, Stephen Schlaikjer, and David Chen. *China’s Program for Science and Technology Modernization: Implications for American Competitiveness*, n.d.
71. Lens.org. “Support Center » What Are Claims?” Accessed May 21, 2019. https://support.lens.org/help-resources/claims/what-are-claims/.

72. "The Lens - Free & Open Patent and Scholarly Search." Accessed December 5, 2018. https://www.lens.org/lens.
73. "Types of Grant Programs | Grants.Nih.Gov," May 16, 2018. https://grants.nih.gov/grants/funding/funding_program.htm.
74. Vertex Pharmaceuticals. "Vertex and CRISPR Therapeutics to Co-Develop and Co-Commercialize CTX001 as CRISPR/Cas9 Gene Edited Treatment for Sickle Cell Disease and β-Thalassemia," December 12, 2017. https://investors.vrtx.com/news-releases/news-release-details/vertex-and-crispr-therapeutics-co-develop-and-co-commercialize.
75. Wang, Xinyi, Guangqiang Sun, Teng Feng, Jing Zhang, Xun Huang, Tao Wang, Zuoquan Xie, et al. "Sodium Oligomannate Therapeutically Remodels Gut Microbiota and Suppresses Gut Bacterial Amino Acids-Shaped Neuroinflammation to Inhibit Alzheimer's Disease Progression." *Cell Research* 29, no. 10 (October 2019): 787–803. doi:https://doi.org/10.1038/s41422-019-0216-x.
76. Clarivate Analytics. "Web of Science." Accessed December 7, 2018. https://clarivate.com/products/web-of-science/.
77. Weede, Erich, and Sebastian Kämpf. "The Impact of Intelligence and Institutional Improvements on Economic Growth." *Kyklos* 55, no. 3 (August 2002): 361–80. doi:https://doi.org/10.1111/1467-6435.00191.
78. "What Are Clinical Trials? | ECMC." Accessed April 15, 2019. https://www.ecmcnetwork.org.uk/what-are-clinical-trials.
79. National Institutes of Health (NIH). "Who We Are." Accessed April 8, 2019. https://www.nih.gov/about-nih/who-we-are.
80. Crunchbase. "Who We Are." Accessed February 10, 2019. https://about.crunchbase.com/about-us/.
81. Zou, Qingjian, Xiaomin Wang, Yunzhong Liu, Zhen Ouyang, Haibin Long, Shu Wei, Jige Xin, et al. "Generation of Gene-Target Dogs Using CRISPR/Cas9 System." *Journal of Molecular Cell Biology* 7, no. 6 (October 12, 2015): 580–83. doi:https://doi.org/10.1093/jmcb/mjv061.

Chapter 6
Raising the Alarm on CRISPR

In an interview in 2019, Steven Walker, then the director of the Defense Advanced Research Projects Agency (DARPA), forecasted that genome editing is one of two biotech areas that could change the world—the other major biotech area is human-machine symbiosis. Walker made this prediction when he was asked: What is the next "big moonshot?" Walker responded that it would have to comparable to the internet,[1] and he pointed to the rapid development of technology in the biotech field. The potential for a technology to have as great of an impact on society as the internet, however, does not come without its risks. For genome editing technologies, Walker warned, "there is so much potential and so much opportunity for misuse."[2]

Genome editing technologies not only have the potential to be used by nefarious actors to modify traditional agents of concern to be more dangerous, but nefarious actors could also apply these technologies to create new types of biological weapons. Given the democratization of biology, which is allowing biology to become more accessible to everyone including nonexperts, there is growing concern over the number and type of actors who could potentially misuse genome editing.[3] It is important to raise awareness over the potential misuse of genome editing technologies, especially CRISPR, but this should be done within reason. By considering the technical capabilities enabled by genome editing technologies within the social context in which they are used, this study offers insight into the different types of genetically engineered biological weapons that different types of actors could reasonably achieve. This chapter explores the findings from the risk of misuse

[1] The DOD Advanced Research Projects Agency, now known as DARPA, invented the technology that paved the way for the internet.

[2] Jill Aitoro, "DARPA Head Resigns, Moving on to Industry," Defense News, December 17, 2019, accessed March 21, 2022, https://www.defensenews.com/breaking-news/2019/12/17/darpa-head-resigns-moving-on-to-industry/

[3] Kirkpatrick et al., "Editing Biosecurity: Needs and Strategies for Governing Genome Editing," 2; Tucker, "Review of the Literature on Dual Use," 25.

K. Paris, *Genome Editing and Biological Weapons*,
https://doi.org/10.1007/978-3-031-21820-0_6

assessment that was carried out in Chaps. 4 and 5 and explains the implications of these findings on the applications of misuse of genome editing technologies.

6.1 Overview of the Risk of Misuse Assessment

The purpose of this study was to determine the risk that genome editing technologies could be misused to develop biological weapons. The data gathered and analyzed for this assessment suggest that the risk of misuse of CRISPR is higher than the risk of misuse of ZFN and TALEN. Compared to ZFN and TALEN, CRISPR is highly accessible and easier to use. This is worrisome because it inevitably increases the pool of potential misusers of the technology. Tacit knowledge and the skill to use CRISPR, which provide a barrier to use, can be acquired through traditional means where students learn through direct, hands-on instruction in the lab (in-person classes) or through untraditional, remote ways (online videos). Furthermore, actors can apply kits to CRISPR protocols to further reduce knowledge and skill requirements; however, actors would need to know how to select and troubleshoot kits. Nevertheless, there are limits to how helpful DIY classes and videos on CRISPR, as well as kits that facilitate CRISPR protocols, are in developing biological weapons because actors must know how to apply these resources.

It is critical to keep in mind that, even with advancements in genome editing technologies, traditional technical obstacles still pose a barrier to biological weapons. While this is the case, CRISPR is maturing very quickly, which is concerning because it becomes more apparent how technologies could be misused as they mature. While ZFN has been around for over 20 years and is mature as well, the use of ZFN has been severely restricted. ZFN is hard to use, and a single company held most of the zinc-finger patents from an early point, which limited its spread and contributed to the inaccessibility of the technology.

6.1.1 Relative Risk of Misuse

The assessment for the risk of misuse for ZFN and TALEN is medium-to-low, and the assessment for the risk of misuse for CRISPR is high. Figure 6.1 illustrates the relative level of concern over the potential misuse of ZFN, TALEN, and CRISPR based on the four parameters that were studied in Chaps. 4 and 5. First, the risk of misuse based on accessibility (i.e., parameter 1) is higher for CRISPR than ZFN and TALEN. CRISPR reagents are inexpensive and globally available from many more providers. Acquisition of the CRISPR reagents is within the financial means of actors ranging from individuals-to-groups and from state-to-nonstate. More software is available, and continues to emerge, to ease the design process for CRISPR. There are many affordable and accessible options for sequencing to validate experimental

Parameter	Operationalizing the Parameter	Level of Concern		
		ZFN	TALEN	CRISPR
Accessibility	Ease of acquiring hardware and software			
	Cost			
	Dependence on other technologies			
Ease of Misuse	Skill and tacit knowledge needed to use technology			
	Ways to acquire skill and knowledge and skills			
	Transferability of knowledge and skill			
Magnitude of Potential Harm	Making pathogens more dangerous			
	Weaponizing the human microbiome			
	Weaponizing gene drives			
	Augmenting humans			
Imminence of potential misuse	Ability to be exploited based on maturity			

Fig. 6.1 Overview of the Level of Concern of ZFN, TALEN, and CRISPR. Regarding misuse, parameters that are of highest concern (i.e., are the most worrisome) are black, medium concern (i.e., are some cause for worry) are gray, and low concern (i.e., are less worrisome) are white

results. Even so, genome editing technologies need to be delivered to the target to work. Delivery remains a challenge for all technologies.

Second, the risk of misuse based on ease of misuse (i.e., parameter 2) is higher for CRISPR than ZFN and TALEN. The skill and tacit knowledge needed to use CRISPR are lower than the other technologies. It mainly takes basic molecular cloning skills and knowledge to build CRISPR plasmids. Additionally, there are ample resources available for help using CRISPR. There are an abundance of providers of services and forums, such as online videos and in-person classes, that are available to acquire skill and knowledge. Studies indicate that classes and videos are an effective way to transfer the types of knowledge and skill needed to use CRISPR.

Third, the risk of misuse based on magnitude of potential harm (i.e., parameter 3) is highest for CRISPR and lower for TALEN and ZFN. The findings of the assessment on magnitude of potential harm generally agree with the analysis from the 2018 *Editing Biosecurity* study. More traditional capabilities enabled by CRISPR, including making existing pathogens more dangerous, poses a high level of concern; however, these types of capabilities are enabled by other genetic engineering tools. Thus, they are less worrisome because they "do not present a fundamental change in the threat landscape."[4] On the other hand, new capabilities that are enabled by CRISPR—such as modifying the human microbiome, immune system, and genome—may pose a medium-to-low level of concern; however, they are new ways to cause harm—thus, they are more worrisome.[5]

Fourth, the risk of misuse based on imminence of potential misuse (i.e., parameter 4) is higher for CRISPR than TALEN. Compared to TALEN, CRISPR is widely accessible, used, and already moving into more advanced phases of clinical trials. While the imminence of potential misuse of ZFN is high given the technology has been maturing for over two decades, scientific publications, clinical studies, patent documents, and companies springing up based on CRISPR have quickly outnumbered those for ZFN.

6.1.2 Expanding the Pool of Potential Misusers

CRISPR is highly accessible and easier to use than ZFN and TALEN, which increases the pool of potential misusers of the technology. Given the affordability and accessibility, as well as the technical advantages, of CRISPR, the 2018 *Editing Biosecurity* report advised that CRISPR has been rapidly adopted by a range of actors. These include "established practitioners of molecular biology such as those working in government laboratories, large corporations, and universities," and "non-traditional citizens or do-it-yourself (DIY) scientists working in community labs and start-ups."[6] Data from this assessment support the finding that CRISPR is highly accessible and easier to use than ZFN and TALEN, which is concerning because it increases the pool of potential misusers of the technology. The results of this assessment confirm that genome editing reagents are no longer available from only a few big commercial players (e.g., Sigma-Aldrich, Thermo Fisher) for tens of thousands of dollars. Instead, CRISPR can now be acquired from many other providers located throughout the world and applied in experiments to edit a gene for as little as $20. This puts CRISPR within the reach of less well-endowed actors.

Providers, like Addgene, are making it easier to share materials. Per the 2018 *Editing Biosecurity* report: "Globalization and the emergence of centralized repositories such as Addgene that facilities sharing technology and know-how further reduces geographic and financial obstacles to accessing the technology."[7] This assessment provides new insight into the amount of time and money that Addgene has singlehandedly saved genome editing users. In September 2018, Addgene

[4] Kirkpatrick et al., "Editing Biosecurity: Needs and Strategies for Governing Genome Editing," 48–49.

[5] Kirkpatrick et al., 48–49.

[6] Kirkpatrick et al., 43.

[7] Kirkpatrick et al., 43.

reported that it costs at least $115 and five hours to make one plasmid.[8] This means that Addgene has saved academic and nonprofit labs that have purchased CRISPR plasmids at least $7.3 million and 83.5 years of cloning work versus only about $0.7 million and 8.0 years of cloning work for ZFN and TALEN plasmids, combined.[9] Not only does the CRISPR technology itself significantly reduce the amount of time that it takes to achieve modifications, but providers of reagents and services are also saving users time or money, or both. Unlike ZFN and TALEN, CRISPR users can get more done in a shorter amount of time—this accelerates the timeline that it takes to develop a bioweapon.

Moreover, the data from Addgene supports larger trends that labs are opting to use CRISPR in lieu of ZFN and TALEN. Analysis of the Addgene data reveals that the requests for units of CRISPR continue to increase, while those for units of ZFN and TALEN continue to decrease. This suggests that labs will continue to use CRISPR and nearly, or completely, stop the use of ZFN and TALEN in the next few years. It is noted that, given the restriction that most of Addgene's inventory is for academic and nonprofit labs, this trend applies to academic and nonprofit labs and not to commercial labs. Nevertheless, an analysis of data from this assessment indicates that there are many more providers of CRISPR reagents than ZFN and TALEN reagents, and providers of ZFN and TALEN reagents are phasing out the tool.

Another interesting observation is the noticeable increase in the number of academic providers of CRISPR products and services compared to those of ZFN and TALEN products and services (see Chap. 4, Parameter 1). This is perhaps a manifestation of the underlying "permissive intellectual property environment" in which CRISPR emerged that has made the technology readily available across the industry and academia.[10] Although there is a heated battle between the Broad Institute and University of California-Berkeley over foundational CRISPR patent rights, this does not seem to slow down the uptake of CRISPR. Two factors continue to facilitate access to CRISPR: the ease of academic and non-profit organizations acquiring CRISPR plasmids made possible through Addgene, and the policy of the Broad Institute—which holds foundational CRISPR patents in the United States—to share CRISPR.[11] It is noted that the Broad Institute makes "CRISPR tools, knowledge, methods and other IP for genome-editing freely available for research" by the

[8] Jennifer Tsang, "The Time and Cost Required to Make a Plasmid," Addgene, accessed October 24, 2018, https://blog.addgene.org/the-time-and-cost-required-to-make-a-plasmid

[9] Addgene distributed 146,341 units of CRISPR between 2012 and 2018. This calculation assumes the cost to purchase each plasmid from Addgene is $65.

[10] Kirkpatrick et al., "Editing Biosecurity: Needs and Strategies for Governing Genome Editing," 29.

[11] Kirkpatrick et al., 29.

academic and non-profit communities, and the Broad Institute licenses CRISPR-IP non-exclusively to companies for research.[12]

CRISPR has not suffered from the same fate as ZFN whereby a single company, Sangamo, pursued an aggressive strategy early in the development of the technology to hold most of the zinc-finger intellectual property.[13] As a result, Sangamo ended up owning or exclusively licensing most of the zinc-finger patents. Unfortunately, due to Sangamo's patent monopoly, many scientists, particularly those in academia, who wanted to use zinc fingers were unable to access Sangamo's closely held proprietary information. Scientists came together and formed the Zinc Finger Consortium in 2005, which began to publish open-access protocols on zinc-finger protein engineering and make reagents available through Addgene.[14] As revealed in Chap. 4, it was generally observed that the build-it-yourself ZFN reagents did not work as well as what users could purchase. While Sangamo licensed[15] its intellectual property to Sigma-Aldrich (now, Millipore Sigma) to allow the development and commercialization of zinc-finger products for research applications, these products cost thousands of dollars and were generally too expensive for academic researchers.[16]

Although CRISPR is more globally available than ZFN or TALEN, it is important not to overlook the "complicated ways in which individuals, groups, and states take up and use materials and equipment"—an important hallmark of the biosocial frame.[17] For instance, an actor may have tried and failed to use ZFN or TALEN, or both, which could influence the actor's decision not to use CRISPR. Or, an actor may want to continue to use established methods for engineering agents that have been described in the literature, such as the Cre/*lox* system for modifying *Bacillus anthracis*, the causative agent of anthrax.[18] Since this study assessed the general risk of misuse for all types of actors, it leaves some uncertainty as to whether each actor will be able to acquire and then make the ultimate decision to misuse CRISPR to

[12] "Information about Licensing CRISPR Systems for Clinical Use," Broad Institute, accessed October 24, 2019, https://www.broadinstitute.org/partnerships/office-strategic-alliances-and-partnering/information-about-licensing-crispr-genome-edi

[13] Pearson, "Protein Engineering," 161–62.

[14] Pearson, 164.

[15] ZFN licensing agreement for material purchased through Millipore Sigma is available on the website. "Zinc Finger Nuclease (ZFN) License Agreement," Sigma-Aldrich, accessed November 15, 2018, https://www.sigmaaldrich.com/life-science/zinc-finger-nuclease-technology/zfn-license-agreement.html

[16] Subhashini Chandrasekharan et al., "Proprietary Science, Open Science and the Role of Patent Disclosure: The Case of Zinc-Finger Proteins," *Nature Biotechnology* 27, no. 2 (February 2009): 140–42, https://doi.org/10.1038/nbt0209-140; DeFrancesco, "Move over ZFNs," 682; Jai Nagarkatti, "Sangamo Lands $20 M in Expanded Sigma Deal," San Francisco Business Times, October 5, 2009, accessed November 15, 2018, https://www.bizjournals.com/sanfrancisco/stories/2009/10/05/daily2.html

[17] Vogel, *Phantom Menace or Looming Danger? A New Framework for Assessing Bioweapons Threats*, 28.

[18] See Chap. 5, Implications of Misuse for Specific Applications of Genome Editing, Altering Microbes.

develop bioweapons. Thus, when applying the findings of this assessment to a specific actor, additional data on local factors (e.g., the actor's ideology, objectives, and attitude toward genome editing) that may affect acquisition and use should be gathered and analyzed.

6.1.3 Overcoming Tacit Knowledge and Skill Barriers

Tacit knowledge and skill to use CRISPR, which provide a barrier to use, can be acquired through traditional means where students learn through direct, hands-on instruction in the lab (in-person classes) or through untraditional, remote ways (online videos). For actors who want to acquire skill and knowledge to use CRISPR, there are resources, including classes and videos, to help. A significant difference between CRISPR and its predecessor technologies is that CRISPR is used and taught in spaces where actors have low, or even no, understanding of biology and lab skills. Research from this assessment revealed that the workshops where ZFN and TALEN were taught were geared toward professionals and PhD-level researchers with at least a basic understanding of molecular biology and genetics as well as intermediate laboratory skills. While actors can still learn CRISPR in these professional-level workshops, they can also acquire skill and knowledge in DIY bio classes and undergraduate classes.

The research from this assessment indicates that DIY bio classes and undergraduate classes are an effective way to learn how to use CRISPR. Approximately half of the DIY bio class at Genspace and half of the undergraduate class at the University of New Mexico were successful in using CRISPR to achieve knockouts. However, this statistic, at least in the case of the DIY bio class, only tells half the story. Analysis of the Genspace lab in Chap. 4, Parameter 2, Step 3, reveals that the efficiency for ligating the gRNA target sequence to the plasmid was expected to be about 50%. For Step 3, 40% of students in the Genspace class successfully built the CRISPR plasmid. Furthermore, analysis of the Genspace lab in Chap. 4, Parameter 2, Extra Step, reveals that only 60% of the Genspace class successfully built the repair template. Given that this was a class, the instructor ensured 100% of students had the correctly assembled CRISPR plasmid and repair template before moving on to Step 4.

If actors were performing the Genspace lab outside of a classroom then they would have needed to troubleshoot the failed steps. A critical aspect of troubleshooting is having the experience to figure out what went wrong and fix it. Troubleshooting these steps would have been very challenging for students who lacked lab experience, knowledge of biology, and skills. As noted in Chap. 4, Parameter 2, Extra Step, students who did not successfully prepare the repair template tended to have the least amount of lab skills and experience. In this case, students were inexperienced with the micropipette and did not know it was broken. It took a highly skilled and experienced student to identify the equipment had malfunctioned.

This is not to say, however, that these students could not have figured out the issue—it just would have taken more time and effort than an actor who is skilled and experienced. There are many videos and guides online for how to use a micropipette. For instance, JoVE offers a video that explains how the micropipette works.[19] Additionally, many micropipette videos are available on YouTube that have hundreds of thousands of views. A popular YouTube micropipette video from the University of Leicester has over 439,000 views, and another popular YouTube video from Eppendorf, a biotech company that manufacturers micropipettes, has over 215,000 views.[20] There are also many YouTube videos available on micropipette troubleshooting and maintenance such as a video from the Arizona Department of Health Services on calibration and cleaning, which has over 131,000 views.[21] These videos are likely an effective way to transfer knowledge and skill given the data from this assessment that indicated JoVE videos could transfer lab knowledge and skill (see Chap. 4, Parameter 2, Transferability of skill and knowledge).

There are also videos that demonstrate how to use CRISPR in protocols. Data on JoVE video article protocols reveals that there are more CRISPR videos, with more overall views, than ZFN and TALEN videos (see Chap. 4, Parameter 2, Ways to acquire skill and knowledge, Forums). While this may be the case, an argument might claim that the videos are not an effective way to transfer skill and knowledge since an instructor is not present and not able to give feedback to either reinforce or correct the viewer's actions. The research from this assessment on students from DeSales University and Clemson University who watched *JoVE* videos, however, indicates that the videos are effective in transferring the types of skill and knowledge needed to use CRISPR. This is concerning because it does not fit the traditional mold of how laboratory skill and knowledge are transferred through long-term, in-person, master-apprentice relationships.

6.1.4 Benefits and Drawbacks of Kits

Kits further reduce knowledge and skill requirements to use CRISPR; however, kits have specific functions (i.e., not a one size fits all), and it takes knowledge to select and troubleshoot kits. Kits that are available to purchase from commercial providers reduce knowledge and skill requirements during the CRISPR workflow. Kits are

---

[19] JoVE Science Education Database, *General Laboratory Techniques. An Introduction to the Micropipettor* (Cambridge, MA, 2017), accessed November 17, 2017, https://www-jove-com.mutex.gmu.edu/science-education/5033/an-introduction-to-the-micropipettor

[20] YouTube, *Using a Micropipette - University of Leicester*, 2009, accessed November 17, 2017, https://www.youtube.com/watch?v=uEy_NGDfo_8; *How to Pipette Correctly – a Short Step-by-Step Introduction into Proper Pipetting I Eppendorf*, 2017, accessed October 26, 2019, https://www.youtube.com/watch?v=QGX490kuKjg

[21] *Pipette Calibration and Cleaning I Arizona Department of Health Services*, 2013, accessed October 26, 2019, https://www.youtube.com/watch?v=MBq55FtOzN4

typically thought of as a black box, which means that users do not have to have any knowledge of the inner workings of the device for it to work. The black boxing of biotechnology is a hallmark of the biotech revolution frame. The trouble with black boxes, however, is that they place the focus on the material end product instead of the messy and contingent processes that it takes to work through problems to arrive at the result.[22]

Bitesize Bio is one platform that is dedicated to helping scientists share the know-how that they acquire in the lab. Per Bitesize Bio, every molecular biologist ought to know the following: "How specific techniques work, what actually *is* in those kits we use and why we use certain approaches rather than others."[23] Furthermore, Bitesize Bio suggests, "these nuggets of knowledge can mean the difference between understanding where an experiment is going wrong and not understanding, between getting a result and not getting one."[24]

Kits have specific functions and are not a one size fits all. An analysis of the Genspace lab CRISPR experimental workflow in Chap. 4, Parameter 2 demonstrates that three different kits were used (see Table 4.4). The Monarch® Plasmid Miniprep Kit (NEB T1010) was used to lyse *E. coli* and isolate the plasmids from the cells. The Monarch® PCR & DNA Cleanup Kit (NEB T1030) was used to purify the repair template (DNA) from the PCR. Lastly, the Frozen-EZ Yeast Transformation II Kit™ (Zymo T2001) facilitated the uptake of the CRISPR plasmid and repair template by yeast.

While the simplicity of using kits is often praised, it takes knowledge to know which kit to select. For instance, when selecting a DNA isolation kit, Bitesize Bio advises that the following factors influence which kit to choose: the sample the user is working with (e.g., soil sample, blood sample, clinical sample, PCR or agarose gel), whether the user is isolating genomic DNA or plasmid DNA, the cell line or organism (e.g., bacteria, animal, plant, or blood), and the amount needed to process the sample. Kits can remove hemoglobin and humic acid contaminates from blood samples and soil samples, respectively, that would interfere with downstream procedures such as PCR. Kits that isolate plasmid DNA are gentler than those that isolate genomic DNA. The lysis step for kits used on animal cells, which lack a cell wall, is typical gentler than bacteria cells, which can be harder to break open.[25] New

---

[22] Vogel, *Phantom Menace or Looming Danger? A New Framework for Assessing Bioweapons Threats*, 23–24.

[23] Nick Oswald et al., *10 Things Every Molecular Biologist Should Know* (UK: Science Squared Ltd, 2012), ii, http://bitesizebio.s3.amazonaws.com/wp-content/uploads/2017/07/10124646/10-Things-Every-Molecular-Biologist-Should-Know.pdf

[24] Oswald et al., ii.

[25] Adrienne Huntress, "Picking the Right DNA Isolation Kit for Your Application," Bitesize Bio, February 7, 2019, accessed October 25, 2019, https://bitesizebio.com/42930/picking-the-right-dna-isolation-kit-for-your-application/

England BioLabs, which was used to source kits for the Genspace lab, offers a variety of nucleic acid purification kits for a range of sample types and organisms.[26]

In addition to knowing how to select kits, users should know how to troubleshoot kits if things go wrong. There are resources available online to help with troubleshooting. For instance, New England BioLabs provides several technical guides and tips for troubleshooting the nucleic acid purification kits used for the Genspace lab (NEB T1010 and NEBT1030).[27] Even so, the availability of troubleshooting guides does not guarantee that all issues are covered. For instance, the UltraClean Microbial DNA Isolation Kit from Mo Bio Laboratories, Inc., which has been used to isolate DNA from *B. anthracis*, has a troubleshooting guide that seems to offer limited assistance.[28] There are complaints that the manual for the UltraClean Kit is too short and simple, the troubleshooting guide is too basic, and suggestions for alternative methods may not work.[29] Actors who have come across these types of issues before and are more experienced, as well as those who can consult with experts, are more likely to successfully troubleshoot, especially when issues are not covered in troubleshooting guides.

6.1.5 Constraints on Applying Learning Tools and Experimental Aids

There are limits to how helpful DIY classes and videos on CRISPR, as well as kits that facilitate CRISPR protocols, are in developing biological weapons because actors must know how to apply these resources. The DIY bio class and videos on CRISPR, as well as kits that can be used to facilitate protocols with CRISPR, only help to a certain extent. As Chap. 4, Parameter 2, revealed, the DIY bio class offered

[26] "Protocols for Nucleic Acid Purification," New England BioLabs, accessed October 25, 2019, https://www.neb.com/products/nucleic-acid-purification/nucleic-acid-purification#tabselect2

[27] "Troubleshooting Guide for Cloning | NEB," accessed March 14, 2019, https://www.neb.com/tools-and-resources/troubleshooting-guides/troubleshooting-guide-for-cloning; "Troubleshooting Guide for DNA Cleanup and Plasmid Purification," New England BioLabs, accessed October 26, 2019, https://www.neb.com/tools-and-resources/troubleshooting-guides/troubleshooting-guide-for-dna-cleanup-and-plasmid-purification; "Monarch® Plasmid Miniprep Kit (T1010): Troubleshooting," New England BioLabs, n.d., accessed October 26, 2019, https://www.neb.com/products/t1010-monarch-plasmid-miniprep-kit#Product%20Information; New England Biolabs, "Monarch® PCR & DNA Cleanup Kit (5 Mg) (T1030): Troubleshooting," n.d., accessed October 26, 2019, https://www.neb.com/products/t1030-monarch-pcr-dna-cleanup-kit-5-ug#FAQs%20&%20Troubleshooting_Troubleshooting

[28] "MO BIO's UltraClean Microbial DNA Kit Handbook - QIAGEN," accessed October 26, 2019, https://www.qiagen.com/us/resources/resourcedetail?id=cda265c5-c1f0-4c99-acee-7282fc6aa371&lang=en

[29] Katrina Williams, "MoBio Laboratories' UltraClean Microbial DNA Isolation Kit," Biocompare, September 17, 2004, http://www.biocompare.com/Product-Reviews/41081-MoBio-Laboratories-UltraClean-Microbial-DNA-isolation-kit/

by Genspace taught students how to use CRISPR to knockout a well-known gene (*ade2*) in a well-studied model organism (yeast). It is widely known in the literature that the knockout of *ade2* will result in a pigment change from white to red. Furthermore, it is widely known how to get started with yeast, grow cultures from scratch, and maintain the organism.[30] In short, CRISPR was applied to knockout a specific gene in a specific organism. In order to apply this to develop biological weapons, actors would have to know how to acquire, isolate, and culture the agent; what part of the genetic code to modify and in what way to modify it using CRISPR; and how to produce and disseminate the weapon. This can be a challenge if the actor is not experienced with the agent; if the information does not exist in the literature; and if the possession, use, and transfer of the agent is controlled by the United States Department of Health and Human Services (HHS) and United States Department of Agriculture (USDA) Select Agents and Toxins List.[31]

Even culturing an agent can be difficult for inexperienced actors. In the context of microbiology, culture describes "the propagation of microorganisms in growth medium" in the laboratory.[32] An extensive handbook on cell culture from Thermo Fisher Scientific covers the basics of cell culture, but it advises how "cell culture conditions vary widely for each cell type" and straying from the ideal culture conditions for that cell type can lead to the expression of different phenotypes.[33] The handbook suggests that it is merely a starting point, and users should review the literature, books, manuals, and product information sheets for more in-depth information.[34] The American Type Culture Collection, ATCC, a nonprofit organization in Manassas, Virginia that sells cell cultures, microorganisms, and other biological products, also has extensive guides on culturing (e.g., animal cell culture, bacterial culture, mycology culture, virology culture, protistology culture (protozoa and algae), stem cell culture, and others).[35] These guides demonstrate that culturing agents is not a one size fits all.

Even with this material available, there are still uncertainties in getting started with culturing agents that scientists have voiced. For example, a member on ResearchGate, a social networking tool for researchers, asked the community for help in establishing a viral culture in July 2017. Another member replied with advice on equipment and offered the following insight: "Unfortunately it would take hours to list reagents (and would depend on the cell line), but you will burn through cell

[30]"Growth and Maintenance of Yeast in the Lab | Protocol," accessed October 17, 2019, https://www-jove-com.mutex.gmu.edu/science-education/5095/yeast-maintenance

[31]"HHS and USDA Select Agents and Toxins 7CFR Part 331, 9 CFR Part 121, and 42 CFR Part 73," accessed April 25, 2018, https://www.selectagents.gov/selectagentsandtoxinslist.html

[32]William C Shiel, "Medical Definition of Culture," MedicineNet, accessed October 18, 2019, https://www.medicinenet.com/script/main/art.asp?articlekey=2879

[33]Thermo Fisher Scientific, "Cell Culture Basics Handbook," 2016, 1, accessed October 29, 2019, https://www.thermofisher.com/content/dam/LifeTech/Documents/PDFs/PG1563-PJT1267-COL31122-Gibco-Cell-Culture-Basics-Handbook-Global-FLR.pdf

[34]Thermo Fisher Scientific, 18.

[35]"Culture Guides," ATCC, n.d., accessed October 29, 2019, https://www.atcc.org/en/Guides.aspx

culture flasks, cell culture plates, tips, stripettes and Falcon tubes (none of which are cheap)."[36]

It is also important to point out that the DIY bio class did not teach students how to modify agents of concern or containment strategies for working with highly pathogenic agents. Containment refers to the "safe methods, facilities and equipment for managing infectious materials in the laboratory environment where they are being handled or maintained biosafety levels."[37] The four levels of containment are referred to as biosafety levels. There are different practices, safety equipment, and other safeguards at laboratory facilities for handing agents at each biosafety level.[38] The Genspace lab follows guidelines for biosafety level 1, which is the most basic level of containment and suitable for handling agents that do not cause human diseases.[39] This level of containment, however, is too low to handle any of the agents of concern.

Furthermore, the plasmid that was used for the Genspace study worked for doing CRISPR in yeast, but it only worked in yeast. If an actor wanted to engineer another organism, such as a plant, then the actor would need to do research to determine the techniques used in the organism. The actor could take apart the plasmid pmL107, which was used to modify yeast during the Genspace lab, and then use elements of it to build other plasmids. For instance, there is a working Cas9 gene on the plasmid pmL107. Although the promoter and terminator of the plasmid pmL107 are meant for yeast, actors could take Cas9 out of the plasmid using advanced techniques and clone it onto a different plasmid that contained the promoter and terminator for the new agent. This type of work, however, is not easy for actors with limited laboratory skills, molecular biology knowledge, and experience. Actors would need proficient lab skills and knowledge as well as troubleshooting expertise.

Additionally, JoVE videos only help to an extent. As James Revill and Catherine Jefferson point out, "tacit knowledge transfer through visualization and the process of viewing only gets you so far."[40] Videos are limited by what the editors decide to include. While videos can transfer some forms of tacit knowledge, they cannot help viewers to transfer somatic tacit knowledge that comes with the development of "physical skills and muscle memory required to execute certain actions."[41] Lastly, and perhaps most importantly for this discussion, there are no JoVE videos on how to use CRISPR to develop a biological weapon. While different parts of video demonstrations may help in biological weapons development, it would take "a

---

36 "What Are the Basic Needs to Establish a Viral Cell Culture Lab?," ResearchGate, accessed October 17, 2019, https://www.researchgate.net/post/What_are_the_basic_needs_to_establish_a_viral_cell_culture_lab

37 Centers for Disease Control and Prevention and National Institutes of Health, *Biosafety in Microbiological and Biomedical Laboratories*, 5th ed. (US Department of Health and Human Services, 2009), 22.

38 Centers for Disease Control and Prevention and National Institutes of Health, 4, 24.

39 Centers for Disease Control and Prevention and National Institutes of Health, 25.

40 Revill and Jefferson, "Tacit Knowledge and the Biological Weapons Regime," 605.

41 Revill and Jefferson, 605.

great deal of dedicated effort to identify and assimilate all the necessary steps required to do something serious."[42] An actor could learn the basic process of how to use CRISPR, but the user would then have to apply this knowledge, which is difficult.

Kits, like the ones used for the Genspace lab, can be used to facilitate experiments involving agents of concern. Actors would need to consult the literature and manufacturers' websites to figure out what kit to select. For instance, when using CRISPR in *B. anthracis*, as described later in this chapter, Wang and colleagues isolated DNA using the Wizard Genomic Purification Kit that is available for purchase from Promega in Madison, Wisconsin.[43] The Wizard Genomic Purification Kit is a solution-based method to isolate DNA from a variety of animals, plants, yeast (including *S. cerevisiae*), and bacteria. In another example, Dale and colleagues used the UltraClean Microbial DNA Isolation Kit purchased from Mo Bio Laboratories, Inc. to isolate DNA from *B. anthracis*.[44]

6.1.6 Persistence of Technical Barriers to Biological Weapons

Despite advancements in genome editing technologies, traditional technical obstacles still pose a barrier to biological weapons. The application of CRISPR to develop biological weapons is still limited by several technical barriers including a lack of scientific understanding, off-target effects, delivery challenges, and low rates of homologous recombination. In spite of these limitations, the multiplexibility of CRISPR is superior to its predecessor technologies, which allows scientists to now make many simultaneous modifications.

First, as revealed in Chap. 3, it is important not to overlook the barriers presented by the state of scientific knowledge. Scientists may just not know the genetic determinants underlying phenotypes. As Ellen Jorgensen, a leader in DIY biology and co-founder of Genspace, advised, CRISPR is only easy "once we figure out all the details for a given system."[45] Unfortunately, many of these details may just not be known.

Additionally, cell lines may have trouble tolerating Cas9. Leigh Brody underscores this issue in a webinar on designing sgRNAs for CRISPR. Brody advised that

---

[42] Revill and Jefferson, 605.

[43] Yanchun Wang et al., "Highly Efficient Genome Engineering in Bacillus Anthracis and Bacillus Cereus Using the CRISPR/Cas9 System," *Frontiers in Microbiology* 10 (August 27, 2019), https://doi.org/10.3389/fmicb.2019.01932

[44] Jennifer L. Dale et al., "A Dual Role for the Bacillus Anthracis Master Virulence Regulator AtxA: Control of Sporulation and Anthrax Toxin Production," *Frontiers in Microbiology* 9 (2018), https://doi.org/10.3389/fmicb.2018.00482

[45] Jorgensen, *What You Need to Know about CRISPR.*

CRISPR experiments may seem easy based on the literature, but some troubleshooting is needed. Scientists should check to make sure the cell line can tolerate Cas9, and they should make sure that the delivery method will work that that particular cell line.[46] In a September 2019 publication, Tycko and colleagues recognize that off-target effects can "result in a major fitness effect due to cellular toxicity as a result of activation of the DNA damage response and apoptosis."[47] Unfortunately, off-target effects are a known issue with CRISPR. This issue, however, is likely to decline as scientists gain more experience and expertise in developing techniques to reduce off-target effects.

If users are not just knocking out a gene using CRISPR, but are trying to replace a gene via HDR, then the rate of homologous recombination can be low. As the Genspace study demonstrated, the rate of the final integration of the *gfp* gene from the repair template was low. Low success rates of gene integration are observed by Caixia Gao, a plant scientist at the Chinese Academy of Sciences who is using CRISPR to engineer crops. In an interview with Gao, she reported that CRISPR could efficiently knock out genes, but her team is "not so good at" efficiently adding new genes—adding that the success rate is low (about 1%) and no one is good at it.[48] Nevertheless, as Gao's lab and others continue to work to improve this issue, it is likely that the success rates of adding new genes will improve.

Despite these technical obstacles, CRISPR does allow actors to modify many genes more easily at one time. A hallmark study that demonstrated the multiplexibility of CRISPR was carried out by Yang and colleagues in 2015. By using two CRISPR-Cas9 gRNAs, the researchers knocked out 62 copies of porcine endogenous retroviruses from porcine kidney epithelial cell line (PK15) all at one time.[49] Prior to this study from Yang and colleagues, researchers only knocked out a few genes at a time—62 genes seemed beyond what was possible.[50]

The study from Yang and colleagues was significant because the researchers knocked out all the porcine endogenous retrovirus genes from the pigs, so these genes would not be expressed if pig organs are transplanted into humans. Unfortunately, viruses are a significant barrier to xenotransplantation, which the FDA describes as: "any procedure that involves the transplantation, implantation or infusion into a human recipient of either (a) live cells, tissues, or organs from a

---

[46]Desktop Genetics, *Custom Designed CRISPR SgRNA Libraries - Webinar*, 2016, accessed March 14, 2017, https://www.youtube.com/watch?v=ghEai9Vtcoc

[47]Josh Tycko et al., "Mitigation of Off-Target Toxicity in CRISPR-Cas9 Screens for Essential Non-Coding Elements," *Nature Communications* 10, no. 1 (September 6, 2019): 1–14, https://doi.org/10.1038/s41467-019-11955-7

[48]Jon Cohen, "To Feed Its 1.4 Billion, China Bets Big on Genome Editing of Crops," Science | AAAS, July 29, 2019, accessed September 29, 2019, https://www-sciencemag-org.mutex.gmu.edu/news/2019/07/feed-its-14-billion-china-bets-big-genome-editing-crops

[49]Luhan Yang et al., "Genome-Wide Inactivation of Porcine Endogenous Retroviruses (PERVs)," *Science* 350, no. 6264 (November 27, 2015): 1101–4, https://doi.org/10.1126/science.aad1191

[50]Church, "The Future of the Genomic Editing Revolution" (UC Berkeley, January 26, 2017), accessed June 28, 2019, https://www.youtube.com/watch?v=Wswvf8Nrubg

nonhuman animal source, or (b) human body fluids, cells, tissues or organs that have had ex vivo contact with live nonhuman animal cells, tissues or organs."[51] George Church explains there were issues with xenotransplantation in the early 2000s because "pig organs were producing viruses that could infect the immune compromised recipient of those organs ... you don't want to have swine flu, or the equivalent, evolving in your immune compromised patient."[52] As this study shows, scientists can now achieve modifications that just could not be done before.

6.1.7 Technology Maturation and Risk of Misuse

Although CRISPR is younger than ZFN and TALEN, CRISPR is maturing very quickly. The maturation of CRISPR concerning because the more a technology matures, the higher its risk of misuse. Data from Chap. 5 revealed that there are many more scientific publications, patent documents, clinical studies, and startup companies for CRISPR than its predecessors. Furthermore, CRISPR projects are receiving more Federal and private funding. The NIH, a lead funder of CRISPR-related projects, has already spent more on CRISPR research than ZFN or TALEN research. Moreover, the NIH is providing more funding for SBIR and STTR programs, which are programs that fund initiatives with strong commercialization potential, that involve CRISPR. The 2018 *Editing Biosecurity* report found that there is a lot of private funding for CRISPR as well as all biotechnologies, which is concerning because there are loopholes in oversight of research with dual use potential that is privately funded. Recommendations for closing oversight loopholes are considered later in Chap. 7.

A major concern of a technology like CRISPR that is advancing so rapidly is that, as the technology matures, the higher its potential risk of misuse. In contrast to a mature technology, a less mature technology that has not yet fully formed has a lower risk of misuse.[53] Less mature technologies can be more difficult to use because they need more expertise, and the expertise may be more specialized, arcane, and limited. Furthermore, less mature technologies may have more kinks to work out, the capabilities and limitations of the technologies may not yet be well understood, and the technologies may not be as widely available. As the end of Chap. 3 reveals, there are already many concerns over the ways that CRISPR could be misused (i.e., alter microbes, weaponize gene drives, and augment humans). As scientists continue to develop and use CRISPR as a tool for genome editing, it becomes clearer how likely these applications of misuse could be achieved using CRISPR. Below, in the second

[51] Center for Biologics Evaluation and Research, "Xenotransplantation," FDA, March 28, 2019, http://www.fda.gov/vaccines-blood-biologics/xenotransplantation

[52] Church, "The Future of the Genomic Editing Revolution," secs. 47th–50th minute.

[53] Tucker, "The Decision Framework," 70.

half of this chapter, the likelihood that CRISPR could achieve these applications of misuse is evaluated.

6.2 Implications of Misuse for Specific Applications of Genome Editing

In addition to being misused in a traditional way to make pathogens more dangerous, CRISPR could be misused in new ways to cause harm (e.g., targeting the human microbiome, gene drives, and human augmentation). The findings from the risk of misuse assessment are applied here to evaluate the implications of misuse for specific applications of genome editing, including making pathogens more dangerous, weaponizing the human microbiome, weaponizing gene drives, and augmenting humans. Each section includes an evaluation as to whether these applications of misuse are near-term, mid-term, or long-term threats.

6.2.1 Making Pathogens More Dangerous

There are several challenges that actors will have to overcome to make pathogens more dangerous using genome editing. There are properties that are inherent to viruses and bacteria that make them difficult to modify; gaps in our knowledge of the genetic determinants that underlie phenotypes; difficulties in transferring knowledge learned in one microbe to another microbe; and challenges acquiring agents that pose a severe threat to humans, animals, and plants because the transfer of such agents is controlled. Given these constraints, this section explores how (1) CRISPR is more likely to be applied to make bacteria, not viruses, more dangerous, (2) CRISPR is likely to target genotypic modifications for phenotypes that are well-understood due to knowledge gaps, (3) knowledge used to modify one microorganism with CRISPR does not necessarily translate to another even closely related microorganism, (4) online design tools are likely to be of limited help, and (5) acquisition challenges for traditional agents of concern are likely to pose a barrier especially for non-state actors.

Given the inherent properties of viruses and bacteria, CRISPR is more likely to be used to modify bacteria—and even large DNA viruses, but not RNA viruses—to be more dangerous. Compared to bacterial genomes, the size of viral genomes is much smaller: "viruses have more fitness pressures and constraints on genome packaging to keep the genome smaller."[54] Due to the restrictions on viral genome organization, mutations that are introduced to a virus often result in a less pathogenic (i.e.,

[54] National Academies of Sciences, Engineering, and Medicine, *Biodefense in the Age of Synthetic Biology*, 52.

attenuated) virus.[55] This is not as much of a problem in bacteria. The genomes of bacteria are genetically more stable, so genetic modifications generally do not attenuate over time.[56] Moreover, most alterations to viruses can be accomplished through technologies that already exist (i.e., standard recombinant DNA technology methods), and, given advancements in synthetic biology, it may make more sense to engineer small viral genome from scratch rather than to use CRISPR.[57] RNA viruses also do not have the "genetic repair mechanisms that CRISPR takes advantage of for genome editing."[58]

The potential for CRISPR to modify viruses and bacteria has been demonstrated. In 2015, the vaccinia virus (VACV), a vector for vaccines that is closely related to the smallpox virus, was rapidly and efficiently modified using CRISPR. Using CRISPR, the homologous recombination in a region of the VACV genome, which plays a role in virulence and host response to the virus, increased approximately 12–60 times.[59] While off-target effects are problematic for CRISPR, Yuan and colleagues reported that there were no "evident off-target effects on the VACV genome."[60] Furthermore, Yuan and colleagues demonstrated the multiplexibility (i.e., making multiple modifications at one time) of CRISPR in VACV. Using CRISPR, Yuan and colleagues were able to simultaneously and rapidly delete two genes in 60% of VACV.[61]

In addition to VACV, the potential to use CRISPR in the bacterial genomes has been demonstrated. In 2017, Zhang and colleagues inserted antibiotic resistance into the bacterium *Escherichia coli*.[62] In 2019, Wang and colleagues used CRISPR to efficiently generate precise point mutations as well as large deletions in *Bacillus anthracis*, the causative agent of anthrax. Wang and colleagues advised that it is far simpler to generate large deletions using CRISPR than the previously reported method that used the Cre/*lox* system.[63] The study from Wang and colleagues underscores a point made in the 2018 *Editing Biosecurity* report that the knowledge used to modify one microorganism does not necessarily translate to another even

[55] National Academies of Sciences, Engineering, and Medicine, 48.

[56] National Academies of Sciences, Engineering, and Medicine, 52.

[57] National Academies of Sciences, Engineering, and Medicine, 48; Kirkpatrick et al., "Editing Biosecurity: Needs and Strategies for Governing Genome Editing," 51.

[58] Kirkpatrick et al., "Editing Biosecurity: Needs and Strategies for Governing Genome Editing," 51.

[59] Ming Yuan et al., "Efficiently Editing the Vaccinia Virus Genome by Using the CRISPR-Cas9 System," *Journal of Virology* 89, no. 9 (May 1, 2015): 5176, https://doi.org/10.1128/JVI.00339-15

[60] Yuan et al., 5179.

[61] Yuan et al., 5177.

[62] Hong Zhang et al., "A Novel and Efficient Method for Bacteria Genome Editing Employing Both CRISPR/Cas9 and an Antibiotic Resistance Cassette," *Frontiers in Microbiology* 8 (May 5, 2017), https://doi.org/10.3389/fmicb.2017.00812

[63] Wang et al., "Highly Efficient Genome Engineering in Bacillus Anthracis and Bacillus Cereus Using the CRISPR/Cas9 System," 7.

closely related microorganism.[64] In this case, conditions used to induce expression of Cas9 in *B. anthracis* and the closely related *B. cereus* were noticeably different.[65]

Given the limited understanding of genotypes that underlie viral and bacterial phenotypes, CRISPR is more likely to be used to modify well-understood traits (e.g., resistance to antibiotics and antivirals, or toxin production) and less likely to be used to modify poorly understood, complex traits (e.g., tropism, transmissibility, production, and delivery). As the 2018 *Editing Biosecurity* report advises, the genotypic modifications needed to confer resistance to antibiotics and antivirals are "well-known" and the "genetic selection schemes are easy to design and execute."[66] Other traits, such as tropism and transmissibility, may be the result of a poorly understood interaction of many genes, or from an interaction of pathogen and host factors that are not easy to change through modifying the pathogen alone. Moreover, modifying one trait may influence other traits and lead to the reduced fitness (i.e., the ability to survive and reproduce) of an organism.[67] There is still a lack of knowledge on traits relating to the scaling-up the production and delivery of agents, so these traits are less likely to be targeted by CRISPR.[68]

As studied in Chap. 4, computational tools are available, and becoming increasingly more available, to help users design gRNA for CRISPR. The 2018 *Editing Biosecurity* report advises that "there is no well-developed standard for evaluating CRISPR sgRNA design for bacteria and archaea, which complicates the design process and the use of CRISPR for even conventional knockout and knock-in applications for these organisms."[69] Furthermore, it would take time and energy to create such a tool for each specific strain of agent.[70] Even so, online design tools from companies that offer services to build gRNA may help to some degree. In a publication from August 2019, Wang and colleagues used the sgRNA design tool that is available online from Integrated DNA Technologies (IDT) and modified *Bacillus anthracis*, the causative agent of anthrax, and *Bacillus cereus*, a closely related bacterium, with CRISPR.[71] Recommendations for mitigating gaps in screening are discussed in Chap. 7.

---

[64] Kirkpatrick et al., "Editing Biosecurity: Needs and Strategies for Governing Genome Editing," 51.

[65] Wang et al., "Highly Efficient Genome Engineering in Bacillus Anthracis and Bacillus Cereus Using the CRISPR/Cas9 System," 3–4, 7.

[66] Kirkpatrick et al., "Editing Biosecurity: Needs and Strategies for Governing Genome Editing," 51.

[67] Kirkpatrick et al., 51.

[68] National Academies of Sciences, Engineering, and Medicine, *Biodefense in the Age of Synthetic Biology*, 53.

[69] Kirkpatrick et al., "Editing Biosecurity: Needs and Strategies for Governing Genome Editing," 51.

[70] Kirkpatrick et al., 51.

[71] Wang et al., "Highly Efficient Genome Engineering in Bacillus Anthracis and Bacillus Cereus Using the CRISPR/Cas9 System," 3.

Additionally, actors who want to apply CRISPR to modify traditional agents of concern, such as *B. anthracis*, will have barriers in acquiring the agent. As mentioned earlier in this chapter, the Federal Select Agent Program, which is made up of the Centers for Disease Control and Prevention/Division of Select Agents and Toxins and the Animal and Plant Health Inspection Service/Agriculture Select Agent Services, oversees the possession, use, and transfer of agents that are a severe threat to humans, animals, and plants. These agents and toxins are listed on the HHS and USDA Select Agents and Toxins list.[72] Thus, these agents are tightly controlled and difficult to acquire. While actors could try to isolate the agent from nature, doing so is difficult. In one example, the Japanese cult Aum Shinrikyo failed many times to disseminate biological weapons. In part, the cult failed because it could only acquire a vaccine strain of *B. anthracis* and could not "cultivate a lethal strain of botulinum toxin from the wild."[73] Overall, the modification of existing pathogens to make them more dangerous is a near-term threat since work has been conducted for decades on how this could be done; however, it is not as worrisome because there are other genetic tools that could modify these pathogens—thus, presenting no significant change to the threat landscape.

6.2.2 Weaponizing the Human Microbiome

Altering the human microbiome via CRISPR poses a longer-term threat since there are still many holes in our understanding of the genetic determinants that underlie desired phenotypes, which only adds to the complexity of knowing how to apply CRIPSR. Still, this type of threat is highly worrisome because it changes the threat landscape. It is clear from this assessment that modification of the human microbiome is unlikely to be carried out by actors with little-to-no laboratory skills, understanding of molecular biology, and expertise working with the human microbiome. This section applies insights from the 2018 NASEM report on synthetic biology with findings from this assessment to comment on the type of barriers that exist to editing the human microbiome and how CRISPR might be applied to overcome these challenges.

Although the human microbiome is an attractive target, a significant barrier that actors will need to overcome is a lack of scientific knowledge of what to edit to produce desired phenotypic changes. Per the 2018 NASEM report on synthetic biology, "Given the current level of understanding of the microbiome, the genetic modification(s) required to effect desired phenotypic changes are not yet certain."[74]

[72] "HHS and USDA Select Agents and Toxins 7CFR Part 331, 9 CFR Part 121, and 42 CFR Part 73."

[73] Koblentz, *Living Weapons*, 205, 213.

[74] National Academies of Sciences, Engineering, and Medicine, *Biodefense in the Age of Synthetic Biology*, 73.

Furthermore, the need to make multiple changes at once or edit multiple symbiotic microbiome species will likely make editing the microbiome more complex. CRISPR has the potential to alleviate these technical challenges. As discussed earlier in this chapter in the section on traditional obstacles that are a barrier to biological weapons, scientists have demonstrated how CRISPR can now achieve modifications that just could not be done before because CRISPR has the powerful ability to make multiple modifications at one time.

Furthermore, CRISPR could be used to increase our general understanding of the microbiome. Studies on CRISPR-based deletion libraries of pathogens might provide give us knowledge about pathogenesis that could potentially be misused. For instance, CRISPR may be used to figure out what genes interact with endogenous flora (i.e., the microbes that naturally reside in the human microbiome) to help a pathogen become established in a host.[75] Even so, the usefulness of CRISPR in targeting the microbiome might be limited. In 2019, Michelle O'Malley, a professor at the University of California Santa Barbara who works on microbes, advised, "CRISPR-Cas cannot be deployed equally well in all microbes."[76] As researchers continue to apply CRISPR to the human microbiome, it will become more apparent what limitations exist.

Experts are cautiously optimistic and warn that there are limits to the control of the microbiome over the human body.[77] Still, a significant investment is being made on the human microbiome because of the potential to manipulate it to improve human health and to treat diseases, such as diabetes, cancer, and autoimmune diseases. Our understanding of the human microbiome, especially its association with diseases, is likely to improve in the coming years. NIH has already funded over $1 billion of research in support of the human microbiome—about 20% of which went to supporting HMP.[78] In 2019, the global market for human-microbiome-based products, diagnostics, and therapeutics was valued at approximately $275 to $400 million and expected to grow to $750 million to $1.9 billion within 5 years.[79] An attack on the human microbiome is likely to generate skepticism about this industry and impede the growth of this market.

CRIPSR has already been used by researchers for microbiome gene editing in mammals. In a breakthrough November 2021 study, researchers used CRISPR-Cas9 to delete genes from *Escherichia coli* bacteria that were already living in the gut microbiome of mice. This research demonstrated that it was possible to directly

[75] National Academies of Sciences, Engineering, and Medicine, 72.

[76] Vivien Marx, "Engineers Embrace Microbiome Messiness," *Nature Methods* 16, no. 7 (July 2019): 582, https://doi.org/10.1038/s41592-019-0460-5

[77] Anand Jagatia et al., "The Future of Research on Human-Associated Microbiota," Nature.com, June 17, 2019, https://www-nature-com.mutex.gmu.edu/immersive/d42859-019-00041-z/index.html

[78] Lita Proctor, "What's Next for the Human Microbiome?," 624.

[79] Lita Proctor, 642.

modify the DNA of microbes in the gut.[80] While this approach could be used in the future to legitimately research and treat human diseases that are associated with the microbiome, it could be misused to directly target microbes of the gut to make them pathogenic. One day, it may be easier for bioweaponeers to modify the microbes of the human microbiome rather than humans themselves.

6.2.3 Weaponizing Gene Drives

Gene drives are worrisome because they alter the threat landscape. Nevertheless, this threat could be countered. Kevin Esvelt, a pioneer of CRISPR-based gene drive technology, advises how "major security risks are likely avoidable if at-risk populations are actively monitored by defense establishments."[81] Additionally, Esvelt identifies the many limitations to gene drive systems: they are "slow, obvious, and easily countered."[82] As this section shows, there are many technical and social obstacles that pose a barrier to gene drive technology. Weaponizing gene drives poses a mid-term threat because there are still due to technical and social barriers, as well as constraints to this application, that limit the usability of gene drives as a weapon. While CRISPR may be applied to engineer gene drives, it is important to point out that there are many constraints and barriers to using this technology that may not make it an attractive option to bioweaponeers.

It takes knowledge, skill, and the ability to troubleshoot to engineer gene drives. Ben Ouagrham-Gormley and Vogel advise that "gene drives would seem to be beyond the capabilities of terrorists or biohackers with limited scientific knowledge and skills."[83] Given the "skills, expertise, resources, and long-term commitment to work out technical problems and develop a long-term research agenda," Ben Ouagrham-Gormley and Vogel suggest that state-level use is a "more serious concern."[84] Research from this assessment also indicates that engineered gene drives would be difficult for actors with little-to-no laboratory skills, knowledge, and experience.

In addition to knowledge, skills, and experience, there is also a need to be able to troubleshoot gene drives. The need for troubleshooting is illustrated by a group of students from the University of Minnesota who worked on a project to build a gene drive and a reversal drive for the 2016 iGEM. The students worked for 5 months

---

[80] Kathy N. Lam et al., "Phage-Delivered CRISPR-Cas9 for Strain-Specific Depletion and Genomic Deletions in the Gut Microbiome," *Cell Reports* 37, no. 5 (November 2, 2021): 109930, https://doi.org/10.1016/j.celrep.2021.109930

[81] Kevin Esvelt, "Gene Drive Technology: The Thing to Fear Is Fear Itself," 2018, 7, accessed October 31, 2019, https://editingbiosecurity.org/new-page

[82] Esvelt, 7.

[83] Ouagrham-Gormley and Vogel, "Gene Drives."

[84] Ouagrham-Gormley and Vogel.

following a protocol from George Church, Kevin Esvelt, and other colleagues.[85] Students were unable to assemble the gene drive and only assembled two-thirds of the reversal drive due to "contaminated lab equipment and a fridge that wasn't cold enough."[86] If given more time to complete the competition, students would have had to work to troubleshoot their mishaps. It is noted that iGEM has its own policy on gene drives that generally does not allow gene drives in the competition unless special permission is granted by its Safety Committee. Additionally, iGEM does not allow participants to release gene drive into the wild.[87]

The knowledge to develop a gene drive in one species does not necessarily translate to another species. David Gurwitz offers more insight into the difficulty of gene drive methodology: "Fortunately, learning how to insert a gene drive into the germline and optimize its function will have to be done anew for each species, and may be far trickier in organisms with larger genomes."[88] As Ben Ouagrham-Gormley and Vogel suggest, only a few members of teams that conduct genome editing experiments may have the skills to deliver gene drives. Furthermore, specialized laboratory skills used on an animal are not necessarily transferable to other species.[89]

6.2.4 Augmenting Humans

Augmenting humans via somatic gene-therapies is a mid-term threat, whereas germline gene-therapies is a longer-term threat. There are already CRISPR-based somatic gene-therapies in clinical trials, so it is not farfetched to believe that humans could be augmented in this manner; however, there are barriers in our understanding of what genes to change (i.e., genetic determinants), off-target effects, and delivery challenges. For germline gene-therapies, however, there is already a norm that is becoming established against using CRISPR to modify the germline. Studies that want to use CRISPR on the human germline will likely be subject to a high level of scrutiny by the international scientific community, so the legitimate use of germline gene-therapies is unlikely to takeoff anytime soon. This will likely impede progress in CRIPSR-based germline gene-therapy, which will limit the available of research with dual use potential. Again, this type of threat is highly worrisome because it changes the threat landscape.

[85] James E. DiCarlo et al., "Safeguarding CRISPR-Cas9 Gene Drives in Yeast," *Nature Biotechnology* 33, no. 12 (December 2015): 1250–55, https://doi.org/10.1038/nbt.3412

[86] "College Students Try to Hack a Gene Drive, and Set a Science Fair Abuzz," *STAT* (blog), December 14, 2016, accessed October 11, 2019, https://www.statnews.com/2016/12/14/gene-drive-students-igem/

[87] iGEM, "Safety Policy," n.d., http://2018.igem.org/Safety/Policies#drives

[88] Gurwitz, "Gene Drives Raise Dual-Use Concerns," 1010.

[89] Ouagrham-Gormley and Vogel, "Gene Drives."

A bioweaponeer would have to overcome technical, social, operational, logistical, and ethical challenges to use CRISPR to engineer super soldiers or to augment other desirable traits in the general population such as intelligence, longevity, and immunity. The constraints posed by these challenges as well as the research from this assessment suggest the following: (1) misuse of gene therapies to enhance human traits are more likely to occur in somatic cells than the germline, (2) misuse would likely occur among users who know how to use CRISPR and have an advanced understanding of the trait targeted for augmentation, and (3) traits controlled by one or a few genes are more likely to be targeted for augmentation than complex traits.

First, bioweaponeers are more likely to use CRISPR-based somatic gene therapies than germline gene therapies. A military program that wanted to conduct germline gene therapy to create super soldiers would encounter operational and logistical issues in recruiting females to be implanted with genetically engineered embryos, carrying the pregnancy to term, and then raising, training, and indoctrinating the individual until an eligible age to become a soldier. However, this is not to say that a military program could not overcome these challenges through financial compensation (e.g. by paying for surrogate mothers) or through its ideology (e.g., the Islamic State recruited young women as brides). Nevertheless, a military program could eliminate these challenges just by carrying out somatic gene therapy on individuals who are already soldiers.

Furthermore, users who wanted to engineer traits in the general population would likely encounter more ethical and safety concerns for CRISPR-based germline gene therapies than somatic gene therapies. He Jiankui, the Chinese scientist who first claimed to use CRISPR to edit the human germline to provide immunity to HIV, received significant backlash from the scientific community when he debuted his research at the Second International Summit on Human Genome Editing in November 2018.[90] Shortly after He Jiankui's presentation at the summit, the NIH, which funds billions of dollars of CRISPR-related research (see Chap. 5, Parameter 4), reaffirmed that it "does not support the use of gene-editing technologies in human embryos."[91] There were serious ethical concerns over the study, including issues of informed consent, as well as safety concerns, such as potential off-target effects caused by CRISPR. Note that He Jiankui's study has since been withdrawn because

[90]He Jiankui, "International Summit on Human Genome Editing - He Jiankui Presentation and Q&A" (The University of Hong Kong, November 28, 2018), accessed October 1, 2019, https://www.youtube.com/watch?v=tLZufCrjrN0&list=PLPuptUEHTSjDGWQ16svQQuOXf78g4z4Bn&index=51&t=0s

[91]Francis S. Collins, "Statement on Claim of First Gene-Edited Babies by Chinese Researcher," National Institutes of Health (NIH), November 28, 2018, accessed October 2, 2019, https://www.nih.gov/about-nih/who-we-are/nih-director/statements/statement-claim-first-gene-edited-babies-chinese-researcher

data could not be provided on the individual participants to evaluate the safety and validity of the work.[92]

Currently, there are already several CRISPR-based somatic gene therapies in clinical trials. In January 2018, the NIH announced a major initiative to invest $190 million over 6 years into somatic genome editing research that makes technologies, such as CRISPR, safe to use in humans.[93] In October 2018, the first research grants for this initiative—21 grants worth $86 million over 5 years—were awarded by the Somatic Cell Genome Editing program.[94] These studies should help to open the door for even more CRISPR-based somatic gene therapy solutions by mitigating challenges with delivery of the therapy into patients and improving methods to test for the safety and efficacy of the tool once delivered into cells.

Second, users who want to apply CRISPR to modify the human genome would need to be able to use CRISPR and have an advanced understanding of human gene therapy and the trait targeted for modification. This type of work is unlikely to occur at the DIY biology level. As discussed previously, there is a difference between using CRISPR to cleave a well-known gene in a well-studied and relatively simple organism (i.e., cleaving *ade2* in yeast during the Genspace class) and using CRISPR to target a trait in a more complex organism. To put it into perspective, the yeast genome is approximately 13 million base pairs, 6000 genes, and 6000 proteins, whereas the human genome is approximately 3 billion base pairs and each of its estimated 30,000 genes encodes approximately 3 proteins.[95]

While students may learn basic molecular cloning skills and how to use CRISPR in yeast in a DIY biology class, it is much harder to apply this to the human genome. Josiah Zayner, CEO Founder of a company called The ODIN that sells biology supplies such as DIY CRISPR kits, offers a CRISPR-Cas9 plasmid that targets the human myostatin gene for knockout and provides instruction for how DIY biologists can do this in their own cells.[96] The objective of knocking out the myostatin gene is to increase muscle mass. This is seemingly simple; after all, scientists in China had reported using CRISPR to engineer a pair of beagles with increased muscle mass by

[92] "Evaluation of the Safety and Efficacy of Gene Editing with Human Embryo CCR5 Gene," Chinese Clinical Trial Register (ChiCTR), accessed October 1, 2019, http://www.chictr.org.cn/showprojen.aspx?proj=32758

[93] "NIH to Launch Genome Editing Research Program."

[94] "NIH Awards $86 Million to Improve Genome Editing Techniques," National Center for Advancing Translational Sciences, October 8, 2018, accessed December 15, 2018, https://ncats.nih.gov/news/releases/2018/somatic-cells

[95] Pevsner, *Bioinformatics and Functional Genomics*, 15, 466; National Human Genome Research Institute, "Human Genome Project FAQ," Genome.gov, accessed October 8, 2019, https://www.genome.gov/human-genome-project/Completion-FAQ

[96] "Human Myostatin Knock-Out Targeting CRISPR-Cas9 Plasmid," The ODIN, accessed October 9, 2019, http://www.the-odin.com/human-myostatin-knock-out-targeting-crispr-cas9-plasmid/; Josiah Zayner, "DIY Human CRISPR Myostatin Knock-Out," YouTube, October 6, 2017, accessed October 9, 2019, https://www.youtube.com/watch?v=o6A9bbDI6fo&t=1438s

knocking out myostatin in 2015.[97] However, Zayner has voiced his frustrations with getting the DIY CRISPR knockout of myostatin to work on himself. He said, "I didn't see any big results in my muscle changes" and suggests that he may not have injected enough DNA or the efficiency of CRISPR may have been too low.[98]

JoVE videos, as assessed in Chap. 4, Parameter 2, are likely to only help a little in teaching users the tacit knowledge and skills needed for gene therapy in humans. JoVE videos from Delerue and Ittner, Li and colleagues, and Yao and colleagues demonstrate how to microinject CRISPR into the oocytes (egg precursor cells) and embryos of mice and hamsters to create animal models.[99] While these videos help transfer knowledge and skill for animal models, users who want to conduct germline gene therapy in humans would still have to know how to perform the procedure for egg retrieval and then how to culture, select, and transfer embryos. Other videos, such as the video from Xu and colleagues that demonstrates how to use adeno-associated virus-mediated delivery of CRISPR in mice to modify a cardiac gene, could be used to gain a basic understanding of the steps involved in somatic gene therapy.[100] However, to apply this to humans, users would still have to overcome significant barriers in off-target effects and delivery.

One unique challenge associated with targeting the brain is that genome-editing components need to be able to traverse the blood-brain barrier. However, this challenge might be alleviated through progress in adeno-associated viral (AAV) vectors. The 2018 *Editing Biosecurity* report warns that "AAVs have been developed that can breach the blood-brain barrier and reach the entire central nervous system which support the development of a neuroweapon."[101] AAV is an attractive option for delivery of CRISPR and has benefits over other options of viruses for gene delivery. The use of AAVs decreases the chances of Cas9 off-target effects because AAVs do not incorporate themselves into the host genome, which prevents the prolonged expression of Cas9. Additionally, AAVs are generally less difficult to

---

[97] Zou et al., "Generation of Gene-Target Dogs Using CRISPR/Cas9 System"; Antonio Regalado, "First Gene-Edited Dogs Reported in China," MIT Technology Review, October 19, 2015, October 8, 2019, https://www.technologyreview.com/s/542616/first-gene-edited-dogs-reported-in-china/

[98] Josiah Zayner, "CRISPR Sucks," YouTube, January 31, 2019, accessed October 9, 2019, https://www.youtube.com/watch?v=ZrEweAhlU4g

[99] Fabien Delerue and Lars M. Ittner, "Generation of Genetically Modified Mice through the Microinjection of Oocytes," *JoVE (Journal of Visualized Experiments)*, no. 124 (June 15, 2017): e55765, https://doi.org/10.3791/55765; Rong Li et al., "Production of Genetically Engineered Golden Syrian Hamsters by Pronuclear Injection of the CRISPR/Cas9 Complex," *JoVE (Journal of Visualized Experiments)*, no. 131 (January 9, 2018): e56263, https://doi.org/10.3791/56263; Xuan Yao et al., "CRISPR/Cas9-Mediated Targeted Integration In Vivo Using a Homology-Mediated End Joining-Based Strategy," *JoVE (Journal of Visualized Experiments)*, no. 133 (March 12, 2018): e56844, https://doi.org/10.3791/56844

[100] Li Xu et al., "Adeno-Associated Virus-Mediated Delivery of CRISPR for Cardiac Gene Editing in Mice," *JoVE (Journal of Visualized Experiments)*, no. 138 (August 2, 2018): e57560, https://doi.org/10.3791/57560.

[101] Kirkpatrick et al., "Editing Biosecurity: Needs and Strategies for Governing Genome Editing," 56.

manipulate because they have a reduced immunogenicity and smaller genomes compared to other non-integrating viruses.[102] Even so, the packaging capacity of AAV is quite low (4.7 kb) relative to other viral delivery systems, such as adenoviral vectors (35 kb). Nevertheless, scientists are devising ways to improve the packaging capacity of AAV. For instance, Xu and colleagues describe an approach using dual AAV systems that increase packaging capacity to 9 kb.[103]

The misuse of CRISPR to augment human traits is limited by the state of scientific knowledge. Jennifer Doudna, a pioneer of CRISPR, explained that we do not know how to modify many traits because "we do not know which genes to change, many traits in us, meaning humans, are multigenic, meaning there are multiple parts of the DNA that contribute to those properties . . . we don't understand yet the way that genes function in cells and interact."[104] Even the work that He Jiankui undertook on using CRISPR to edit the germline, as described above, seemed premature. Although there had been decades of research on CCR5, there was still uncertainty about whether enough was known about CCR5, the role of CCR5 in immunity, and the function of CCR5 throughout the body (namely the brain and potential for enhanced cognitive function).[105] Furthermore, the Committee on Human Gene Editing: Scientific, Medical, and Ethical Considerations, a committee convened by the National Academy of Sciences and the National Academy of Medicine, advised that the current state of understanding of human genetics may only allow some, but not all, gene targets to be appropriately selected.[106]

Not only are there knowledge gaps, but there are still uncertainties in the outcome of genome editing. As Luhan Yang, Chief Scientific Officer at the CRISPR-focused startup company called eGenesis, said when questioned on genome editing in humans, "sometimes I think the constraint is [that] we don't know what to modify, and we don't know after modification whether there is some unexpected consequence."[107] The Committee on Human Gene Editing echoed this concern by advising how the human gene pool may unexpectedly be affected by genome editing. More specifically, the Committee on Human Gene Editing suggested that issues with

---

[102] "Improved CRISPR/Cas9 Genome Editing in Hard-to-Transfect Mammalian Cells Using AAV," Takara, n.d., accessed October 31, 2019, https://www.takarabio.com/learning-centers/gene-function/gene-editing/crispr/cas9-delivery-methods/crispr/cas9-gene-editing-with-aav

[103] Christine L Xu et al., "Viral Delivery Systems for CRISPR," *Viruses* 11, no. 28 (2019): 2, https://doi.org/doi:10.3390/v11010028.

[104] "Conversations in Science with Dan Rather & Jennifer Doudna: CRISPR - YouTube," 2016, accessed October 5, 2019, https://www.youtube.com/watch?v=GMndqLvTqhA&list=PLPuptUEHTSjDGWQ16svQQuOXf78g4z4Bn&index=52&t=0s

[105] Jiankui, "International Summit on Human Genome Editing - He Jiankui Presentation and Q&A," pt. 27th minute.

[106] National Academies of Sciences, Engineering, and Medicine, *Human Genome Editing: Science, Ethics, and Governance*, 116–18.

[107] *Issue Briefing: Editing Humans*, Annual Meeting of the New Champions 2017 (Dalian, People's Republic of China, 2017), accessed April 28, 2018, https://www.weforum.org/events/annual-meeting-of-the-new-champions-2017/sessions/cloning-humans/

mosaicism can result if a zygote or early embryo is edited and only a portion of cells in the resulting embryo have the modification. Issues with mosaicism were observed in He Jiankui's study on CCR5 knockout and the Chinese study on myostatin knockout in dogs (both studies were described above).

Even when using CRISPR, it would still take a skilled healthcare team of medical geneticists—such as clinical geneticists, laboratory geneticists, nurses, and other professionals—to target a gene of interest, target the correct cells, determine a way to deliver the genetic material into the cells, and avoid unwanted side effects. Gene therapy is difficult and comes with many risks, such as an unintentional overreaction from the immune system that could lead to organ failure, or the insertion of genes in the wrong location that could lead to tumor formation. Given the difficulties of genetic therapeutics, state programs are more likely than nonstate actors to achieve desired gains in human augmentation using genome editing.

Third, given the state of scientific knowledge and difficulty of editing complex pathways, it is likely that CRISPR would be misused to augment well-studied, well-understood traits that are controlled by only one gene or a few genes as well as multigenic traits in which single gene variants have been observed to have a big impact.[108] In his work on editing the human germline, He Jiankui reasoned that CCR5 was a "simple, well understood single gene"—it was the target of the first clinical trial that used ZFN in the United States (NCT00842634).[109] He Jiankui suggested that targeting CCR5 would serve as a "conservative first model so that maybe [in] the future [scientists] can move on to multiple genes [and] more complicated genotypes."[110]

One way that militaries could gain a strategic advantage is by treating soldiers with a CRISPR-based gene therapy to diminish pain. In August 2019, the MIT Technology reported on the first use of CRISPR in mice to treat pain.[111] Scientists from the University of California, San Diego had used dCas9 to non-permanently repress a single gene, *SCN9A*, which encodes $Na_v1.7$, a sodium ion channel that has a key role in transmitting pain signals.[112] Given that much is known about this gene, it is a single gene that influences pain, and preclinical work has already demonstrated

[108] For a list of traits, see "Protective Alleles," accessed October 11, 2019, http://arep.med.harvard.edu/gmc/protect.html

[109] "Autologous T-Cells Genetically Modified at the CCR5 Gene by Zinc Finger Nucleases SB-728 for HIV - Full Text View," ClinicalTrials.gov, accessed October 2, 2019, https://clinicaltrials.gov/ct2/show/NCT00842634; The He Lab, "Why We Chose HIV and CCR5 First," November 25, 2018, accessed October 1, 2019, https://www.youtube.com/watch?v=aezxaOn0efE; Jiankui, "International Summit on Human Genome Editing - He Jiankui Presentation and Q&A," pt. 28th minute.

[110] Jiankui, "International Summit on Human Genome Editing - He Jiankui Presentation and Q&A," pt. 28th minute.

[111] Antonio Regalado, "The next Trick for CRISPR Is Gene-Editing Pain Away," MIT Technology Review, accessed October 5, 2019, https://www.technologyreview.com/s/614210/the-next-trick-for-crispr-is-gene-editing-pain-away/

[112] Ana M. Moreno et al., "Long-Lasting Analgesia via Targeted in Vivo Epigenetic Repression of Nav1.7," *BioRxiv*, July 24, 2019, 711812, https://doi.org/10.1101/711812

that targeting it leads to diminished pain, it is likely that a CRISPR-based somatic gene therapy could be developed to target this trait in soldiers in the near future. Note that it is generally ill-advisable to permanently knockout pain signals, since doing so might cause the body to might miss injuries that could lead to disability or death.

Based on the interest from actors to augment humans, the ability to target the gene of interest via CRISPR, and the ability to successfully deliver the gene therapy, this assessment supports the prediction that genetic engineering could likely be used to augment human performance as early as 2047.[113] Even so, it is highly unlikely that every trait will be augmented to the desired level by 2047, but it is possible to see moderate improvement in some qualities. Genetic engineers will have to work within the gene pool. In other words, there are limits to how fast or how strong scientists can make humans because scientists must work with the genes that are in front of them. The genetic enhancement of traits that go beyond correcting a diseased back to normal is still ethically questionable. It is likely that non-Western nations would take more liberties in genetically enhancing traits that cross ethical "red lines." Furthermore, state-level actors that elect to pursue human augmentation are likely to undertake these programs covertly given the ethical stigma.

6.3 Conclusion

This assessment provides valuable insight into the likelihood of misuse of CRISPR compared to its predecessor technologies. By analyzing and evaluating the accessibility, ease of misuse, magnitude of potential harm, and imminence of potential misuse, this assessment concluded that the risk of misuse of CRISPR is high, while the risk of misuse of ZFN and TALEN is only medium to low. CRISPR is highly accessible and easier to use than its predecessor technologies, which increases the pool of potential misusers. While this is concerning, data from this study suggest that CRISPR is not so easy to misuse that anyone could misuse it.

Although the skill and knowledge needed to use CRISPR could be transferred through forums such as DIY bio classes and JoVE videos, there are limits in the ability of actors to apply this knowledge. Moreover, kits that help to reduce knowledge and skill requirements during CRISPR experiences still necessitate the actor be able to select the appropriate kit and troubleshoot as needed. Additionally, there are still traditional technical obstacles that pose a barrier to applying CRISPR to develop biological weapons. It is concerning, however, that CRISPR is maturing very quickly because it becomes more apparent how CRISPR could be misused as it develops more fully.

This assessment also provides insight into the likelihood of using CRISPR to achieve specific applications of misuse. Traditional concerns over making pathogens more dangerous are less worrisome because there are already technologies that could

[113] "Emerging Science and Technology Trends," 13.

do this; however, this is a near-term threat. Weaponizing gene drives is assessed as a mid-term since there are still technical and social barriers to overcome and constraints to this application, and, while they are worrisome, investments into defensive monitoring systems could counter this threat. Lastly, augmenting humans via somatic gene-therapies is a near- to mid-term threat given that it is already in clinical trials and there is high interest from some nation-states, whereas germline gene-therapies is a longer-term threat given the norm against germline genome editing. Applying CRISPR to either type of gene therapy is highly worrisome because it changes the threat landscape.

References

1. Aitoro, Jill. "DARPA Head Resigns, Moving on to Industry." Defense News, December 17, 2019. https://www.defensenews.com/breaking-news/2019/12/17/darpa-head-resigns-moving-on-to-industry/.
2. Anand Jagatia, Rochellys Heijtz, Jennifer Wargo, and Eran Elinav. "The Future of Research on Human-Associated Microbiota." Nature.com, June 17, 2019. https://www-nature-com.mutex.gmu.edu/immersive/d42859-019-00041-z/index.html.
3. ClinicalTrials.gov. "Autologous T-Cells Genetically Modified at the CCR5 Gene by Zinc Finger Nucleases SB-728 for HIV - Full Text View." Accessed October 2, 2019. https://clinicaltrials.gov/ct2/show/NCT00842634.
4. Centers for Disease Control and Prevention and National Institutes of Health. *Biosafety in Microbiological and Biomedical Laboratories*. 5th ed. US Department of Health and Human Services, 2009.
5. Chandrasekharan, Subhashini, Sapna Kumar, Cory M. Valley, and Arti Rai. "Proprietary Science, Open Science and the Role of Patent Disclosure: The Case of Zinc-Finger Proteins." *Nature Biotechnology* 27, no. 2 (February 2009): 140–44. doi:https://doi.org/10.1038/nbt0209-140.
6. Church, George. "The Future of the Genomic Editing Revolution." UC Berkeley, January 26, 2017. https://www.youtube.com/watch?v=Wswvf8Nrubg.
7. Cohen, Jon. "To Feed Its 1.4 Billion, China Bets Big on Genome Editing of Crops." Science | AAAS, July 29, 2019. https://www-sciencemag-org.mutex.gmu.edu/news/2019/07/feed-its-14-billion-china-bets-big-genome-editing-crops.
8. STAT. "College Students Try to Hack a Gene Drive, and Set a Science Fair Abuzz," December 14, 2016. https://www.statnews.com/2016/12/14/gene-drive-students-igem/.
9. Collins, Francis S. "Statement on Claim of First Gene-Edited Babies by Chinese Researcher." National Institutes of Health (NIH), November 28, 2018. https://www.nih.gov/about-nih/who-we-are/nih-director/statements/statement-claim-first-gene-edited-babies-chinese-researcher.
10. "Conversations in Science with Dan Rather & Jennifer Doudna: CRISPR - YouTube," 2016. https://www.youtube.com/watch?v=GMndqLvTqhA&list=PLPuptUEHTSjDGWQ16svQQuOXf78g4z4Bn&index=52&t=0s.
11. ATCC. "Culture Guides," n.d. https://www.atcc.org/en/Guides.aspx.
12. Dale, Jennifer L., Malik J. Raynor, Maureen C. Ty, Maria Hadjifrangiskou, and Theresa M. Koehler. "A Dual Role for the Bacillus Anthracis Master Virulence Regulator AtxA: Control of Sporulation and Anthrax Toxin Production." *Frontiers in Microbiology* 9 (2018). doi:https://doi.org/10.3389/fmicb.2018.00482.
13. DeFrancesco, Laura. "Move over ZFNs." *Nature Biotechnology* 29 (August 5, 2011): 681–84. doi:https://doi.org/10.1038/nbt.1935.

14. Delerue, Fabien, and Lars M. Ittner. "Generation of Genetically Modified Mice through the Microinjection of Oocytes." *JoVE (Journal of Visualized Experiments)*, no. 124 (June 15, 2017): e55765. doi:https://doi.org/10.3791/55765.
15. Desktop Genetics. *Custom Designed CRISPR SgRNA Libraries - Webinar*, 2016. https://www.youtube.com/watch?v=ghEai9Vtcoc.
16. DiCarlo, James E., Alejandro Chavez, Sven L. Dietz, Kevin M. Esvelt, and George M. Church. "Safeguarding CRISPR-Cas9 Gene Drives in Yeast." *Nature Biotechnology* 33, no. 12 (December 2015): 1250–55. doi:https://doi.org/10.1038/nbt.3412.
17. "Emerging Science and Technology Trends: 2017-2047." Office of the Deputy Assistant Secretary of the Army (Research & Technology), November 21, 2017.
18. Esvelt, Kevin. "Gene Drive Technology: The Thing to Fear Is Fear Itself," 2018. https://editingbiosecurity.org/new-page.
19. Chinese Clinical Trial Register (ChiCTR). "Evaluation of the Safety and Efficacy of Gene Editing with Human Embryo CCR5 Gene." Accessed October 1, 2019. http://www.chictr.org.cn/showprojen.aspx?proj=32758.
20. "Growth and Maintenance of Yeast in the Lab | Protocol." Accessed October 17, 2019. https://www-jove-com.mutex.gmu.edu/science-education/5095/yeast-maintenance.
21. Gurwitz, David. "Gene Drives Raise Dual-Use Concerns." *Science* 345, no. 6200 (August 29, 2014): 1010–1010. https://doi.org/https://doi.org/10.1126/science.345.6200.1010-b.
22. "HHS and USDA Select Agents and Toxins 7CFR Part 331, 9 CFR Part 121, and 42 CFR Part 73." Accessed April 25, 2018. https://www.selectagents.gov/selectagentsandtoxinslist.html.
23. How to Pipette Correctly – a Short Step-by-Step Introduction into Proper Pipetting I Eppendorf, 2017. https://www.youtube.com/watch?v=QGX490kuKjg.
24. The ODIN. "Human Myostatin Knock-Out Targeting CRISPR-Cas9 Plasmid." Accessed October 9, 2019. http://www.the-odin.com/human-myostatin-knock-out-targeting-crispr-cas9-plasmid/.
25. Huntress, Adrienne. "Picking the Right DNA Isolation Kit for Your Application." Bitesize Bio, February 7, 2019. https://bitesizebio.com/42930/picking-the-right-dna-isolation-kit-for-your-application/.
26. iGEM. "Safety Policy," n.d. http://2018.igem.org/Safety/Policies#drives.
27. Takara. "Improved CRISPR/Cas9 Genome Editing in Hard-to-Transfect Mammalian Cells Using AAV," n.d. https://www.takarabio.com/learning-centers/gene-function/gene-editing/crispr/cas9-delivery-methods/crispr/cas9-gene-editing-with-aav.
28. Broad Institute. "Information about Licensing CRISPR Systems for Clinical Use." Accessed October 24, 2019. https://www.broadinstitute.org/partnerships/office-strategic-alliances-and-partnering/information-about-licensing-crispr-genome-edi.
29. *Issue Briefing: Editing Humans*. Annual Meeting of the New Champions 2017. Dalian, People's Republic of China, 2017. https://www.weforum.org/events/annual-meeting-of-the-new-champions-2017/sessions/cloning-humans/.
30. Jiankui, He. "International Summit on Human Genome Editing - He Jiankui Presentation and Q&A." The University of Hong Kong, November 28, 2018. https://www.youtube.com/watch?v=tLZufCrjrN0&list=PLPuptUEHTSjDGWQ16svQQuOXf78g4z4Bn&index=51&t=0s.
31. Jorgensen, Ellen. *What You Need to Know about CRISPR*. Banff, Canada: TEDSummit, 2016.
32. JoVE Science Education Database. *General Laboratory Techniques. An Introduction to the Micropipettor*. Cambridge, MA, 2017. https://www-jove-com.mutex.gmu.edu/science-education/5033/an-introduction-to-the-micropipettor.
33. Kirkpatrick, Jesse, Gregory D. Koblentz, Megan J. Palmer, Edward Perello, David A. Relman, and Sarah W. Denton. "Editing Biosecurity: Needs and Strategies for Governing Genome Editing," December 2018.
34. Koblentz, Gregory D. *Living Weapons: Biological Warfare and International Security*. Ithaca: Cornell University Press, 2009.
35. Lam, Kathy N., Peter Spanogiannopoulos, Paola Soto-Perez, Margaret Alexander, Matthew J. Nalley, Jordan E. Bisanz, Renuka R. Nayak, Allison M. Weakley, Feiqiao B. Yu, and Peter

J. Turnbaugh. "Phage-Delivered CRISPR-Cas9 for Strain-Specific Depletion and Genomic Deletions in the Gut Microbiome." *Cell Reports* 37, no. 5 (November 2, 2021): 109930. doi: https://doi.org/10.1016/j.celrep.2021.109930.
36. Li, Rong, Jinxin Miao, Zhiqiang Fan, SeokHwan Song, Il-keun Kong, Yaohe Wang, and Zhongde Wang. "Production of Genetically Engineered Golden Syrian Hamsters by Pronuclear Injection of the CRISPR/Cas9 Complex." *JoVE (Journal of Visualized Experiments)*, no. 131 (January 9, 2018): e56263. https://doi.org/https://doi.org/10.3791/56263.
37. Lita Proctor. "What's Next for the Human Microbiome?" *Nature* 569 (May 30, 2019): 623–25.
38. Marx, Vivien. "Engineers Embrace Microbiome Messiness." *Nature Methods* 16, no. 7 (July 2019): 581–84. https://doi.org/https://doi.org/10.1038/s41592-019-0460-5.
39. "MO BIO's UltraClean Microbial DNA Kit Handbook - QIAGEN." Accessed October 26, 2019. https://www.qiagen.com/us/resources/resourcedetail?id=cda265c5-c1f0-4c99-acee-7282fc6aa371&lang=en.
40. New England BioLabs. "Monarch® Plasmid Miniprep Kit (T1010): Troubleshooting," n.d. https://www.neb.com/products/t1010-monarch-plasmid-miniprep-kit#Product%20Information.
41. Moreno, Ana M., Glaucilene F. Catroli, Fernando Alemán, Andrew Pla, Sarah A. Woller, Michael Hu, Tony Yaksh, and Prashant Mali. "Long-Lasting Analgesia via Targeted in Vivo Epigenetic Repression of Nav1.7." *BioRxiv*, July 24, 2019, 711812. https://doi.org/https://doi.org/10.1101/711812.
42. Nagarkatti, Jai. "Sangamo Lands $20M in Expanded Sigma Deal." San Francisco Business Times, October 5, 2009. https://www.bizjournals.com/sanfrancisco/stories/2009/10/05/daily2.html.
43. National Academies of Sciences, Engineering, and Medicine. *Biodefense in the Age of Synthetic Biology*. Washington, DC: The National Academies Press, 2018. https://doi.org/10.17226/24890.
44. National Academies of Sciences, . *Human Genome Editing: Science, Ethics, and Governance*. Washington, DC: The National Academies Press, 2017. https://doi.org/10.17226/24623.
45. National Human Genome Research Institute. "Human Genome Project FAQ." Genome.gov. Accessed October 8, 2019. https://www.genome.gov/human-genome-project/Completion-FAQ.
46. New England Biolabs. "Monarch® PCR & DNA Cleanup Kit (5 Mg) (T1030): Troubleshooting," n.d. https://www.neb.com/products/t1030-monarch-pcr-dna-cleanup-kit-5-ug#FAQs%20&%20Troubleshooting_Troubleshooting.
47. National Center for Advancing Translational Sciences. "NIH Awards $86 Million to Improve Genome Editing Techniques," October 8, 2018. https://ncats.nih.gov/news/releases/2018/somatic-cells.
48. National Institutes of Health (NIH). "NIH to Launch Genome Editing Research Program," January 23, 2018. https://www.nih.gov/news-events/news-releases/nih-launch-genome-editing-research-program.
49. Oswald, Nick, Suzanne Kennedy, Megan Hogan, and Megan Cartwright. *10 Things Every Molecular Biologist Should Know*. UK: Science Squared Ltd, 2012. http://bitesizebio.s3.amazonaws.com/wp-content/uploads/2017/07/10124646/10-Things-Every-Molecular-Biologist-Should-Know.pdf.
50. Ouagrham-Gormley, Sonia Ben, and Kathleen M. Vogel. "Gene Drives: The Good, the Bad, and the Hype." Bulletin of the Atomic Scientists, October 14, 2016. https://thebulletin.org/gene-drives-good-bad-and-hype10027.
51. Pearson, Helen. "Protein Engineering: The Fate of Fingers." *Nature* 455, no. 7210 (September 11, 2008): 160–64. https://doi.org/https://doi.org/10.1038/455160a.
52. Pevsner, Jonathan. *Bioinformatics and Functional Genomics*. 2nd ed. New Jersey: John Wiley & Sons, Inc., 2009.
53. Pipette Calibration and Cleaning I Arizona Department of Health Services, 2013. https://www.youtube.com/watch?v=MBq55FtOzN4.

54. "Protective Alleles." Accessed October 11, 2019. http://arep.med.harvard.edu/gmc/protect.html.
55. New England BioLabs. "Protocols for Nucleic Acid Purification." Accessed October 25, 2019. https://www.neb.com/products/nucleic-acid-purification/nucleic-acid-purification#tabselect2.
56. Regalado, Antonio. "First Gene-Edited Dogs Reported in China." MIT Technology Review, October 19, 2015. https://www.technologyreview.com/s/542616/first-gene-edited-dogs-reported-in-china/.
57. ———. "The next Trick for CRISPR Is Gene-Editing Pain Away." MIT Technology Review. Accessed October 5, 2019. https://www.technologyreview.com/s/614210/the-next-trick-for-crispr-is-gene-editing-pain-away/.
58. Research, Center for Biologics Evaluation and. "Xenotransplantation." FDA, March 28, 2019. http://www.fda.gov/vaccines-blood-biologics/xenotransplantation.
59. Revill, James, and Catherine Jefferson. "Tacit Knowledge and the Biological Weapons Regime." *Science and Public Policy* 41, no. 5 (October 1, 2014): 597–610. https://doi.org/https://doi.org/10.1093/scipol/sct090.
60. Shiel, William C. "Medical Definition of Culture." MedicineNet. Accessed October 18, 2019. https://www.medicinenet.com/script/main/art.asp?articlekey=2879.
61. The He Lab. "Why We Chose HIV and CCR5 First," November 25, 2018. https://www.youtube.com/watch?v=aezxaOn0efE.
62. Thermo Fisher Scientific. "Cell Culture Basics Handbook," 2016. https://www.thermofisher.com/content/dam/LifeTech/Documents/PDFs/PG1563-PJT1267-COL31122-Gibco-Cell-Culture-Basics-Handbook-Global-FLR.pdf.
63. "Troubleshooting Guide for Cloning | NEB." Accessed March 14, 2019. https://www.neb.com/tools-and-resources/troubleshooting-guides/troubleshooting-guide-for-cloning.
64. New England BioLabs. "Troubleshooting Guide for DNA Cleanup and Plasmid Purification." Accessed October 26, 2019. https://www.neb.com/tools-and-resources/troubleshooting-guides/troubleshooting-guide-for-dna-cleanup-and-plasmid-purification.
65. Tsang, Jennifer. "The Time and Cost Required to Make a Plasmid." Addgene. Accessed October 24, 2018. https://blog.addgene.org/the-time-and-cost-required-to-make-a-plasmid.
66. Tucker, Jonathan B. "Review of the Literature on Dual Use." In *Innovation, Dual Use, and Security: Managing the Risks of Emerging Biological and Chemical Technologies*, edited by Jonathan B. Tucker, 19–44. Cambridge, Massachusetts: The MIT Press, 2012.
67. ———. "The Decision Framework." In *Innovation, Dual Use, and Security: Managing the Risks of Emerging Biological and Chemical Technologies*, edited by Jonathan B. Tucker, 67–82. Cambridge, Massachusetts: The MIT Press, 2012.
68. Tycko, Josh, Michael Wainberg, Georgi K. Marinov, Oana Ursu, Gaelen T. Hess, Braeden K. Ego, Aradhana, et al. "Mitigation of Off-Target Toxicity in CRISPR-Cas9 Screens for Essential Non-Coding Elements." *Nature Communications* 10, no. 1 (September 6, 2019): 1–14. https://doi.org/https://doi.org/10.1038/s41467-019-11955-7.
69. Vogel, Kathleen M. *Phantom Menace or Looming Danger? A New Framework for Assessing Bioweapons Threats*. Baltimore: Johns Hopkins University Press, 2013.
70. Wang, Yanchun, Dongshu Wang, Xiaojing Wang, Haoxia Tao, Erling Feng, Li Zhu, Chao Pan, et al. "Highly Efficient Genome Engineering in Bacillus Anthracis and Bacillus Cereus Using the CRISPR/Cas9 System." *Frontiers in Microbiology* 10 (August 27, 2019). https://doi.org/https://doi.org/10.3389/fmicb.2019.01932.
71. ResearchGate. "What Are the Basic Needs to Establish a Viral Cell Culture Lab?" Accessed October 17, 2019. https://www.researchgate.net/post/What_are_the_basic_needs_to_establish_a_viral_cell_culture_lab.
72. Williams, Katrina. "MoBio Laboratories' UltraClean Microbial DNA Isolation Kit." Biocompare, September 17, 2004. http://www.biocompare.com/Product-Reviews/41081-MoBio-Laboratories-UltraClean-Microbial-DNA-isolation-kit/.

73. Xu, Christine L, Merry Z C Ruan, Vinit B Mahajan, and Stephen H Tsang. "Viral Delivery Systems for CRISPR." *Viruses* 11, no. 28 (2019). https://doi.org/https://doi.org/10.3390/v11010028.
74. Xu, Li, Yandi Gao, Yeh Siang Lau, and Renzhi Han. "Adeno-Associated Virus-Mediated Delivery of CRISPR for Cardiac Gene Editing in Mice." *JoVE (Journal of Visualized Experiments)*, no. 138 (August 2, 2018): e57560. https://doi.org/https://doi.org/10.3791/57560.
75. Yang, Luhan, Marc Güell, Dong Niu, Haydy George, Emal Lesha, Dennis Grishin, John Aach, et al. "Genome-Wide Inactivation of Porcine Endogenous Retroviruses (PERVs)." *Science* 350, no. 6264 (November 27, 2015): 1101–4. https://doi.org/https://doi.org/10.1126/science.aad1191.
76. Yao, Xuan, Xing Wang, Junlai Liu, Linyu Shi, Pengyu Huang, and Hui Yang. "CRISPR/Cas9-Mediated Targeted Integration In Vivo Using a Homology-Mediated End Joining-Based Strategy." *JoVE (Journal of Visualized Experiments)*, no. 133 (March 12, 2018): e56844. https://doi.org/https://doi.org/10.3791/56844.
77. YouTube. *Using a Micropipette - University of Leicester*, 2009. https://www.youtube.com/watch?v=uEy_NGDfo_8.
78. Yuan, Ming, Wensheng Zhang, Jun Wang, Chadwan Al Yaghchi, Jahangir Ahmed, Louisa Chard, Nick R. Lemoine, and Yaohe Wang. "Efficiently Editing the Vaccinia Virus Genome by Using the CRISPR-Cas9 System." *Journal of Virology* 89, no. 9 (May 1, 2015): 5176–79. https://doi.org/https://doi.org/10.1128/JVI.00339-15.
79. Zayner, Josiah. "CRISPR Sucks." YouTube, January 31, 2019. https://www.youtube.com/watch?v=ZrEweAhlU4g.
80. ———. "DIY Human CRISPR Myostatin Knock-Out." YouTube, October 6, 2017. https://www.youtube.com/watch?v=o6A9bbDI6fo&t=1438s.
81. Zhang, Hong, Qiu-Xiang Cheng, Ai-Min Liu, Guo-Ping Zhao, and Jin Wang. "A Novel and Efficient Method for Bacteria Genome Editing Employing Both CRISPR/Cas9 and an Antibiotic Resistance Cassette." *Frontiers in Microbiology* 8 (May 5, 2017). https://doi.org/https://doi.org/10.3389/fmicb.2017.00812.
82. Sigma-Aldrich. "Zinc Finger Nuclease (ZFN) License Agreement." Accessed November 15, 2018. https://www.sigmaaldrich.com/life-science/zinc-finger-nuclease-technology/zfn-license-agreement.html.
83. Zou, Qingjian, Xiaomin Wang, Yunzhong Liu, Zhen Ouyang, Haibin Long, Shu Wei, Jige Xin, et al. "Generation of Gene-Target Dogs Using CRISPR/Cas9 System." *Journal of Molecular Cell Biology* 7, no. 6 (October 12, 2015): 580–83. https://doi.org/10.1093/jmcb/mjv061.

Chapter 7
Conclusion: Reducing the Perils from the Misuse of Genome Editing

Rudolf Jaenisch, Professor of Biology at the Massachusetts Institute of Technology and creator of the first transgenic animal, said CRISPR is so easy that "any idiot can do it."[1] This assessment adds an important qualifier to this statement: CRISPR is so easy that anyone can use it; *however*, CRISPR is not so easy that anyone can use it to develop a biological weapon. This book applied Tucker's *Decision Framework for Governance of Dual-Use Technologies* to determine the risk of misuse of genome editing technologies and provide policymakers recommendations for governance strategies to prevent misuse. A key feature of this risk of misuse assessment is that it is conducted within a biosocial frame, instead of the traditional biotech revolution frame, to better account for the contingencies and complexities involved in applying genome editing to develop biological weapons. If this assessment had used a biotech framing that mainly focused on what genome editing technologies could technically do and theoretically achieve, it would have likely overestimated the abilities of nefarious actors.

Ultimately, the results from the risk of misuse assessment suggest that the relative risk of misuse of CRISPR is much higher than its predecessor technologies (i.e., ZFN and TALEN) with a few important caveats. Although CRISPR is more accessible and easier to use than its predecessor technologies, the deliberate misuse of CRISPR to develop biological weapons is likely still beyond what actors with no-to-low laboratory skills, knowledge, and experience can achieve. Even actors who are proficient working in a molecular biology laboratory, but who lack specialized biological weapons knowledge, would likely still have trouble developing a biological weapon. Additionally, applications of misuse still have several barriers (i.e., technical, social, operational, logistical, and ethical) that even well-funded, well-organized state-level actors will have to overcome. These obstacles would likely

[1] Jon Cohen, "'Any Idiot Can Do It.' Genome Editor CRISPR Could Put Mutant Mice in Everyone's Reach," *Science*, November 2, 2016, https://doi.org/10.1126/science.aal0334.

K. Paris, *Genome Editing and Biological Weapons*,
https://doi.org/10.1007/978-3-031-21820-0_7

pose a near-term barrier for state actors, a medium-term barrier for skilled groups of actors, and a longer-term barrier for skilled individuals.[2]

A comprehensive strategy to reduce the potential dangers from genome editing should include efforts that range from lessening the possibility of an attack to those that reduce the negative outcomes of an attack that has taken place. This should include measures for deterrence and prevention, recognition and attribution of an attack, and consequence management.[3] There are many hard laws, soft laws, and informal measures that could be applied to deter and prevent the misuse of enabling technologies. While it is important to stop misuse before it happens, it is also important not to impede the use of genome editing technologies by scientists who rely on these technologies for legitimate research and other applications. Therefore, strategies that address security concerns of genome editing technologies need to carefully balance the cost-benefit trade-off of regulating the technologies.

As is the challenge with other emerging technologies, advancements in biodefense and biosecurity are not keeping up with innovations in genome editing. There are new applications of misuse, such as gene drives, that do not fit within the traditional mold of biological weapons. It is important to continue, reinforce, and expand on traditional approaches to nonproliferation and counterproliferation. It is also important to consider new approaches that leverage automation, artificial intelligence (AI), and other advancements in computer science and biotechnology to support a proactive approach to biodefense and biosecurity in the era of synthetic biology. This chapter first includes an assessment of the governability of genome editing technologies, specifically CRISPR. Based on the results of the governability assessment, recommendations on select strategies that could be applied to address security concerns and prevent misuse are provided to help guide policymakers.

7.1 Assessment of Governability

The assessment of governability of genome editing technologies, specifically CRISPR, is conducted here in accordance with Tucker's *Decision Framework for Governance of Dual-Use Technologies*. According to Tucker's framework, the following parameters are used to assess governability: embodiment, maturity, convergence, rate of advance, and international diffusion. The relationships between these parameters and the level of governability (scored as either high, medium, or low) of dual-use technologies are presented in Fig. 7.1.

First, embodiment considers the extent that technologies are based on hardware, intangible information, or a combination of both. In general, hardware is easier to govern through measures such as export controls, while intangible information that

[2] Kirkpatrick et al., "Editing Biosecurity: Needs and Strategies for Governing Genome Editing," 9.

[3] National Academies of Sciences, Engineering, and Medicine, *Biodefense in the Age of Synthetic Biology*, 95.

	Embodiment	Maturity	Convergence	Rate of advance	International diffusion
Low Governability	Intangible information	Not mature	High level of convergence	High rate of advance	High level of international diffusion
Medium Governability	Hybrid (combines intangible information and hardware)	Very mature	Moderate level of convergence	Moderate rate of advance	Moderate level of international diffusion
High Governability	Hardware	Moderately mature	Low level of convergence	Low rate of advance	Low level of international diffusion

Fig. 7.1 Tucker's relationships between parameters and level of governability of dual-use technologies. Tucker uses five parameters to determine governability: embodiment, maturity, convergence, rate of advance, and international diffusion. The overall governability is determined by averaging the governability score (high, medium, or low) for each parameter (Jonathan B. Tucker, ed. Foreword by Richard Danzig., *Innovation, Dual Use, and Security*, Table 4.1 (p. 74), © 2012 Massachusetts Institute of Technology, by permission of The MIT Press)

can be transmitted without detection is harder to govern.[4] CRISPR is comprised of a hybrid of both, which includes general laboratory equipment (e.g., micropipettes), reagents (e.g., CRISPR plasmids), software (e.g., gRNA design software), and techniques (e.g., molecular cloning). While a medium level of governability would typically work for technologies that are a hybrid of hardware and intangible information, this level of governability is too high for CRISPR. General laboratory equipment and reagents to use CRISPR are ubiquitous in the molecular biology community. Although the software and techniques to use CRISPR are more specialized, they are still widely available. Based on its embodiment and these stipulations, CRISPR is assessed here as having a low level of governability.

Second, maturity refers to how far the technology has moved on the developmental pipeline, which ranges from basic research to commercialization.[5] Data from this assessment on clinical trials indicate that CRISPR has moved beyond basic research and is in the stage of product development. Those working on CRISPR-based therapeutics will have to continue to overcome delivery and manufacturing hurdles. For instance, more work is needed to make viral vectors easier to manufacture and scale up.[6] Still, the use of CRISPR diagnostics is on the rise. In 2020, the

[4] Tucker, "The Decision Framework," 72.

[5] Tucker, 72.

[6] Jim Burns, "Getting CRISPR to Market: Solve Delivery, Manufacturing Issues," STAT, February 1, 2019, accessed September 26, 2019, https://www.statnews.com/2019/02/01/crispr-cas9-commercialization-delivery-manufacturing/.

FDA allowed the emergency use of a CRIPSR test for coronavirus.[7] Moreover, crops developed using genome editing technologies are already available to consumers. In February 2019, a soybean oil that was developed using TALEN by the Minnesota-based company called Calyxt was the first genome-edited food product in the United States.[8] The first food product developed using CRISPR, a tomato that was genetically edited for high levels of γ-aminobutyric acid (GABA), entered the market 2 years later.[9] Thus, CRISPR still has a medium level of governability due to its high maturity.

Third, convergence refers to how many disciplines have been pulled together to bring about the technology. It is typically harder to govern technologies that have converged from many disciplines given that "each discipline has its own professional community, culture, lexicon, and level of awareness of dual-use issues."[10] As demonstrated by this assessment, CRISPR has converged from the disciplines of bioinformatics, genetic engineering, and molecular biology. CRISPR has a moderate level of convergence given that the technology draws from a few fields. Thus, CRISPR has a medium level of governability based on its moderate convergence.

Fourth, rate of advance measures the effectiveness of the technology over time (i.e., whether it is increasing, decreasing, or plateauing) based on indicators such as "reliability, speed, throughput, accuracy, or cost."[11] It is generally tough to keep up with the governance of technologies that are advancing very fast.[12] For most of these indicators, the rate of advance of CRISPR is much faster than its predecessor technologies (ZFN and TALEN)—CRISPR is significantly more affordable and can generate genetically engineered organisms in fewer generations (e.g., cuts breeding time in crops and mice). CRISPR is a reliable technology; however, there are known issues with off-target effects. Nevertheless, the overall rate of advance of CRISPR is high, which suggests a low level of governability.

Lastly, international diffusion, the fifth parameter to assess governability, considers the extent that the technologies are accessible around the world. Typically, as technologies become available in more countries, it becomes harder to govern because of the increased need to coordinate and harmonize.[13] Data from this assessment indicate that CRISPR is widely available. As data from Addgene revealed, units of CRISPR are deposited from and shipped to countries around the world. Additionally, many software design tools that help to guide the use of genome editing technologies are available online. Furthermore, the number of CRISPR

[7] Giorgia Guglielmi, "First CRISPR Test for the Coronavirus Approved in the United States," *Nature*, May 8, 2020, https://doi.org/10.1038/d41586-020-01402-9.

[8] Cohen, "To Feed Its 1.4 Billion, China Bets Big on Genome Editing of Crops."

[9] Emily Waltz, "GABA-Enriched Tomato Is First CRISPR-Edited Food to Enter Market," *Nature Biotechnology* 40, no. 1 (December 14, 2021): 9–11, https://doi.org/10.1038/d41587-021-00026-2.

[10] Tucker, "The Decision Framework," 72.

[11] Tucker, 73.

[12] Tucker, 73.

[13] Tucker, 73.

patent documents published each year is exponentially increasing, and data from this assessment suggest that patent applicants seeking invention rights for CRISPR are pursuing a more international approach. Thus, the high level of international diffusion of CRISPR suggests it has a low level of governability.

Overall, the average governability score of CRISPR is medium to low. It is advisable to mainly use soft law (e.g., voluntary guidelines and self-governance mechanisms) and informal measures (e.g., codes of conduct, education, and other training) on technologies with medium to low governability.[14] In this case, hard law (e.g., civil and criminal statues and binding regulations) would not be as effective in regulating the technology because it is better to apply hard law to technologies with a high level of governability.[15] The assessment of the governability of CRISPR is applied later in this chapter to provide recommendations for strategies to address security concerns.

7.2 Strategies to Address Security Concerns

Based on the assessment of governability, soft law and informal measures are key to protect against the misuse of genome editing technologies. A one-size-fits-all approach will not work to govern genome editing technologies, especially CRISPR. A variety of laboratories all over the world use CRISPR for a number of purposes. This includes users who work out of their homes, community-based DIY bio laboratories, universities, government laboratories, and other public and private companies. Additionally, CRISPR is used in many areas from biomedical research to human health, agriculture, and industrial biotechnology. A comprehensive plan should include strategies to address misuse across these areas from the local, community level to the international level.

It is important for this plan to keep in mind the changing nature of biology, and the current and future of biodefense and biosecurity in the era of synthetic biology. Since the turn of the twenty-first century, biology has been moving from a descriptive science to a design science. Advancements in genetic engineering, and eventually synthetic biology, have enabled scientists to go beyond basic observing, recording, describing, and cataloging. The future of biology is oriented toward programmability and design. Where possible, the following suggestions have incorporated how advancements in automation, AI, computer science, and biotechnology can be leveraged for defense against the misuse of genome editing technologies. These suggestions include strategies to prevent and deter misuse, to bolster capabilities to recognize and attribute an attack, and to strengthen capabilities for consequence management.

[14] Tucker, 75, 77.

[15] Tucker, 77.

7.2.1 Norms Against Misuse

There should be deep-seated norms against the misuse of genome editing technologies, especially CRISPR, such that the idea of misusing the technologies is unthinkable. Norms against misuse need to address all applications of CRISPR. For some of these applications, red lines already exist. For instance, the BWC maintains a red line against the development, production, stockpiling, and use of all biological weapons. Even though the BWC is weakened by its lack of effective enforcement mechanisms, international agreements like the BWC still help to foster global cooperation and reinforce norms. One way to strengthen the language of the BWC to prevent the misuse of genome editing technologies is to clarify the definition of "other biological agents" to include "organisms modified with gene drives and genetic constructs."[16] Certain cases, such as applying genome editing to make traditional agents of concern more pathogenic, are clearly offensive in nature and are prohibited by the BWC. Other cases, such as augmenting human traits, are more ambiguous.

Bioethics committees and leaders in the defense and scientific communities can help to fill in these gaps and provide clear guidance to scientists on what constitutes appropriate research conduct. In the United States, Steven Walker, director of DARPA, expressed DAPRA's interest in genetic modifications that could protect soldiers from biological and chemical warfare agents. Walker set a precedent by clarifying that this type of work is defensive in nature since the objective is keep the soldier safe and not to engineer a super soldier.[17] This precedent translates into DARPA's programs. DARPA's PReemptive Expression of Protective Alleles and Response Elements (PREPARE) program applies insights from genome editing to develop novel medical countermeasures and technologies with effects that are temporary and reversible. Teams that work on the PREPARE program are instructed: "do not alter primary DNA sequence or create DNA lesions in a host genome in any way."[18] In France, the Minister of the Armed Forces Florence Parly created the Defense Ethics Committee in January 2020, which investigated issues on several topics including soldier enhancement. Later that year, the French Defense Ethics Committee issued a report that established a framework on limits for handling military human enhancement and drew a red line on unacceptable practices for

---

[16] Jesse Kirkpatrick et al., "Biosecurity in the Age of Genome Editing – YouTube," "Biosecurity in the Age of Genome Editing-YouTube," accessed October 13, 2019, https://www.youtube.com/watch?v=jCTDuy5pE_I&list=PLPuptUEHTSjDGWQ16svQQuOXf78g4z4Bn&index=54&t=0s; Kirkpatrick et al., "Editing Biosecurity: Needs and Strategies for Governing Genome Editing," 68.

[17] Todd Lopez, "DARPA Director Talks Promise of Life Sciences Research," U.S. DEPARTMENT OF DEFENSE, September 24, 2019, accessed October 5, 2019, https://www.defense.gov/explore/story/Article/1969741/darpa-director-talks-promise-of-life-sciences-research/.

[18] "PReemptive Expression of Protective Alleles and Response Elements Program Information," DARPA: PREPARE Proposers Day, June 12, 2018, accessed May 11, 2022, https://events.sa-meetings.com/ehome/306665/661233/.

genetic engineering.[19] To prevent distrust and an international arms race, governments and scientists should promote transparency of their work.

There will be challenges in developing norms that are shared by governments and scientists around the world because there are divergent viewpoints on applications of CRISPR both within and among countries. For instance, country A might think one way about gene drives, while country B might think another way; and these countries might have different opinions about germline and somatic gene therapy, too.[20] This concept is illustrated by the different viewpoints that exist in China and the United States on deafness in humans: "deaf Americans welcome more deaf American children into the deaf community, in China, that's considered a less viable human."[21] Given these differences, it is essential for security experts, scientists, and policymakers from different countries to come together to develop norms.[22] Ultimately, the goal is for everyone who works with genome editing to share a common view that the technology should be used for peaceful applications and to do their best to safeguard against intentional, accidental, and reckless misuse.[23]

7.2.2 *Responsible Research Culture*

A responsible research culture will reinforce norms against misuse. One of the most important lines of defense against the misuse of CRISPR is the scientists, themselves, who witness the day in and day out work. Scientists need to be encouraged to speak up if they suspect suspicious activity. Already, the incident of He Jiankui using CRISPR for germline gene therapy has illustrated the need to encourage a culture of responsibility. Since He Jiankui debuted his research, more details have emerged including how five-dozen people either knew of or suspected the study before it was publicized. The incident has been called a "failure of self-regulation by the scientific community."[24] While George Church advises there should be a

---

[19] Pierre Bourgois, "Strategic Brief No. 18 – 2021 – 'Yes to Iron Man, No to Spiderman!' A New Framework for the Enhanced Soldier Brought by the Report from the Defense Ethics Committee in France," IRSEM, February 24, 2021, accessed March 29, 2022, https://www.irsem.fr/publications-de-l-irsem/breves-strategiques/strategic-brief-no-18-2021.html.

[20] Kirkpatrick et al., "Biosecurity in the Age of Genome Editing – YouTube."

[21] "The Future of Genetic Engineering – George Church – YouTube," March 18, 2019, accessed October 14, 2019, https://www.youtube.com/watch?v=FGCFIED815Y&t=1s.

[22] Kirkpatrick et al., "Biosecurity in the Age of Genome Editing – YouTube," "Biosecurity in the Age of Genome Editing-YouTube."

[23] Kirkpatrick et al., "Biosecurity in the Age of Genome Editing-YouTube."

[24] Jon Cohen, "The Untold Story of the 'Circle of Trust' behind the World's First Gene-Edited Babies," Science | AAAS, August 1, 2019, accessed October 1, 2019, https://www-sciencemag-org.mutex.gmu.edu/news/2019/08/untold-story-circle-trust-behind-world-s-first-gene-edited-babies.

balanced discussion on this incident, he suggests that no one who was aware of the study felt a "desire or need to report" and that it was a "failure of the system."[25]

The scientific community should learn from the incident mentioned above and promote responsible conduct. A 2011 report by the NSABB on how to strengthen the culture of responsibility in the laboratory recommended that institutions have programs or mechanisms that support the reporting of concerning behavior. Additionally, the 2011 NSABB report recommended education and training to promote research ethics and responsible conduct. Trainings should also cover warning signs of concerning behaviors. Furthermore, leadership at the institution plays a key role in supporting the organization's culture and promoting trust, integrity, responsibility, and biosecurity. Thus, it is critical for leadership to communicate the expectations of responsible CRISPR behavior and to ensure the institution is abiding by all laws, regulations, and policies of the institution.[26]

Institutional biosafety offices and committees also are a part of promoting responsible conduct. For instance, biosafety guidelines in 2018 at the University of Florida indicated that all CRISPR-Cas research at the university should be registered with the Biosafety Office and reviewed by the Institutional Biosafety Committee.[27] The University of Florida Biological Safety Program provided researchers with guidance on how to assess and mitigate potential risk with work that involves CRISPR.[28] Responsible conduct also means implementing alert systems so that accidents and near misses are reported. In a discussion on oversight for research and development that involves gene drives, the 2018 *Editing Biosecurity* report suggested that alert systems should be developed to make sure "rapid responses and information are fed into networks that adapt policies and practices."[29]

In having adopted their own practices to support biosafety and biosecurity, two non-traditional research communities—the DIY bio community labs and the International Genetically Engineered Machine (iGEM) competition—serve as a model for promoting responsibility. While the DIY bio community has been depicted in the popular media as reckless actors who could develop doomsday pathogens, Todd Kuiken advised how the DIY bio community sought to disprove this inaccurate idea

[25] "The Future of Genetic Engineering – George Church – YouTube."

[26] A Report of the National Science Advisory Board for Biosecurity, "Guidance for Enhancing Personnel Reliability and Strengthening the Culture of Responsibility," September 2011, 20–24.

[27] "Biosafety Guidelines for CRISPR-Cas Research » Environmental Health & Safety » University of Florida," accessed December 18, 2018, http://www.ehs.ufl.edu/programs/bio/crispr-cas-research/.

[28] "Biosafety Guidelines for CRISPR-Cas Research » Environmental Health & Safety » University of Florida I Risk Assessment Questions and Risk Mitigation," accessed October 16, 2019, http://www.ehs.ufl.edu/programs/bio/crispr-cas-research/.

[29] Kirkpatrick et al., "Editing Biosecurity: Needs and Strategies for Governing Genome Editing," 74.

and worked to promote a culture of responsibility.[30] This is evident in the DIY bio code of ethics that the community created for itself.[31] Additionally, at iGEM, which is an international synthetic biology competition with student teams that use standard biological parts to genetically engineer organisms, there is a Safety Committee as well as a Safety and Security Hub to support participants. Furthermore, parts that are used in the competition are screened by a commercial partner. iGEM instills a sense of responsibility for safety for all participants, including supervisors and principal investigators.[32] iGEM uses an adaptive risk management approach to biosafety and biosecurity that encompasses traditional agent-based risks and emerging risks from new technologies as well as new risks.[33] As a case study on the iGEM competition reveals, this approach could be especially useful for overseeing gene drive-related research.[34] The scientific community at large could adopt similar practices as these nontraditional communities to promote responsible CRISPR conduct.

Scientists should be encouraged and trained to use state-of-the-art safeguards. In 2017, DARPA announced that it would invest $65 million over 4 years in a program, called SAFE Genes, to protect against genome editing technologies.[35] Initially, SAFE gene performers worked to understand fundamentals of genome editing to provide a foundation for the safe development of the technology and to learn how to defeat its potential misuse. One of its research teams is working to improve control over when and where genome editing function, and the team has developed open-source software for scientists to analyze genome editing sequences quickly and accurately. Other teams are working on anti-CRISPR proteins that would block or diminish unwanted edits in cells. Additionally, teams are working to understand gene drive performance, so they can develop tools to control, counter, or reverse the effects of gene drives.[36]

---

[30] Piers Millet and Todd Kuiken, "Genome Editing and Security: Governance of Non-Traditional Research Communities" (Webinar, U.S. National Academies of Sciences, Engineering, and Medicine, June 18, 2018), http://nas-sites.org/dels/events/ge-security/.

[31] "Codes," *DIYbio* (blog), March 29, 2011, accessed October 27, 2019, https://diybio.org/codes/.

[32] Millet and Kuiken, "Genome Editing and Security: Governance of Non-Traditional Research Communities"; "Safety & Security at IGEM," iGEM, accessed October 17, 2019, https://igem.org/Safety; "Safety and Security Hub," iGEM, accessed October 17, 2019, http://2018.igem.org/Safety.

[33] Piers Millett et al., "Developing a Comprehensive, Adaptive, and International Biosafety and Biosecurity Program for Advanced Biotechnology: The IGEM Experience," *Applied Biosafety* 24, no. 2 (June 2019): 64–71, https://doi.org/10.1177/1535676019838075.

[34] Piers Millett et al., "IGEM and Gene Drives: A Case Study for Governance," *Health Security* 20, no. 1 (February 1, 2022): 26–34, https://doi.org/10.1089/hs.2021.0157.

[35] "Building the Safe Genes Toolkit," DARPA, July 19, 2017, accessed September 8, 2021, https://www.darpa.mil/news-events/2017-07-19.

[36] "Safe Genes Tool Kit Takes Shape," DARPA, October 15, 2019, accessed September 13, 2021, https://www.darpa.mil/news-events/2019-10-15.

7.2.3 Product Screening and Sales Training

In addition to the users of CRISPR, the providers of goods and services should also have a role in ensuring CRISPR is not misused by screening product orders and providing sales awareness training. As this assessment revealed, there are many providers of CRISPR products and services across the academic, non-profit, and commercial sectors. The U.S. government should work with these providers to establish voluntary guidelines that incorporate "know your customer" standards.[37] These standards could be modeled after the voluntary guidelines that the International Gene Synthesis Consortium, an international group of DNA synthesis firms, have adopted to screen customer orders. Federal incentives could be provided to encourage the industry to make purchases from companies that follow customer screening standards.[38]

Sales personnel should be trained on red flags in CRISPR customers as well as how to report suspicious interactions with customers. The 2010 voluntary guidelines published by the Department of Health and Human Services, titled *Screening Framework for Guidance for Providers of Synthetic Double-Stranded DNA*, provides a list of indicators of suspicious orders, such as when a customer's identity is unclear and cannot be confirmed, the customer has an odd form of payment (e.g., cash, non-bank third party), or the customer requests transaction records be destroyed. Additionally, the Guidance provides instruction on how to report suspicious interactions to the Federal Bureau of Investigation, Centers for Disease Control and Prevention, Animal and Plant Health Inspection Service, and Department of Commerce.[39] A similar set of guidelines could be used to help providers of CRISPR flag suspicious orders.

Gaps in screening should be closed. The 2010 guidance from the Department of Health and Human Services, which provided voluntary guidelines for U.S.-based providers of synthetic DNA, only covered dsDNA greater than 200 base pairs (i.e., not short oligonucleotides) and did not have a mechanism to assess if companies are in compliance.[40] While there are fears that screening shorter DNA sequences might lead to an excessive number of false positives, there may be ways to improve screening for short oligonucleotides and make access to DNA synthesis safer and

---

[37] Kirkpatrick et al., "Editing Biosecurity: Needs and Strategies for Governing Genome Editing," 70.

[38] Kirkpatrick et al., 70–71.

[39] Department of Health and Human Services, *Screening Framework Guidance for Providers of Synthetic Double-Stranged DNA* (Washington, DC: Department of Health and Human Services, 2010), 7, 19.

[40] Kirkpatrick et al., "Editing Biosecurity: Needs and Strategies for Governing Genome Editing," 71; Department of Health and Human Services, *Screening Framework Guidance for Providers of Synthetic Double-Stranged DNA*.

more secure.[41] An Intelligence Advanced Research Projects Activity program on Functional Genomic and Computational Assessment of Threats (Fun GCAT) launched in June 2017 to improve DNA synthesis order screening tools. The program works on threat screening software that is powered by AI.[42] It is important to continue to fund this type of research and develop better tools for screening.

7.2.4 Financial Loophole Closure

Governments should ensure oversight loopholes on privately funded research are closed. This assessment demonstrates that not only have governments invested billions of dollars into CRISPR research, but private investors have also pumped over a billion dollars into CRISPR-focused startup companies. Currently, there is a lack of oversight of privately funded dual-use research of concern (DURC). DURC is described as the following:

> life sciences research that, based on current understanding, can be reasonably anticipated to provide knowledge, information, products, or technologies that could be directly misapplied to pose a significant threat with broad potential consequences to public health and safety, agricultural crops and other plants, animals, the environment, materiel, or national security.[43]

Per the 2018 *Editing Biosecurity* report, "DURC oversight only applies to public or private research institutions that receive Federal funding for life science research" as well as "foreign research institutions that receive U.S. funding for DURC research.[44] This is worrisome because CRISPR research that is privately funded and falls within the definition of DURC is exempt from oversight.

One way to close the DURC loophole in the United States is through legislation that mandates a nationwide DURC oversight system for research that is publicly and privately funded.[45] A policy such as this would be effective if there is compliance. A 2016 report by the NSABB recommended, "The effectiveness of policies can be enhanced by a commitment to proper implementation and enforcement at the federal,

[41] James Diggans and Emily Leproust, "Next Steps for Access to Safe, Secure DNA Synthesis," *Frontiers in Bioengineering and Biotechnology* 7, no. 86 (April 24, 2019): 3, https://doi.org/10.3389/fbioe.2019.00086.

[42] "IARPA – Fun GCAT," Intelligence Advanced Research Projects Activity, accessed April 6, 2022, https://www.iarpa.gov/research-programs/fun-gcat.

[43] "Dual Use Research of Concern," National Institutes of Health Office of Science Policy, accessed September 25, 2019, https://osp.od.nih.gov/biotechnology/dual-use-research-of-concern/.

[44] Kirkpatrick et al., "Editing Biosecurity: Needs and Strategies for Governing Genome Editing," 65; U.S. Department of Health and Human Services, "Dual Use Research of Concern," accessed September 25, 2019, https://www.phe.gov/s3/dualuse/Pages/default.aspx.

[45] Kirkpatrick et al., "Editing Biosecurity: Needs and Strategies for Governing Genome Editing," 64.

institutional, and individual investigator levels."[46] Ways to support the implementation and enforcement of policy could be through "training, education, codes of conduct, and other mechanisms for continuing to build a culture of responsibility."[47] While the assessment of the governability of CRISPR at the beginning of this chapter did not find hard law to be as effective of a governance measure, it is, nonetheless, still critical to close known gaps in oversight.

7.2.5 Bioengineered Attack Indicators

Governments should work to establish indicators of bioengineered attacks to facilitate the prompt detection, response, and mitigation of attacks. Current capabilities for detecting biological engineering are slow, expensive, and limited in their ability to detect only a portion of all possible genetic modifications. The Finding Engineering-Linked Indicators (FELIX) is an Intelligence Advanced Research Projects Activity program that began in 2018 to create new computational and experimental tools for detecting engineered organisms. Early detection of engineered organisms will help to speed up appropriate responses and reduce downstream consequences from an attack.[48]

Furthermore, it would be helpful to have the capability to forensically trace and attribute bioengineered attacks to perpetrators. One way to approach this problem is to have registries that track known expertise and materials that could be used to engineer a biological weapon. Additionally, it may be possible to profile designs that come from these laboratories, personnel, and materials and develop digital signatures of typical bioengineering designs. Deviation from typical signatures could indicate nefarious actions. It would be hard, however, to obtain access to the data needed to develop profile designs and to differentiate between legitimate and illegitimate work.[49]

7.2.6 Diagnostics, Vaccines, and Other Medical Countermeasures

The current U.S. biodefense strategy that is based on lists, such as the Federal Select Agent Program Select Agent and Toxins list, along with access controls on physical

---

[46] A Report of the National, Science Advisory Board for, and Biosecurity, "Recommendations for the Evaluation and Oversight Proposed Gain-of-Function Research," May 2016, 37.

[47] A Report of the National, Science Advisory Board for, and Biosecurity, 37.

[48] "IARPA – FELIX," accessed April 7, 2022, https://www.iarpa.gov/research-programs/felix.

[49] National Academies of Sciences, Engineering, and Medicine, *Biodefense in the Age of Synthetic Biology*, 105.

material are not sufficient to manage all the possible threats that are enabled by genetic engineering and synthetic biology.[50] As this assessment demonstrated, genome editing could be applied by nefarious actors to circumvent agent-specific therapeutics. One approach to overcome this challenge is to pursue more generalized therapeutic strategies. For instance, it may be possible to invest in research on vaccines that more generally boost the immune system response.[51] The teams of the DARPA PREPARE program are applying insights from genome editing and working on a more general approach to influenza defense and treatment, which could possibly protect against nearly all influenza strains.[52] Additionally, investments should be made in CRISPR-based diagnostics, such as Specific High-Sensitivity Enzymatic Reporter UnLOCKing (SHERLOCK), which have already demonstrated their ability to detect biological agents quickly and affordably with high specificity and high sensitivity.[53] In May 2020, the FDA approved emergency use of SHERLOCK, the CRIPSR-based diagnostic tool, to detect coronavirus.[54] In the future, the fast, reliable, and deployable detection platforms, like SHERLOCK, may be preferred over gold standard PCR-based diagnostics.

7.3 The Future of Biodefense and Biosecurity

Investments into biotechnology are critical to secure the future technological leadership of the United States. According to a report to Congress from the National Security Commission on Artificial Intelligence (NSCAI), biotechnology is among the top eight critical technologies that the United States needs to invest in to ensure its competitiveness. Advancements in genetic engineering and synthetic biology enable the programmability of biology, which can be optimized by AI to make biotechnology breakthroughs and offer solutions to pressing global challenges in human health, agriculture, and the environment.[55] On the other hand, AI could also be used by adversaries to develop precisely engineered biological agents or to

---

[50] Gregory Koblentz, "The Impact of Synthetic Biology and Genome Editing on Biosecurity – YouTube," accessed October 14, 2019, https://www.youtube.com/watch?v=O8yM6UKM0Ew&list=PLPuptUEHTSjDGWQ16svQQuOXf78g4z4Bn&index=46&t=178s; National Academies of Sciences, Engineering, and Medicine, *Biodefense in the Age of Synthetic Biology*, 126.

[51] Petro, Plasse, and McNulty, "Biotechnology," 166.

[52] "A Dose of Inner Strength to Survive and Recover from Potentially Lethal Health Threats," DARPA, accessed May 11, 2022, https://www.darpa.mil/news-events/2019-06-27.

[53] "CRISPR • Sherlock Biosciences," accessed April 7, 2022, https://sherlock.bio/platforms/crispr/.

[54] Kyle E. Watters et al., "The CRISPR Revolution and Its Potential Impact on Global Health Security," *Pathogens and Global Health* 115, no. 2 (n.d.): 80–92, https://doi.org/10.1080/20477724.2021.1880202.

[55] *Final Report National Security Commission on Artificial Intelligence*, 2021, 260–61, 439, accessed April 3, 2022, https://www.nscai.gov/wp-content/uploads/2021/03/Full-Report-Digital-1.pdf.

optimize the enhancement of humans. Overall, the NSCAI report warned, "The U.S. government is not prepared to defend the United States in the coming artificial intelligence (AI) era."[56]

Moving forward, the United States and its partners should not be caught off guard by the rise of technological advancements, especially in China. The world was surprised in 2018 when Chinese scientists used genome editing on human embryos. Especially in Western nations, this practice is unethical. This assessment on the risk of misuse of genome editing illuminated how China hopes to develop biologically enhanced soldiers and has already undertaken human testing on members of its armed forces.[57] Countries, like China, could be motivated to more broadly use genome editing on their population to improve IQ and subsequently gain a global economic advantage. It is no secret that China hopes to surpass the United States and soon become the global frontrunner in AI.[58] Already, the invasive use of AI and biometric data for repression and surveillance in China is unsettling. We should not underestimate how state and non-state actors will use AI and other technologies to advance the threat from biotechnologies, like genome editing.

To reduce this threat, it is important for scientists, security experts, and policy makers to continue to engage in international forums and find common ground for biodefense and biosecurity measures. In the summer of 2019, over 100 representatives from China, Europe, and the United States came together in Washington, D.C. to discuss biosafety and biosecurity challenges, opportunities, and governance options for synthetic biology. Participants agreed that there was a deficiency of international governance structures for synthetic biology, and they called on China and the United States as leaders in synthetic biology to help establish norms and standards. Participants advised that they needed to meet again to continue discussions.[59]

Biological weapons have been criticized in the past for their perceived lack of utility. We need to continue to make them useless. It is imperative for governments to double down on their efforts to commit to using genome editing for ethical and peaceful applications. Governments need to also continue to invest in genome editing defensive systems. The goal is to dissuade people from even trying to make a genetically engineered biological weapon because it would be so easily detected and countered that it would be pointless. Governments alone, however, will not stop misuse. It is up to the institutions to implement policies, for leaders to reinforce them, and for scientists to carry them out. At the end of the day, scientists who are working with CRISPR at the lab bench will make the ultimate decision.

---

[56] *NSCAI Report*, 45.

[57] John Ratcliffe, "China Is National Security Threat No. 1."

[58] *NSCAI Report*, 2.

[59] Thomas Inglesby et al., "Biosafety and Biosecurity in the Era of Synthetic Biology: Meeting the Challenges in China and the U.S.," *Journal of Biosafety and Biosecurity* 1, no. 2 (September 1, 2019): 73–74, https://doi.org/10.1016/j.jobb.2019.09.003.

References

1. DARPA. “A Dose of Inner Strength to Survive and Recover from Potentially Lethal Health Threats.” Accessed May 11, 2022. https://www.darpa.mil/news-events/2019-06-27.
2. A Report of the National Science Advisory Board for Biosecurity. “Guidance for Enhancing Personnel Reliability and Strengthening the Culture of Responsibility,” September 2011.
3. A Report of the National, Science Advisory Board for, and Biosecurity. “Recommendations for the Evaluation and Oversight Proposed Gain-of-Function Research,” May 2016.
4. “Biosafety Guidelines for CRISPR-Cas Research » Environmental Health & Safety » University of Florida.” Accessed December 18, 2018. http://www.ehs.ufl.edu/programs/bio/crispr-cas-research/.
5. “Biosafety Guidelines for CRISPR-Cas Research » Environmental Health & Safety » University of Florida I Risk Assessment Questions and Risk Mitigation.” Accessed October 16, 2019. http://www.ehs.ufl.edu/programs/bio/crispr-cas-research/.
6. DARPA. “Building the Safe Genes Toolkit,” July 19, 2017. https://www.darpa.mil/news-events/2017-07-19.
7. Burns, Jim. “Getting CRISPR to Market: Solve Delivery, Manufacturing Issues.” STAT, February 1, 2019. https://www.statnews.com/2019/02/01/crispr-cas9-commercialization-delivery-manufacturing/.
8. DIYbio. “Codes,” March 29, 2011. https://diybio.org/codes/.
9. Cohen, Jon. “‘Any Idiot Can Do It.’ Genome Editor CRISPR Could Put Mutant Mice in Everyone’s Reach.” *Science*, November 2, 2016. https://doi.org/10.1126/science.aal0334.
10. ———. “The Untold Story of the ‘Circle of Trust’ behind the World’s First Gene-Edited Babies.” Science I AAAS, August 1, 2019. https://www-sciencemag-org.mutex.gmu.edu/news/2019/08/untold-story-circle-trust-behind-world-s-first-gene-edited-babies.
11. ———. “To Feed Its 1.4 Billion, China Bets Big on Genome Editing of Crops.” Science I AAAS, July 29, 2019. https://www-sciencemag-org.mutex.gmu.edu/news/2019/07/feed-its-14-billion-china-bets-big-genome-editing-crops.
12. “CRISPR • Sherlock Biosciences.” Accessed April 7, 2022. https://sherlock.bio/platforms/crispr/.
13. Department of Health and Human Services. *Screening Framework Guidance for Providers of Synthetic Double-Stranged DNA*. Washington, DC: Department of Health and Human Services, 2010.
14. Diggans, James, and Emily Leproust. “Next Steps for Access to Safe, Secure DNA Synthesis.” *Frontiers in Bioengineering and Biotechnology* 7, no. 86 (April 24, 2019). https://doi.org/10.3389/fbioe.2019.00086.
15. National Institutes of Health Office of Science Policy. “Dual Use Research of Concern.” Accessed September 25, 2019. https://osp.od.nih.gov/biotechnology/dual-use-research-of-concern/.
16. *Final Report National Security Commission on Artificial Intelligence*, 2021. https://www.nscai.gov/wp-content/uploads/2021/03/Full-Report-Digital-1.pdf.
17. Guglielmi, Giorgia. “First CRISPR Test for the Coronavirus Approved in the United States.” *Nature*, May 8, 2020. https://doi.org/10.1038/d41586-020-01402-9.
18. “IARPA – FELIX.” Accessed April 7, 2022. https://www.iarpa.gov/research-programs/felix.
19. Intelligence Advanced Research Projects Activity. “IARPA – Fun GCAT.” Accessed April 6, 2022. https://www.iarpa.gov/research-programs/fun-gcat.
20. Inglesby, Thomas, Anita Cicero, Caitlin Rivers, and Weiwen Zhang. “Biosafety and Biosecurity in the Era of Synthetic Biology: Meeting the Challenges in China and the U.S.” *Journal of Biosafety and Biosecurity* 1, no. 2 (September 1, 2019): 73–74. https://doi.org/10.1016/j.jobb.2019.09.003.
21. John Ratcliffe. “China Is National Security Threat No. 1.” The Wall Street Journal, December 3, 2020. https://www.wsj.com/articles/china-is-national-security-threat-no-1-11607019599.

22. Kirkpatrick, Jesse, Gregory D. Koblentz, Megan J. Palmer, Edward Perello, David A. Relman, and Sarah W. Denton. "Editing Biosecurity: Needs and Strategies for Governing Genome Editing," December 2018.
23. Kirkpatrick, Jesse, Gregory Koblentz, Megan Palmer, and Edward Perello. "Biosecurity in the Age of Genome Editing – YouTube." Accessed October 13, 2019. https://www.youtube.com/watch?v=jCTDuy5pE_I&list=PLPuptUEHTSjDGWQ16svQQuOXf78g4z4Bn&index=54&t=0s.
24. Koblentz, Gregory. "The Impact of Synthetic Biology and Genome Editing on Biosecurity – YouTube." Accessed October 14, 2019. https://www.youtube.com/watch?v=O8yM6UKM0Ew&list=PLPuptUEHTSjDGWQ16svQQuOXf78g4z4Bn&index=46&t=178s.
25. Lopez, Todd. "DARPA Director Talks Promise of Life Sciences Research." U.-S. DEPARTMENT OF DEFENSE, September 24, 2019. https://www.defense.gov/explore/story/Article/1969741/darpa-director-talks-promise-of-life-sciences-research/.
26. Millet, Piers, and Todd Kuiken. "Genome Editing and Security: Governance of Non-Traditional Research Communities." Webinar, U.S. National Academies of Sciences, Engineering, and Medicine, June 18, 2018. http://nas-sites.org/dels/events/ge-security/.
27. Millett, Piers, Tessa Alexanian, Megan J. Palmer, Sam Weiss Evans, Todd Kuiken, and Kenneth Oye. "IGEM and Gene Drives: A Case Study for Governance." *Health Security* 20, no. 1 (February 1, 2022): 26–34. https://doi.org/10.1089/hs.2021.0157.
28. Millett, Piers, Thomas Binz, Sam Weiss Evans, Todd Kuiken, Ken Oye, Megan J. Palmer, Cécile van der Vlugt, Kathrina Yambao, and Samuel Yu. "Developing a Comprehensive, Adaptive, and International Biosafety and Biosecurity Program for Advanced Biotechnology: The IGEM Experience." *Applied Biosafety* 24, no. 2 (June 2019): 64–71. https://doi.org/10.1177/1535676019838075.
29. National Academies of Sciences, Engineering, and Medicine. *Biodefense in the Age of Synthetic Biology*. Washington, DC: The National Academies Press, 2018. https://doi.org/10.17226/24890.
30. Petro, James B., Theodore R. Plasse, and Jack A. McNulty. "Biotechnology: Impact on Biological Warfare and Biodefense." *Biosecurity and Bioterrorism: Biodefense Strategy, Practice, and Science* 1, no. 3 (September 1, 2003): 161–68. https://doi.org/10.1089/153871303769201815.
31. Pierre Bourgois. "Strategic Brief No. 18 – 2021 – 'Yes to Iron Man, No to Spiderman!' A New Framework for the Enhanced Soldier Brought by the Report from the Defense Ethics Committee in France." IRSEM, February 24, 2021. https://www.irsem.fr/publications-de-l-irsem/breves-strategiques/strategic-brief-no-18-2021.html.
32. DARPA: PREPARE Proposers Day. "PReemptive Expression of Protective Alleles and Response Elements Program Information," June 12, 2018. https://events.sa-meetings.com/ehome/306665/661233/.
33. DARPA. "Safe Genes Tool Kit Takes Shape," October 15, 2019. https://www.darpa.mil/news-events/2019-10-15.
34. iGEM. "Safety & Security at IGEM." Accessed October 17, 2019. https://igem.org/Safety.
35. iGEM. "Safety and Security Hub." Accessed October 17, 2019. http://2018.igem.org/Safety.
36. "The Future of Genetic Engineering – George Church – YouTube," March 18, 2019. https://www.youtube.com/watch?v=FGCFIED815Y&t=1s.
37. Tucker, Jonathan B. "The Decision Framework." In *Innovation, Dual Use, and Security: Managing the Risks of Emerging Biological and Chemical Technologies*, edited by Jonathan B. Tucker, 67–82. Cambridge, Massachusetts: The MIT Press, 2012.
38. U.S. Department of Health and Human Services. "Dual Use Research of Concern." Accessed September 25, 2019. https://www.phe.gov/s3/dualuse/Pages/default.aspx.
39. Waltz, Emily. "GABA-Enriched Tomato Is First CRISPR-Edited Food to Enter Market." *Nature Biotechnology* 40, no. 1 (December 14, 2021): 9–11. https://doi.org/10.1038/d41587-021-00026-2.

40. Watters, Kyle E., Jesse Kirkpatrick, Megan J. Palmer, and Gregory D. Koblentz. "The CRISPR Revolution and Its Potential Impact on Global Health Security." *Pathogens and Global Health* 115, no. 2 (n.d.): 80–92. https://doi.org/10.1080/20477724.2021.1880202.

Appendix A: Providers of Genome Editing Technologies

Table A.1 Providers of ZFN, TALEN, and CRISPR (Data for the table were obtained from the Science Exchange on April 18, 2019)

	Sector	Provider name	Location	ZFN	TALEN	CRISPR
1	Academic	Animal Models Core at University of North Carolina at Chapel Hill School of Medicine	North Carolina, USA			X
2	Academic	Animal Models Shared Resource - Zebrafish at Georgetown University Medical Center	Washington, DC, USA			X
3	Academic	Drug Discovery Core at University of Utah	Utah, USA			X
4	Academic	ES Cell Targeting Core at the University of California, San Francisco	California, USA		X	X
5	Academic	Functional Genomics (Sanford Burnham Medical Research Institute) at Sanford Burnham Medical Research Institute	California, USA			X
6	Academic	Gene Targeting and Transgenic Core at University of Rochester Medical Center	New York, USA			X
7	Academic	Genome Engineering and iPSC Center at Washington University in St. Louis	Missouri, USA			X

(continued)

K. Paris, *Genome Editing and Biological Weapons*,
https://doi.org/10.1007/978-3-031-21820-0

Table A.1 (continued)

	Sector	Provider name	Location	ZFN	TALEN	CRISPR
8	Academic	Mouse Gene Modification Facility at Albert Einstein College of Medicine	New York, USA			X
9	Academic	Mouse Genome Engineering Core at UT Southwestern Medical Center	Texas, USA			X
10	Academic	Mutation Generation and Detection Facility at the University of Utah	Utah, USA		X	X
11	Academic	Transgenic & Gene Targeting Core at University of Utah	Utah, USA			X
12	Academic	Translational Genomics Facility	Wisconsin, USA			X
13	Academic	USC Stem Cell Engineering at University of Southern California Keck School of Medicine	California, USA			X
14	Academic	UCSF Laboratory for Cell Analysis—Genome Analysis Core Services at University of California, San Francisco	California, USA			X
15	Academic	Viral Vector Core at La Jolla at Sanford Burnham Medical Research Institute	California, USA			X
16	Nonprofit (nonacademic)	Addgene	Massachusetts, USA	X	X	X
17	Nonprofit (nonacademic)	ATCC	Virginia, USA			X
18	Nonprofit (nonacademic)	Bioneer A/S	Hørsholm, Denmark			X
19	Nonprofit (nonacademic)	Nanjing BioMedical Research Institute of Nanjing University	Jiangsu, China		X	X
20	Commercial	AcceGen Biotechnology	New Jersey, USA			X
21	Commercial	Aelian Biotechnology	Landstraße, Austria			X
22	Commercial	Agilent Technologies, Inc.	California, USA			X
23	Commercial	Aldevron—Protein at Aldevron	Wisconsin, USA			X
24	Commercial	Allele Biotechnology and Pharmaceuticals, Inc.	California, USA			X

(continued)

Table A.1 (continued)

	Sector	Provider name	Location	ZFN	TALEN	CRISPR
25	Commercial	AMRI—Drug Discovery Solutions at AMRI	Location not available			X
26	Commercial	Applied Biological Materials	British Columbia, Canada			X
27	Commercial	Applied StemCell, Inc.	California, USA			X
28	Commercial	Axol Bioscience	Little Chesterford, UK			X
29	Commercial	Axolabs GmbH	Kulmbach, Germany			X
30	Commercial	B-MoGen Biotechnologies	Minnesota, USA			X
31	Commercial	Basepair	New York, USA			X
32	Commercial	BestGene Inc	California, USA			X
33	Commercial	Biobide	Donostia, Spain		X	X
34	Commercial	Biocytogen LLC.	Massachusetts, USA; Beijing, China		X	X
35	Commercial	Blue Heron Biotech, LLC.	Washington, USA			X
36	Commercial	Brooklyn Bio Inc.	New York, USA			X
37	Commercial	Canopy Biosciences	Missouri, USA			X
38	Commercial	Charles River	Massachusetts, USA			X
39	Commercial	Cellecta, Inc.	California, USA			X
40	Commercial	Censo Biotechnologies	Roslin, United Kingdom	X		X
41	Commercial	COBO Technologies Aps	Maaloev, Denmark			X
42	Commercial	Creative Biogene	New York, USA		X	X
43	Commercial	Cyagen Biosciences	California, USA			X
44	Commercial	Desktop Genetics	London, UK			X
45	Commercial	Dharmacon, Inc. (now Dharmacon, a Horizon Discovery Group company)	Colorado, USA			X
46	Commercial	DiaCarta, Inc.	California, USA			X

(continued)

Table A.1 (continued)

	Sector	Provider name	Location	ZFN	TALEN	CRISPR
47	Commercial	EdiGene Biotechnology Inc.	Massachusetts, USA			X
48	Commercial	EverCell Bio, INC.	Massachusetts, USA			X
49	Commercial	Gene Oracle Inc	California, USA			X
50	Commercial	GeneCopoeia, Inc.	Maryland, USA		X	X
51	Commercial	GeneWerk	Heidelberg, Germany	X	X	X
52	Commercial	GENEWIZ, Inc.	New Jersey, USA			X
53	Commercial	GenScript USA Inc	New Jersey, USA			X
54	Commercial	HD Biosciences	Shanghai, China			X
55	Commercial	Horizon Discovery Group plc	Cambridge, UK	X		X
56	Commercial	ingenious targeting laboratory	New York, USA			X
57	Commercial	Innovative Protein Technologies	Ontario, Canada			X
58	Commercial	Integrated DNA Technologies	Illinois, USA			X
59	Commercial	KBioBox LLC	Massachusetts, USA		X	X
60	Commercial	KromaTiD, Inc.	Colorado, USA			X
61	Commercial	Labomics	Nivelles, Belgium	X	X	X
62	Commercial	MedGenome	California, USA			X
63	Commercial	Millipore Sigma (formerly, Sigma-Aldrich)	St. Louis, Missouri and Burlington, Massachusetts	X		X
64	Commercial	NBS Biologicals	Huntingdon, UK			X
65	Commercial	NemaMetrix	Utah, USA			X
66	Commercial	New England Biolabs	Massachusetts, USA			X
67	Commercial	NMI TT Pharmaservices	Reutlingen, Germany			X
68	Commercial	NovaBioAssays LLC	Massachusetts, USA			X
69	Commercial	OnRamp Bioinformatics	California, USA			X

(continued)

Table A.1 (continued)

	Sector	Provider name	Location	ZFN	TALEN	CRISPR
70	Commercial	OriGene Technologies, Inc	Maryland, USA			X
71	Commercial	Oxford Genetics Ltd	New York, USA			X
72	Commercial	PNA BIO INC	California, USA			X
73	Commercial	Quick Biology Inc.	California, USA			X
74	Commercial	SciGenom	Kerala, India			X
75	Commercial	Sunomix Biosciences	California, USA		X	X
76	Commercial	Synbio Technologies	New Jersey, USA			X
77	Commercial	Synthego	California, USA			X
78	Commercial	System Biosciences	California, USA			X
79	Commercial	Taconic Biosciences at Taconic Biosciences—USA	New York, USA			X
80	Commercial	Takara Bio USA, Inc.	California, USA			X
81	Commercial	Thermo Fischer Scientific	Massachusetts, USA		X	X
82	Commercial	Toolgen	Seoul, South Korea	X	X	X
83	Commercial	Transposagen, Inc.	Lexington, Kentucky		X	X
84	Commercial	TransViragen	North Carolina, USA			X
85	Commercial	United BioLabs	Maryland, USA			X
86	Commercial	VectorBuilder, Inc.	California, USA			X
87	Commercial	ZeClinics	Barcelona, Spain			X

Appendix B: Computational Design Tools

Table B.1 ZFN design tools (Data was obtained using omicX on December 29, 2018)

	Name	Restrictions to Use	Computer Skills	Tool in Use	Year	Publication	Web Address
1	ZFN target site algorithm	None	Basic	Yes	Not specified	Not specified	http://mccb.umassmed.edu/ZFPmodularsearch.html
2	Zinc Finger Tools	None	Basic	Yes	2006	Mandell and Barbas (2006)	https://www.scripps.edu/barbas/zfdesign/zfdesignhome.php
3	ZiFiT (Alternative name: Zinc Finger Targeter)	None	Basic	No	2006	Not specified	http://zifit.partners.org/ZiFiT/
4	ZiF-Predict	None	Basic	Yes	2010	Molparia et al. (2010)	http://web.iitd.ac.in/~sundar/zifpredict/
5	ZFN-Site	None	Basic	Yes	2011	Cradick et al. (2011)	https://ccg.vital-it.ch/tagger/targetsearch.html
6	PROGNOS (Alternative Name: Predicted Report of Genome-wide Nuclease Off-target Sites)	None	Advanced	Yes	2014	Fine EJ et al. (2014)	http://bao.rice.edu/Research/BioinformaticTools/prognos.html

(continued)

K. Paris, *Genome Editing and Biological Weapons*,
https://doi.org/10.1007/978-3-031-21820-0

Table B.1 (continued)

	Name	Restrictions to Use	Computer Skills	Tool in Use	Year	Publication	Web Address
7	ZifNN	None	Basic	Yes	2016	Dutta et al. (2016)	https://www.scripps.edu/barbas/zfdesign/zfdesignhome.php

Table B.2 TALEN design tools (Data was obtained using omicX on December 29, 2018)

	Name	Restrictions to use	Computer skills	Tool in use	Year	Publication	Web Address
1	TALE-NT (Alternative Name: TAL Effector-Nucleotide Targeter)	None	Basic	Yes	2011	Cermak et al. (2011)	https://tale-nt.cac.cornell.edu/
2	TALEN Targeter	None	Basic	Yes	2012	Doyle et al. (2012)	https://tale-nt.cac.cornell.edu/node/add/talen
3	TALE Sequence Generator	None	Basic	No	2012	Sanjana et al. (2012)	http://taletoolbox.genome-engineering.org/
4	Paired Target Finder	None	Basic	Yes	2012	Doyle et al. (2012)	https://tale-nt.cac.cornell.edu/node/add/talef-off-paired
5	idTALE	None	Basic	No	2012	Li et al. (2012)	Not in use
6	TAL Effector Targeter	None	Basic	Yes	2012	Doyle et al. (2012)	https://tale-nt.cac.cornell.edu/node/add/single-tale
7	Target Finder	None	Basic	Yes	2012	Doyle et al. (2012)	https://tale-nt.cac.cornell.edu/node/add/talef-off
8	E-TALEN	None	Basic	Yes	2013	Heigwer et al. (2013)	http://www.e-talen.org/E-TALEN/
9	TALgetter (Alternative Name: TAL effector target site finder)	None	Basic	Yes	2013	Grau et al. (2013)	http://galaxy.informatik.uni-halle.de/root?tool_id=TALgetter Server not available
10	Mojo Hand	None	Basic	Yes	2013	Neff et al. (2013)	http://www.talendesign.org/
11	TALENoffer	None	Advanced	No	2013	Grau et al. (2013)	http://galaxy.informatik.uni-halle.de Server not available
12	TALgetterLong (Alternative Name: TAL effector target site finder Long)	None	Basic	No	2013	Grau et al. (2013)	http://galaxy.informatik.uni-halle.de/root?tool_id=TALgetterLong No longer in use
13	CHOPCHOP	None	Basic	Yes	2014	Montague et al. (2014)	http://chopchop.cbu.uib.no/

(continued)

Table B.2 (continued)

	Name	Restrictions to use	Computer skills	Tool in use	Year	Publication	Web Address
14	TALENdesigner	None	Basic	No	2014	Jankele and Svoboda (2014)	Tool no longer used.
15	SAPTA (Alternative Name: Scoring Algorithm for Predicting TALEN Activity)	None	Basic	Yes	2014	Lin et al. (2014)	http://bao.rice.edu/Research/BioinformaticTools/TAL_targeter.html
16	PROGNOS (Alternative Name: Predicted Report Of Genome-wide Nuclease Off-target Sites)	None	Advanced	Yes	2014	Fine et al. (2014)	http://bao.rice.edu/Research/BioinformaticTools/prognos.html
17	QueTAL (Alternative Name: Que Transcription Activator-Like)	None	Advanced	Yes	2015	Pérez-Quintero et al. (2015)	http://bioinfo-web.mpl.ird.fr/cgi-bin2/quetal/quetal.cgi
18	ZiFiT (Alternative Name: Zinc Finger Targeter)	None	Basic	No	Not specified when TALEN incorporated into tool	Not specified	http://zifit.partners.org/ZiFiT/

Table B.3 CRISPR/Cas9 sgRNA design tools (Data was obtained using omicX on December 29, 2018)

	Name	Restrictions to Use	Computer Skills	Tool in Use	Year	Publication	Web Address
1	CRISPR design	None	Basic	Yes	2013	Hsu et al. (2013)	https://zlab.bio/guide-design-resources Tool is no longer available
2	Cas9 Design	None	Basic	No	2013	Ma et al. (2013)	Tool no longer available.
3	CRISPR-P	None	Basic	Yes	2014	Yang et al. (2014)	http://crispr.hzau.edu.cn/CRISPR2/
4	COSMID (Alternative Name: CRISPR Off-target Sites with Mismatches, Insertions, and Deletions)	None	Basic	Yes	2014	Cradick et al. (2014)	https://crispr.bme.gatech.edu/
5	CRISPRseek	None	Advanced	Yes	2014	Zhu et al. (2014)	http://www.bioconductor.org/packages/release/bioc/html/CRISPRseek.html
6	CHOPCHOP	None	Basic	Yes	2014	Montague et al., 2014	http://chopchop.cbu.uib.no/
7	OutKnocker	None	Basic	Yes	2014	Schmid-Burgk et al. (2014)	http://www.outknocker.org/
8	CasFinder	None	Advanced	Yes	2014	Aach et al. (2014)	http://arep.med.harvard.edu/CasFinder/
9	Cas-OFFinder	None	Advanced	Yes	2014	Bae et al. (2014)	http://www.rgenome.net/cas-offinder/
10	GT-Scan (Alternative Name: Genome Target Scan)	None	Basic	Yes	2014	O'Brien and Bailey (2014)	https://gt-scan.csiro.au/
11	Cas-Designer	None	Advanced	Yes	2014	Bae et al. (2014)	http://www.rgenome.net/cas-designer/
12	SSFinder	None	Advanced	Yes	2014	Upadhyay and Sharma (2014)	https://github.com/alaindomissy/ssfinder

(continued)

Table B.3 (continued)

	Name	Restrictions to Use	Computer Skills	Tool in Use	Year	Publication	Web Address
13	E-CRISP	None	Basic	Yes	2014	Heigwer et al. (2016)	http://www.e-crisp.org/E-CRISP/
14	flyCRISPR	None	Basic	Yes	2014	Gratz et al. (2014)	http://tools.flycrispr.molbio.wisc.edu/targetFinder/
15	sgRNAcas9	None	Advanced	Yes	2014	Xie et al. (2014)	http://www.biootools.com/col.jsp?id=103
16	CasOT	None	Advanced	No	2014	Xiao et al. (2014)	Tool no longer available.
17	Azimuth	None	Advanced	Yes	2015	Fusi et al. (2015)	https://www.microsoft.com/en-us/research/project/azimuth/
18	CRISPR multitargeter	None	Basic	Yes	2015	Prykhozhij et al. (2015)	http://www.multicrispr.net/
19	CGAT (Alternative Name: CRISPR Genome Analysis Tool)	None	Basic	Yes	2015	Brazelton et al. (2015)	http://cbc.gdcb.iastate.edu/cgat/
20	CRISPR-ERA (Alternative Name: CRISPR-mediated Editing, Repression, and Activation)	None	Basic	Yes	2015	Liu et al. (2015)	http://crispr-era.stanford.edu/
21	SSC (Alternative Name: Spacer Scoring for CRISPR)	None	Basic	Yes	2015	Xu et al. (2015)	http://crispr.dfci.harvard.edu/SSC/
22	CRISPRscan	None	Basic	Yes	2015	Moreno-Mateos et al. (2015)	http://www.crisprscan.org/
23	CROP-IT (Alternative Name: CRISPR/Cas9 Off-target Prediction and Identification Tool)	None	Basic	Yes	2015	Singh et al. (2015)	http://cheetah.bioch.virginia.edu/AdliLab/CROP-IT/homepage.html

24	CCTOP (Alternative Name: CRISPR/Cas9 target online predictor)	None	Advanced	Yes	2015	Stemmer et al. (2015)	https://crispr.cos.uni-heidelberg.de/
25	CRISPRdirect	None	Basic	Yes	2015	Naito et al. (2015)	http://crispr.dbcls.jp/
26	WU-CRISPR	None	Basic	Yes	2015	Wong et al. (2015)	http://crispr.wustl.edu/
27	ProtospacerWB (Alternative Name: Protospacer Workbench)	None	Medium	Yes	2015	MacPherson and Scherf (2015)	http://www.protospacer.com/
28	CasBLASTR	None	Basic	Yes	2015	Kleinstiver et al. (2015)	http://www.casblastr.org/
29	ZhouEtAl2017	None	Advanced	Yes	2015	Zhou et al. (2015)	Contact the maintainer, Hong Zhou, to obtain tool
30	Off-Spotter	Academic or non-commercial use	Basic	Yes	2015	Pliatsika and Rigoutsos (2015)	https://cm.jefferson.edu/Off-Spotter/
31	Genedata Selector	License purchase required	Medium	Yes	2015	Niklas et al. (2015)	https://www.genedata.com/products/selector/
32	AGEseq	None	Basic	Yes	2015	Xue and Tsai (2015)	http://aspendb.uga.edu:8085/
33	EuPaGDT (Alternative Name: Eukaryotic Pathogen gRNA Design Tool)	None	Basic	Yes	2015	Peng and Tarleton (2015)	http://grna.ctegd.uga.edu/
34	sgRNA Scorer	None	Basic	Yes	2015	Chari et al. (2015)	https://crispr.med.harvard.edu/sgRNAScorer/
35	CLD (Alternative Name: CRISPR Library Designer)	None	Advanced	Yes	2016	Heigwer et al. (2016)	https://github.com/boutroslab/cld
36	CRISPETa (Alternative Name: CRISPR Paired Excision Tool)	Academic or non-commercial use	Basic	Yes	2016	Pulido-Quetglas et al. (2016)	http://crispeta.crg.eu/
37	CRISPR-DO	None	Basic	Yes	2016	Ma et al. (2016)	http://cistrome.org/crispr/
38	ge-CRISPR	None	Basic	Yes	2016	Kaur et al. (2016)	http://bioinfo.imtech.res.in/manojk/gecrispr/

(continued)

Table B.3 (continued)

	Name	Restrictions to Use	Computer Skills	Tool in Use	Year	Publication	Web Address
39	CRISPOR	None	Basic	Yes	2016	Haeussler et al. (2016)	http://crispor.tefor.net/
40	CRISPRer	None	Basic	No	2016	Budhagatapalli et al. (2016)	http://galaxy.informatik.uni-halle.de/ Server is unavailable
41	pgRNADesign	None	Advanced	Yes	2016	Zhu et al. (2016)	https://bitbucket.org/liulab/pgrnadesign
42	CRISPy-web	None	Basic	Yes	2016	Blin et al. (2016)	https://crispy.secondarymetabolites.org/#/input
43	Breaking-Cas	None	Basic	Yes	2016	Oliveros et al. (2016)	http://bioinfogp.cnb.csic.es/tools/breakingcas/index.php
44	CT-Finder	None	Basic	Yes	2016	Zhu et al. (2016)	http://bioinfolab.miamioh.edu/ct-finder/
45	PhytoCRISP-Ex	None	Basic	Yes	2016	Rastogi et al. (2016)	http://www.phytocrispex.biologie.ens.fr/CRISP-Ex/
46	BSNP (Alternative Name: BaySnpper)	None	Advanced	Yes	2017	Hazelbaker et al. (2017)	https://github.com/dat4git/BaySnpper
47	CRISPys	None	Basic	Yes	2017	Hyams et al. (2017)	http://multicrispr.tau.ac.il/
48	CRISPR LifePipe	License purchase required	Basic	No	2017	CRISPR LifePipe User Guide (2018)	https://crispr.lifeandsoft.com/crispr/login/?next=/crispr/
49	CRISPR-FOCUS	None	Basic	Yes	2017	Cao et al. (2017)	http://cistrome.org/crispr-focus/
50	GuideScan	None	Basic	Yes	2017	Perez et al. (2017)	http://www.guidescan.com/
51	CRISPR-DS	None	Advanced	Yes	2017	Nachmanson et al. (2017)	https://github.com/risqueslab/CRISPR-DS

52	Guide Picker	None	Basic	Yes	2017	Hough et al. (2017)	https://www.deskgen.com/landing/
53	FlashFry	None	Advanced	Yes	2017	McKenna and Shendure (2017)	http://aaronmck.github.io/FlashFry/
54	CRISPRpred	None	Advanced	Yes	2017	Rahman and Rahman (2017)	https://github.com/khaled-rahman/CRISPRpred
55	CRISPR-RT (Alternative Name: CRISPR RNA-Targeting)	None	Basic	Yes	2017	Zhu et al. (2017)	http://bioinfolab.miamioh.edu/CRISPR-RT/
56	CRISTA (Alternative Name: CRISPR Target Assessment)	None	Basic	Yes	2017	Abadi et al. (2017)	http://crista.tau.ac.il/
57	CRISPRcleanR	None	Advanced	Yes	2017	Iorio et al. (2017)	https://github.com/francescojm/CRISPRcleanR
58	CASPER (Alternative Name: CRISPR Associated Software for Pathway Engineering and Research)	None	Advanced	Yes	2017	Mendoza and Trinh (2018)	https://github.com/TrinhLab/CASPER
59	gRNA_Tool	None	Advanced	Yes	2017	Boettcher et al. (2017)	https://github.com/quasiben/gRNA_Tool
60	predictSGRNA	None	Advanced	Yes	2017	Kuan et al. (2017)	http://www.ams.sunysb.edu/~pfkuan/softwares.html
61	sgTiler	None	Advanced	Yes	2017	Ahmed and He. (2017)	https://github.com/HansenHeLab/sgTiler
62	PITCh designer (Alternative Name: Precise Integration into Target Chromosome designer)	None	Basic	Yes	2017	Nakamae et al. (2017)	http://www.mls.sci.hiroshima-u.ac.jp/smg/PITChdesigner/index.html
63	pgRNAFinder	None	Basic	No	2017	Xiong et al. (2017)	Tool no longer available.
64	Redkmer	None	Advanced	Yes	2017	Papathanos and Windbichler (2017)	https://github.com/genome-traffic/redkmer-hpc

(continued)

Table B.3 (continued)

	Name	Restrictions to Use	Computer Skills	Tool in Use	Year	Publication	Web Address
65	Hi-TOM (Alternative Name: High-throughput Tracking Of Mutations)	None	Basic	Yes	2017	Liu et al. (2017)	Contact the maintainer, Kejian Wang, to obtain tool.
66	ExcisionFinder	None	Advanced	Yes	2018	Keough et al. (2018)	https://github.com/keoughkath/AlleleAnalyzer
67	CUNE (Alternative Name: Computational Universal Nucleotide Editor)	None	Basic	Yes	2018	O'Brien et al. (2018)	https://gt-scan.csiro.au/cune/Mus_musculus
68	GRIBCG (Alternative Name: Guide RNA Identifier for Balancer Chromosome Generation)	None	Advanced	Yes	2018	Merritt and Cheung (2018)	https://sourceforge.net/projects/gribcg/
69	CRISPR-Local	None	Advanced	Yes	2018	Sun et al. (2018)	http://crispr.hzau.edu.cn/CRISPR-Local/
70	CRISPRMatch	None	Advanced	Yes	2018	You et al. (2018)	https://github.com/zhangtaolab/CRISPRMatch
71	CRISPR-SURF	None	Basic	Yes	2018	Hsu et al. (2018)	Stand-alone: https://github.com/pinellolab/CRISPR-SURF or web application: http://crisprsurf.pinellolab.org/
72	CRISPRtrack	None	Advanced	Yes	2018	Lam and Ye (2018)	https://sourceforge.net/projects/crisprtrack/
73	PAVOOC (Alternative Name: Prediction And Visualization of On- and Off-targets for CRISPR)	None	Basic	Yes	2018	Schaefer et al. (2018)	https://pavooc.me/

74	CRISPRpic (Alternative Name: CRISPR via prefixed index counting)	None	Advanced	Yes	2018	Cho et al. (2018)	https://github.com/compbio/CRISPRpic
75	ICE (Alternative Name: Inference of CRISPR Edits)	None	Advanced	Yes	2018	Hsiau et al. (2018)	https://ice.synthego.com/#/ github.com/synthego-open/ice
76	CRISPR-DT (Alternative Name: CRISPR DNA Targeting)	None	Basic	Yes	2018	Zhu and Liang (2018)	http://bioinfolab.miamioh.edu/CRISPR-DT/
77	scGESTALT	None	Advanced	Yes	2018	Raj et al. (2018)	https://github.com/aaronmck/SC_GESTALT
78	SynergizingCRISPR	None	Advanced	Yes	2018	Zhang et al. (2018)	https://github.com/Alexzsx/CRISPR
79	CASPERpam	None	Advanced	Yes	2018	Mendoza and Trinh (2018)	https://github.com/TrinhLab/CASPERpam
80	MENTHU (Alternative Name: Microhomology-mediated End joining kNock-out Target Heuristic Utility)	None	Basic	Yes	2018	Ata et al. (2018)	http://genesculpt.org/menthu/
81	inDelphi	None	Basic	Yes	2018	Shen et al. (2018)	https://www.crisprindelphi.design/
82	MAGESTIC (Alternative Name: Multiplexed Accurate Genome Editing with Short, Trackable, Integrated Cellular barcodes)	None	Advanced	Yes	2018	Roy et al. (2018)	https://github.com/k-roy/MAGESTIC
83	BE-Designer	None	Basic	Yes	2018	Hwang et al. (2018)	http://www.rgenome.net/be-designer/
84	BE-Analyzer	None	Basic	Yes	2018	Hwang et al. (2018)	http://www.rgenome.net/be-analyzer/#!

(continued)

Table B.3 (continued)

	Name	Restrictions to Use	Computer Skills	Tool in Use	Year	Publication	Web Address
85	DeepCRISPR	None	Advanced	Yes	2018	Chuai et al. (2018)	https://github.com/bm2-lab/DeepCRISPR
86	ZiFiT (Alternative Name: Zinc Finger Targeter)	None	Basic	No	Not specified when CRISPR incorporated into tool	Not specified	http://zifit.partners.org/ZiFiT/
87	CRISPR sgRNA Design Tool	None	Basic	Yes	Not specified	Not specified (Feng Zhang lab, Broad Institute, developed and validated the tool.)	https://www.genscript.com/gRNA-design-tool.html
88	DESKGEN	None	Medium	Yes	Not specified	Deskop CRISPR Libraries, n.d.	https://www.deskgen.com/landing/
89	CRISPR gRNA Design tool	None	Basic	Yes	Not specified	Not specified	https://www.atum.bio/eCommerce/cas9/input
90	Benchling	License purchase required	Basic	Yes	Not specified	Not specified	Registration required to access the tool.

Appendix C: JoVE Video Articles

Table C.1 JoVE Video Articles on ZFN, TALEN, and CRISPR (Data for the table were obtained from *JoVE* on February 19, 2019)

	Video article name	ZFN	TALEN	CRISPR	Reference for *JoVE* video article
1	Genome Editing with CompoZr Custom Zinc Finger Nucleases (ZFNs)	X			Hansen, K., Coussens, M. J., Sago, J., Subramanian, S., Gjoka, M., Briner, D. Genome Editing with CompoZr Custom Zinc Finger Nucleases (ZFNs). J. Vis. Exp. (64), e3304, doi:10.3791/3304 (2012).
2	Substrate Generation for Endonucleases of CRISPR/Cas Systems			X	Zoephel, J., Dwarakanath, S., Richter, H., Plagens, A., Randau, L. Substrate Generation for Endonucleases of CRISPR/Cas Systems. J. Vis. Exp. (67), e4277, doi:10.3791/4277 (2012).
3	Production of Apolipoprotein C-III Knockout Rabbits using Zinc Finger Nucleases	X			Yang, D., Zhang, J., Xu, J., Zhu, T., Fan, Y., Fan, J., Chen, Y. E. Production of Apolipoprotein C-III Knockout Rabbits using Zinc Finger Nucleases. J. Vis. Exp. (81), e50957, doi:10.3791/50957 (2013).
4	Mouse Genome Engineering Using Designer Nucleases		X		Hermann, M., Cermak, T., Voytas, D. F., Pelczar, P. Mouse Genome Engineering Using Designer

(continued)

K. Paris, *Genome Editing and Biological Weapons*,
https://doi.org/10.1007/978-3-031-21820-0

Table C.1 (continued)

	Video article name	ZFN	TALEN	CRISPR	Reference for *JoVE* video article
					Nucleases. J. Vis. Exp. (86), e50930, doi:10.3791/50930 (2014).
5	Zinc-finger Nuclease Enhanced Gene Targeting in Human Embryonic Stem Cells	X	X		Hartley, B. J., Fabb, S. A., Finnin, B. A., Haynes, J. M., Pouton, C. W. Zinc-finger Nuclease Enhanced Gene Targeting in Human Embryonic Stem Cells. J. Vis. Exp. (90), e51764, doi:10.3791/51764 (2014).
6	Generation of Genomic Deletions in Mammalian Cell Lines via CRISPR/Cas9			X	Bauer, D. E., Canver, M. C., Orkin, S. H. Generation of Genomic Deletions in Mammalian Cell Lines via CRISPR/Cas9. J. Vis. Exp. (95), e52118, doi:10.3791/52118 (2015).
7	Transfection, Selection, and Colony-picking of Human Induced Pluripotent Stem Cells TALEN-targeted with a GFP Gene into the AAVS1 Safe Harbor		X		Cerbini, T., Luo, Y., Rao, M. S., Zou, J. Transfection, Selection, and Colony-picking of Human Induced Pluripotent Stem Cells TALEN-targeted with a GFP Gene into the AAVS1 Safe Harbor. J. Vis. Exp. (96), e52504, doi:10.3791/52504 (2015).
8	Efficient Generation of hiPSC Neural Lineage Specific Knockin Reporters Using the CRISPR/Cas9 and Cas9 Double Nickase System			X	Li, S., Xue, H., Long, B., Sun, L., Truong, T., Liu, Y. Efficient Generation of hiPSC Neural Lineage Specific Knockin Reporters Using the CRISPR/Cas9 and Cas9 Double Nickase System. J. Vis. Exp. (99), e52539, doi:10.3791/52539 (2015).
9	Establishment of Genome-edited Human Pluripotent Stem Cell Lines: From Targeting to Isolation	X	X	X	Blair, J. D., Bateup, H. S., Hockemeyer, D. F. Establishment of Genome-edited Human Pluripotent Stem Cell Lines: From Targeting to Isolation. J. Vis. Exp. (108), e53583, doi:10.3791/53583 (2016).

(continued)

Table C.1 (continued)

	Video article name	ZFN	TALEN	CRISPR	Reference for *JoVE* video article
10	Genome Editing in Astyanax mexicanus Using Transcription Activator-like Effector Nucleases (TALENs)		X		Kowalko, J. E., Ma, L., Jeffery, W. R. Genome Editing in Astyanax mexicanus Using Transcription Activator-like Effector Nucleases (TALENs). J. Vis. Exp. (112), e54113, doi:10.3791/54113 (2016).
11	Microinjection for Transgenesis and Genome Editing in Threespine Sticklebacks		X	X	Erickson, P. A., Ellis, N. A., Miller, C. T. Microinjection for Transgenesis and Genome Editing in Threespine Sticklebacks. J. Vis. Exp. (111), e54055, doi:10.3791/54055 (2016).
12	Isolation of Specific Genomic Regions and Identification of Associated Molecules by enChIP		X	X	Fujita, T., Fujii, H. Isolation of Specific Genomic Regions and Identification of Associated Molecules by enChIP. J. Vis. Exp. (107), e53478, doi:10.3791/53478 (2016).
13	Designing, Packaging, and Delivery of High Titer CRISPR Retro and Lentiviruses via Stereotaxic Injection			X	Fricano-Kugler, C. J., Williams, M. R., Salinaro, J. R., Li, M., Luikart, B. Designing, Packaging, and Delivery of High Titer CRISPR Retro and Lentiviruses via Stereotaxic Injection. J. Vis. Exp. (111), e53783, doi:10.3791/53783 (2016).
14	Generating CRISPR/Cas9 Mediated Monoallelic Deletions to Study Enhancer Function in Mouse Embryonic Stem Cells			X	Moorthy, S. D., Mitchell, J. A. Generating CRISPR/Cas9 Mediated Monoallelic Deletions to Study Enhancer Function in Mouse Embryonic Stem Cells. J. Vis. Exp. (110), e53552, doi:10.3791/53552 (2016).
15	A Simple and Efficient Approach to Construct Mutant Vaccinia Virus Vectors			X	Yuan, M., Wang, P., Chard, L. S., Lemoine, N. R., Wang, Y. A Simple and Efficient Approach to Construct Mutant Vaccinia Virus Vectors. J. Vis. Exp. (116), e54171, doi:10.3791/54171 (2016).

(continued)

Table C.1 (continued)

	Video article name	ZFN	TALEN	CRISPR	Reference for *JoVE* video article
16	High-throughput CRISPR Vector Construction and Characterization of DNA Modifications by Generation of Tomato Hairy Roots			X	Jacobs, T. B., Martin, G. B. High-throughput CRISPR Vector Construction and Characterization of DNA Modifications by Generation of Tomato Hairy Roots. J. Vis. Exp. (110), e53843, doi:10.3791/53843 (2016).
17	Generation of Genetically Modified Mice through the Microinjection of Oocytes			X	Delerue, F., Ittner, L. M. Generation of Genetically Modified Mice through the Microinjection of Oocytes. J. Vis. Exp. (124), e55765, doi:10.3791/55765 (2017).
18	Genome Editing and Directed Differentiation of hPSCs for Interrogating Lineage Determinants in Human Pancreatic Development		X	X	Shi, Z. D., Soh, C. L., Zhu, Z., Huangfu, D. Genome Editing and Directed Differentiation of hPSCs for Interrogating Lineage Determinants in Human Pancreatic Development. J. Vis. Exp. (121), e55267, doi:10.3791/55267 (2017).
19	Using a Fluorescent PCR-capillary Gel Electrophoresis Technique to Genotype CRISPR/Cas9-mediated Knockout Mutants in a High-throughput Format			X	Ramlee, M. K., Wang, J., Cheung, A. M., Li, S. Using a Fluorescent PCR-capillary Gel Electrophoresis Technique to Genotype CRISPR/Cas9-mediated Knockout Mutants in a High-throughput Format. J. Vis. Exp. (122), e55586, doi:10.3791/55586 (2017).
20	Selection-dependent and Independent Generation of CRISPR/Cas9-mediated Gene Knockouts in Mammalian Cells			X	Sternburg, E. L., Dias, K. C., Karginov, F. V. Selection-dependent and Independent Generation of CRISPR/Cas9-mediated Gene Knockouts in Mammalian Cells. J. Vis. Exp. (124), e55903, doi:10.3791/55903 (2017).
21	A Protocol for Multiple Gene Knockout in Mouse Small			X	Merenda, A., Andersson-Rolf, A., Mustata, R. C., Li,

(continued)

Table C.1 (continued)

	Video article name	ZFN	TALEN	CRISPR	Reference for *JoVE* video article
	Intestinal Organoids Using a CRISPR-concatemer				T., Kim, H., Koo, B. K. A Protocol for Multiple Gene Knockout in Mouse Small Intestinal Organoids Using a CRISPR-concatemer. J. Vis. Exp. (125), e55916, doi:10.3791/55916 (2017).
22	A Protocol for the Production of Integrase-deficient Lentiviral Vectors for CRISPR/Cas9-mediated Gene Knockout in Dividing Cells			X	Vijayraghavan, S., Kantor, B. A Protocol for the Production of Integrase-deficient Lentiviral Vectors for CRISPR/Cas9-mediated Gene Knockout in Dividing Cells. J. Vis. Exp. (130), e56915, doi:10.3791/56915 (2017).
23	QTL Mapping and CRISPR/Cas9 Editing to Identify a Drug Resistance Gene in Toxoplasma gondii			X	Shen, B., Powell, R. H., Behnke, M. S. QTL Mapping and CRISPR/Cas9 Editing to Identify a Drug Resistance Gene in Toxoplasma gondii. J. Vis. Exp. (124), e55185, doi:10.3791/55185 (2017).
24	Efficient Generation and Editing of Feeder-free IPSCs from Human Pancreatic Cells Using the CRISPR-Cas9 System			X	Nandal, A., Mallon, B., Telugu, B. P. Efficient Generation and Editing of Feeder-free IPSCs from Human Pancreatic Cells Using the CRISPR-Cas9 System. J. Vis. Exp. (129), e56260, doi:10.3791/56260 (2017).
25	Protocols for C-Brick DNA Standard Assembly Using Cpf1			X	Li, S. Y., Zhao, G. P., Wang, J. Protocols for C-Brick DNA Standard Assembly Using Cpf1. J. Vis. Exp. (124), e55775, doi:10.3791/55775 (2017).
26	Embryo Microinjection and Transplantation Technique for Nasonia vitripennis Genome Manipulation			X	Li, M., Bui, M., Akbari, O. S. Embryo Microinjection and Transplantation Technique for Nasonia vitripennis Genome Manipulation. J. Vis. Exp. (130), e56990, doi:10.3791/56990 (2017).

(continued)

Table C.1 (continued)

	Video article name	ZFN	TALEN	CRISPR	Reference for *JoVE* video article
27	CRISPR/Cas9-mediated Targeted Integration In Vivo Using a Homology-mediated End Joining-based Strategy			X	Yao, X., Wang, X., Liu, J., Shi, L., Huang, P., Yang, H. CRISPR/Cas9-mediated Targeted Integration In Vivo Using a Homology-mediated End Joining-based Strategy. J. Vis. Exp. (133), e56844, doi:10.3791/56844 (2018).
28	Using CRISPR/Cas9 Gene Editing to Investigate the Oncogenic Activity of Mutant Calreticulin in Cytokine Dependent Hematopoietic Cells			X	Abdelfattah, N. S., Mullally, A. Using CRISPR/Cas9 Gene Editing to Investigate the Oncogenic Activity of Mutant Calreticulin in Cytokine Dependent Hematopoietic Cells. J. Vis. Exp. (131), e56726, doi:10.3791/56726 (2018).
29	Endogenous Protein Tagging in Human Induced Pluripotent Stem Cells Using CRISPR/Cas9			X	Haupt, A., Grancharova, T., Arakaki, J., Fuqua, M. A., Roberts, B., Gunawardane, R. N. Endogenous Protein Tagging in Human Induced Pluripotent Stem Cells Using CRISPR/Cas9. J. Vis. Exp. (138), e58130, doi:10.3791/58130 (2018).
30	Efficient Production and Identification of CRISPR/Cas9-generated Gene Knockouts in the Model System Danio rerio			X	Sorlien, E. L., Witucki, M. A., Ogas, J. Efficient Production and Identification of CRISPR/Cas9-generated Gene Knockouts in the Model System Danio rerio. J. Vis. Exp. (138), e56969, doi:10.3791/56969 (2018).
31	Microinjection of Western Corn Rootworm, Diabrotica virgifera virgifera, Embryos for Germline Transformation, or CRISPR/Cas9 Genome Editing			X	Chu, F. C., Wu, P. S., Pinzi, S., Grubbs, N., Lorenzen, M. D. Microinjection of Western Corn Rootworm, Diabrotica virgifera virgifera, Embryos for Germline Transformation, or CRISPR/Cas9 Genome Editing. J. Vis. Exp. (134), e57497, doi:10.3791/57497 (2018).

(continued)

Table C.1 (continued)

	Video article name	ZFN	TALEN	CRISPR	Reference for *JoVE* video article
32	Enhanced Genome Editing with Cas9 Ribonucleoprotein in Diverse Cells and Organisms			X	Farboud, B., Jarvis, E., Roth, T. L., Shin, J., Corn, J. E., Marson, A., Meyer, B. J., Patel, N. H., Hochstrasser, M. L. Enhanced Genome Editing with Cas9 Ribonucleoprotein in Diverse Cells and Organisms. J. Vis. Exp. (135), e57350, doi:10.3791/57350 (2018).
33	CRISPR Guide RNA Cloning for Mammalian Systems			X	Nageshwaran, S., Chavez, A., Cher Yeo, N., Guo, X., Lance-Byrne, A., Tung, A., Collins, J. J., Church, G. M. CRISPR Guide RNA Cloning for Mammalian Systems. J. Vis. Exp. (140), e57998, doi:10.3791/57998 (2018).
34	CRISPR/Cas9 Gene Editing to Make Conditional Mutants of Human Malaria Parasite P. falciparum			X	Kudyba, H. M., Cobb, D. W., Florentin, A., Krakowiak, M., Muralidharan, V. CRISPR/Cas9 Gene Editing to Make Conditional Mutants of Human Malaria Parasite P. falciparum. J. Vis. Exp. (139), e57747, doi:10.3791/57747 (2018).
35	An Efficient Strategy for Generating Tissue-specific Binary Transcription Systems in Drosophila by Genome Editing			X	Du, L., Zhou, A., Sohr, A., Roy, S. An Efficient Strategy for Generating Tissue-specific Binary Transcription Systems in Drosophila by Genome Editing. J. Vis. Exp. (139), e58268, doi:10.3791/58268 (2018).
36	Culturing and Manipulation of O9-1 Neural Crest Cells			X	Nguyen, B. H., Ishii, M., Maxson, R. E., Wang, J. Culturing and Manipulation of O9-1 Neural Crest Cells. J. Vis. Exp. (140), e58346, doi:10.3791/58346 (2018).
37	CRISPR-mediated Loss of Function Analysis in Cerebellar Granule Cells Using In			X	Feng, W., Herbst, L., Lichter, P., Pfister, S. M., Liu, H. K., Kawauchi, D. CRISPR-mediated Loss

(continued)

Table C.1 (continued)

	Video article name	ZFN	TALEN	CRISPR	Reference for *JoVE* video article
	Utero Electroporation-based Gene Transfer				of Function Analysis in Cerebellar Granule Cells Using In Utero Electroporation-based Gene Transfer. J. Vis. Exp. (136), e57311, doi:10.3791/57311 (2018).
38	CRISPR-Cas9-based Genome Engineering to Generate Jurkat Reporter Models for HIV-1 Infection with Selected Proviral Integration Sites			X	Bialek, J. K., Walther, T., Hauber, J., Lange, U. C. CRISPR-Cas9-based Genome Engineering to Generate Jurkat Reporter Models for HIV-1 Infection with Selected Proviral Integration Sites. J. Vis. Exp. (141), e58572, doi:10.3791/58572 (2018).
39	Dissection of Enhancer Function Using Multiplex CRISPR-based Enhancer Interference in Cell Lines			X	Carleton, J. B., Berrett, K. C., Gertz, J. Dissection of Enhancer Function Using Multiplex CRISPR-based Enhancer Interference in Cell Lines. J. Vis. Exp. (136), e57883, doi:10.3791/57883 (2018).
40	CRISPR-Mediated Reorganization of Chromatin Loop Structure			X	Morgan, S. L., Chang, E. Y., Mariano, N. C., Bermudez, A., Arruda, N. L., Wu, F., Luo, Y., Shankar, G., Huynh, S. K., Huang, C. C., Pitteri, S. J., Wang, K. C. CRISPR-Mediated Reorganization of Chromatin Loop Structure. J. Vis. Exp. (139), e57457, doi:10.3791/57457 (2018).
41	CRISPR-mediated Genome Editing of the Human Fungal Pathogen Candida albicans			X	Evans, B. A., Pickerill, E. S., Vyas, V. K., Bernstein, D. A. CRISPR-mediated Genome Editing of the Human Fungal Pathogen Candida albicans. J. Vis. Exp. (141), e58764, doi:10.3791/58764 (2018).
42	A Customizable Protocol for String Assembly gRNA Cloning (STAgR)			X	Breunig, C. T., Neuner, A. M., Giehrl-Schwab, J., Wurst, W., Götz, M., Stricker, S. H. A

(continued)

Table C.1 (continued)

	Video article name	ZFN	TALEN	CRISPR	Reference for *JoVE* video article
					Customizable Protocol for String Assembly gRNA Cloning (STAgR). J. Vis. Exp. (142), e58556, doi:10.3791/58556 (2018).
43	Generation of Knock-out Primary and Expanded Human NK Cells Using Cas9 Ribonucleoproteins			X	Naeimi Kararoudi, M., Dolatshad, H., Trikha, P., Hussain, S. R., Elmas, E., Foltz, J. A., Moseman, J. E., Thakkar, A., Nakkula, R. J., Lamb, M., Chakravarti, N., McLaughlin, K. J., Lee, D. A. Generation of Knock-out Primary and Expanded Human NK Cells Using Cas9 Ribonucleoproteins. J. Vis. Exp. (136), e58237, doi:10.3791/58237 (2018).
44	Adeno-Associated Virus-Mediated Delivery of CRISPR for Cardiac Gene Editing in Mice			X	Xu, L., Gao, Y., Lau, Y. S., Han, R. Adeno-Associated Virus-Mediated Delivery of CRISPR for Cardiac Gene Editing in Mice. J. Vis. Exp. (138), e57560, doi:10.3791/57560 (2018).
45	A Rapid and Facile Pipeline for Generating Genomic Point Mutants in C. elegans Using CRISPR/Cas9 Ribonucleoproteins			X	Prior, H., MacConnachie, L., Martinez, J. L., Nicholl, G. C., Beg, A. A. A Rapid and Facile Pipeline for Generating Genomic Point Mutants in C. elegans Using CRISPR/Cas9 Ribonucleoproteins. J. Vis. Exp. (134), e57518, doi:10.3791/57518 (2018).
46	Highly Efficient Gene Disruption of Murine and Human Hematopoietic Progenitor Cells by CRISPR/Cas9			X	Brunetti, L., Gundry, M. C., Kitano, A., Nakada, D., Goodell, M. A. Highly Efficient Gene Disruption of Murine and Human Hematopoietic Progenitor Cells by CRISPR/Cas9. J. Vis. Exp. (134), e57278, doi:10.3791/57278 (2018).

(continued)

Table C.1 (continued)

	Video article name	ZFN	TALEN	CRISPR	Reference for *JoVE* video article
47	Production of Genetically Engineered Golden Syrian Hamsters by Pronuclear Injection of the CRISPR/Cas9 Complex			X	Li, R., Miao, J., Fan, Z., Song, S., Kong, I. k., Wang, Y., Wang, Z. Production of Genetically Engineered Golden Syrian Hamsters by Pronuclear Injection of the CRISPR/Cas9 Complex. J. Vis. Exp. (131), e56263, doi:10.3791/56263 (2018).

Appendix D: Forums to Acquire Skill and Knowledge

Table D.1 Forums to acquire skills and knowledge to use genome editing technologies

Type of forum	Name of forum	Location	Description
Training Workshops	Bio-Trac	Germantown, Maryland	"Gene Editing with CRISPR" is a 3-day workshop led by active researchers that is designed to give scientists an overview of CRISPR through presentations and laboratory exercises. The workshop is offered throughout the year, costs $895, and limited to 24 participants. The workshop is "ideal for basic research and translational biology scientists."[a]
	Thermo Fisher Scientific	Carlsbad, California	"MB-205: Introduction to CRISPR-Cas9 Genome Editing - Design to Analysis" is a 4-day training workshop offered throughout the year to teach CRISPR to life science professionals. The course includes lectures and laboratory exercises, is limited to eight participants, and costs $2995. It is an entry-level CRISPR course; however, attendees should have intermediate lab skills including PCR and mammalian cell culture.[b]

(continued)

K. Paris, *Genome Editing and Biological Weapons*,
https://doi.org/10.1007/978-3-031-21820-0

Table D.1 (continued)

Type of forum	Name of forum	Location	Description
	CNRS Formation Enterprises	Illkirch, France	"In Vivo CRISPR-Cas9 genome editing" is a 2-day training to learn to use CRISPR/Cas9 in rodents for *in vivo* gene editing. The training includes lectures and a hands-on bioinformatics computer session to design a CRISPR/Cas9 experiment. CNRS Formation Enterprises is a training organization. The training is organized by Phenomin in collaboration with Celphedia for 6–10 participants at a cost of 800 €. Participants are researchers, engineers, PhD students, and post-doctoral researchers who have "the basic knowledge in genetics and molecular biology to understand the training content."[c]
	National Institutes of Health	Bethesda, Maryland	"BioTech 55: Engineering with CRISPR, TALENs & ZFNs" is a 5-day, hands-on workshop that costs $1295 and is limited to 24 participants. It covers design and assembly of CRISPR, TALEN, and ZFN as well as "different applications including engineering in mice, disease modeling, generating iPSC reporter lines, and high throughput approaches."[d]
	European Molecular Biology Laboratory (EMBL)	Heidelberg, Germany	EMBL offered an advanced course for ZFN on May 10-12, 2011 in which lectures provided knowledge of ZFN fundamentals and laboratory exercises provided additional training in ZFN.[e]Genome Engineering: CRISPR/Cas course designed for participants to have hands-on training using CRISPR/Cas to generate knock-ins and knockouts. The course was held September 16–21, 2018

(continued)

Table D.1 (continued)

Type of forum	Name of forum	Location	Description
			and will be held again March 17–22, 2019. Thermo Fisher co-organized the course. Participants should have "basic molecular and cell biology techniques."[f]
DIY Bio Laboratories	Genspace	Brooklyn, New York	Class offered on CRISPR. Members conduct CRISPR experiments in the Genspace laboratory.
	BUGSS	Baltimore, Maryland	Introduction to CRISPR classes offered. A class was held August 4–25, 2018, and it cost $60 for members and $260 for non-members.[g]
	Indie Lab	Richmond, Virginia	As of November 2017, did not have gene editing classes or events planned, but can support such things. Have discussed CRISPR class but nothing has come about. In November 2017, a member was interested in CRISPR, but the state of the project did not required genome editing yet.[h]
	BosLab	Somerville, Massachusetts	As of November 2017, BosLab was in the process of developing a CRISPR class.[i] Genspace sent a CRISPR class protocol to BosLab.
	Charlottesville Open Bio Labs	Charlottesville, Virginia	CRISPR class offered. Will be offered again in January/February 2018.[j]
College Classes	University of New Mexico	Albuquerque, New Mexico	A class taught to undergraduate students on how to use CRISPR/Cas9 to knock-out Drosophila genes was first offered in the Spring of 2015. Three out of six students in the class were successful in mutating the genes that were targeted. Overall, the initial class supported the finding that a laboratory genome editing could be taught that gives undergraduate students effective, hands-on laboratory training.[k]

(continued)

Table D.1 (continued)

Type of forum	Name of forum	Location	Description
	University of California, Berkeley	Berkeley, California	During the summer 2018 session (July to August), 30 undergraduate students learned to use CRISPR/Cas9 and carried out experiments in bacteria. There are plans to offer the course again during the summer 2019 session.[l]
	Université Laval	Quebec City, Canada	As of February 2017, undergraduate students in the BSc microbiology program studied CRISPR in a laboratory course for the previous 3 years. Students are using *Streptococcus thermophilus*, a biosafety level 1 bacteria generally considered safe.[m]
	GetSmarter in partnership with Harvard Office of the Vice Provost for Advances in Learning	Online	An 8-week, online, short course is offered on CRISPR gene-editing applications. Students learn about CRISPR basics and applications in treating genetic diseases, agriculture, and ethics. The course costs $3000.[n]
	University of Toronto	Toronto, Ontario, Canada	Undergraduate students in the human biology program used CRISPR in a laboratory class to modify cells that are normally blue to green under fluorescence microscopy.[o]

[a]"CRISPR/Cas9 Workshop: Technology Foundations, Applications Principles," *Bio-Trac®* (blog), accessed November 4, 2018, https://biotrac.com/crispr/

[b]"MB-205: Introduction to CRISPR-Cas9 Genome Editing - Design to Analysis (4 Days)," Thermo Fisher Scientific, accessed November 4, 2018, https://learn.thermofisher.com/courses/search?q=crispr

[c]"In Vivo CRISPR-Cas9 Genome Editing," CNRS Formation Entreprises, accessed November 4, 2018, https://cnrsformation.cnrs.fr/stage-18188-In-vivo-CRISPR-Cas9-genome-editing.html?mc=crispr

[d]"BioTech 55 | Engineering with CRISPR, TALENs & ZFNs | 5-Day Hands On Workshop," FAES@NIH, accessed November 4, 2018, https://faes.org/events/biotech-55-engineering-crispr-talens-zfns-5-day-hands-workshop-1

[e]"ZFN 2011 : EMBL Advanced Course - Targeted Genome Editing Using Zinc Finger Nucleases," accessed January 20, 2019, http://www.wikicfp.com/cfp/servlet/event.showcfp?eventid=13257&copyownerid=3197

[f]"EMBL Course - Genome Engineering: CRISPR/Cas - 17 - 22 March 2019," European Molecular Biology Laboratory, accessed January 20, 2019, https://www.embl.de/training/events/2019/GEE19-01/

(continued)

[g]Baltimore UnderGround Science Space (BUGSS), "Last Chance for You to Try out CRISPR! Gene Editing Class Starts This Sat!," August 1, 2018
[h]"Email Communications with the Indie Lab," November 2017
[i]"Email Communications with the BosLab," November 13, 2017
[j]"Email Communications with Charlottesville Open Bio Labs," November 20, 2017
[k]Adame et al., "An Undergraduate Laboratory Class Using CRISPR/Cas9 Technology to Mutate Drosophila Genes"
[l]Robert S, "Only at Cal: New Class Teaches Undergrads How to 'CRISPR,'" Berkeley News, August 21, 2018, accessed November 4, 2018, https://news.berkeley.edu/story_jump/only-at-cal-new-class-teaches-undergrads-how-to-crispr/
[m]Luc Trudel, Michel Frenette, and Sylvain Moineau, "CRISPR–Cas in the Laboratory Classroom," *Nature Microbiology* 2 (February 17, 2017): 17018, https://doi.org/10.1038/nmicrobiol.2017.18
[n]"CRISPR: Gene-Editing Applications - Harvard's VPAL Online Short Course|GetSmarter," accessed November 4, 2018, https://www.getsmarter.com/courses/us/harvard-gene-editing-applications-online-short-course
[o]"Undergraduate Students Get Rare Chance to Work with Groundbreaking Genetic Engineering Technique CRISPR," University of Toronto News, April 6, 2018, accessed November 4, 2018, https://www.utoronto.ca/news/undergraduate-students-get-rare-chance-work-groundbreaking-genetic-engineering-technique-crispr

Appendix E: Company Startups

Table E.1 Startup Companies Focused on ZFN, TALEN, or CRISPR (Data for this table was obtained through Crunchbase and Owler unless otherwise noted. "Crunchbase: Discover Innovative Companies and the People behind Them," Crunchbase, accessed November 30, 2018, https://www.crunchbase.com; "Owler: Competitive Intelligence to Outsmart Your Competition," Owler, accessed November 30, 2018, https://www.owler.com/)

#	Year founded	Company name	Location	ZFN	TALEN	CRISPR
1	1995	Sangamo Biosciences (now Sangamo Therapeutics)	Richmond, California	X		
2	2000 (closed in 2004)	Phytodyne[a]	Ames, Iowa	X		
3	2010	Calyxt (separate unit in Cellectis)	Roseville, Minnesota		X	
4	2011	Caribou Biosciences	Berkeley, California			X
5	July 1, 2012	Synthego	Redwood City, California			X
6	2012	Benchling	San Francisco, California			X
7	August 7, 2012	Desktop Genetics	London, England			X
8	2013	Ligandal	San Francisco, California			X
9	Sept. 2013	Editas Medicine	Cambridge, Massachusetts			X
10	Oct. 31, 2013	CRISPR Therapeutics	Cambridge, Massachusetts			X
11	2014	eGenesis Bio	Cambridge, Massachusetts			X

(continued)

K. Paris, *Genome Editing and Biological Weapons*,
https://doi.org/10.1007/978-3-031-21820-0

Table E.1 (continued)

#	Year founded	Company name	Location	ZFN	TALEN	CRISPR
12	May 2014	Intellia Therapeutics	Cambridge, Massachusetts			X
13	2014	Agenovir	San Francisco, California			X
14	2014	ERS Genomics	Dublin, Ireland			X
15	2014	Eligo Bioscience	Paris, France			X
16	2015	Poseida Therapeutics	San Diego, California			X
17	2015	Inscripta	Boulder, Colorado			X
18	2015	NTrans Technologies	Utrecht, The Netherlands			X
19	2015	Locus Biosciences	Raleigh, North Carolina			X
20	2015	Hudson River Biotechnology	Wageningen,The Netherlands			X
21	2016	Senti Biosciences	San Francisco, California			X
22	2016	Arbor Biotechnologies	Cambridge, Massachusetts			X
23	2016	GenEdit	Berkeley, California			X
24	2016	Casebia Therapeutics	Cambridge, Massachusetts			X
25	2016	Inari Agriculture	Cambridge, Massachusetts			X
26	November 16, 2016	Folium Science	Cambridge and Bristol, United Kingdom			X
27	April 4, 2017	PLANTeDIT	Cork, Ireland			X
28	June 1, 2017	Mammoth Biosciences	San Francisco, California			X
29	2017	Exonics Therapeutics	Cambridge, Massachusetts			X
30	2017	Beam Therapeutics	Cambridge, Massachusetts			X
31	2017	CasZyme	Vilnius, Lithuania			X
32	2017	LifeEdit	Durham, North Carolina			X
33	July 2017	Nextbiotics	San Francisco, California			X
34	2017	SNIPR Biome	Denmark, Copenhagen			X

[a]Pearson, "Protein Engineering," 163

www.ingramcontent.com/pod-product-compliance
Lightning Source LLC
Chambersburg PA
CBHW071950050326
40576CB00034B/92
9783031218194